D1381066

UNIVERSITY OF GLAMORGAN
LEARNING RESOURCES CENTRE

Pontypridd, Mid Glamorgan, CF37 1DL
Telephone: Pontypridd (01443) 482626

Books are to be returned on or before the last date below

Statistics for Industry and Technology

Series Editor
N. Balakrishnan
McMaster University

Editorial Advisory Board

Advances in Stochastic Models
for Reliability, Quality and Safety

Waltraud Kahle
Elart von Collani
Jürgen Franz
Uwe Jensen
Editors

Birkhäuser
Boston • Basel • Berlin

Waltraud Kahle
Dept. of Mathematical Stochastics
Otto-von-Guericke-
University Magdeburg
D-39016 Magdeburg
Germany

Elart von Collani
Dept. of Applied Mathematics
University of Würzburg
D-97070 Würzburg
Germany

Jürgen Franz
Dept. of Mathematical Stochastics
Technical University Dresden
Zellescher Weg 12-14
D-01069 Dresden
Germany

Uwe Jensen
Dept. of Stochastics
University of Ulm
D-89069 Ulm
Germany

Library of Congress Cataloging-in-Publication Data
Advances in stochastic models for reliability, quality, and safety /
Elart von Collani . . . [et al.], editors.
        p.   cm.
    Papers from a workshop held in Magdeburg, Germany, in 1997.
    Includes bibliographical references and index.
    ISBN 0-8176-4049-5 (alk. paper). -- ISBN 3-7643-4049-5 (alk.
paper)
    1. Reliability (Engineering)--Mathematical models. 2. Stochastic
processes.   I. Von Collani, Elart, 1944-   .
    TA169.A384   1998
    620' .00452'015118--dc21   98-3506

Printed on acid-free paper
© 1998 Birkhäuser Boston   *Birkhäuser*

ISBN 0-8176-4049-5
ISBN 3-7643-4049-5
Typeset by the Editors in LATEX.
Cover design by Vernon Press, Boston, MA.
Printed in the U.S.A.
9 8 7 6 5 4 3 2 1

Learning Resources
Centre

1191888

Contents

PART II: RELIABILITY ANALYSIS

7. Maximum Likelihood Estimation With Different Sequential k-out-of-n Systems
Erhard Cramer and Udo Kamps   **101**

8. Stochastic Models for the Return of Used Devices
Berthold Heiligers and Jürgen Ruf   **113**

9. Some Remarks on Dependent Censoring in Complex Systems
Tina Herberts and Uwe Jensen   **125**

Preface

Fast technological development produces systems of ever-increasing complexity. The demand for reliable functioning of these systems has become more and more important. Thus, there is a need for highly reliable technical devices and systems, for monitoring and controlling their functioning and for planning maintenance and corrective actions to fulfill given requirements considering economic limitations.

These tasks reflect the wide field of engineering activities that are accompanied by and based on a wide range of stochastical models. The book presents the main contributions to a workshop on *Stochastic Models of Reliability, Quality, and Safety* held in Schierke near Magdeburg, Germany. This workshop was part of a series of meetings that take place every two years organized by the Society of Reliability, Quality and Safety.

The basic idea of these workshops is to bring together theorists, applied statisticians, and practitioners to exchange experiences and ideas of common interest. The book contains recent results in reliability and related fields. The presentation aims at making at least a part of the results accessible to engineers. This would be most desirable because many of the relevant textbooks written for engineers are mainly restricted to the simplest models, whereas the more theoretically oriented literature in fields like point processes and survival analysis, being important for reliability, is rarely presented in a way appealing to engineers. Thus, one purpose of this book is to bridge the gap between theory and practice, although we are aware of the fact that this is a very ambitious aim.

Each workshop paper published in this book has been refereed and, if appropriate, revised by the author. The papers are classified in four sections: Lifetime Analysis, Reliability Analysis, Network Analysis, and Process Control.

The section on lifetime analysis deals with certain classes of lifetime distributions and focuses on parameter estimation. New estimators and methods for deriving estimators are proposed and their application is illustrated. The topics in the second section on reliability analysis cover damage processes, first passage times as lifetime models, and statistical methods for certain complex systems including the case of dependent censoring.

The third section on network analysis provides different methods for computing the connectiveness probabilities in large networks. They include a very simple approximate one and a rather sophisticated, exact computation method. The papers of the concluding section on process control investigate different aspects of production processes. Various control chart methods are analyzed, the integration of statistical and engineering process control is investigated, a process model with three different economic states is introduced, and process capability is dealt with.

Editing such a book needs the assistance of many colleagues willing to serve as referees. We are very grateful to all those who supported us in this respect. Finally, we want to thank the contributors to this proceedings volume. The smooth collaboration with the series editor, Professor N. Balakrishnan, is particularly acknowledged.

WALTRAUD KAHLE
Madgeburg, Germany

ELART VON COLLANI
Würzburg, Germany

JÜRGEN FRANZ
Dresden, Germany

UWE JENSEN
Ulm, Germany   FEBRUARY 1998

Contributors

Christoph, Gerd Fakultät für Mathematik, Otto–von–Guericke–Universität Magdeburg, Institut für Mathematische Stochastik, PF 4120, D–39016 Magdeburg, Germany
e-mail: *Gerd.Christoph@Mathematik.Uni-Magdeburg.DE*

Cramer, Erhard Institute of Statistics, Aachen University of Technology, D-52056 Aachen, Germany
e-mail: *cramer@stochastik.rwth-aachen.de*

Dräger, Klaus Institut für Angewandte Mathematik & Statistik, Sanderring 2, D-97070 Würzburg, Germany
e-mail: *draeger@mathematik.uni-wuerzburg.de*

Ferger, Dietmar Department of Mathematics, Dresden University of Technology, Mommsenstr. 13, 01062 Dresden, Germany
e-mail: *dietmar.ferger@math.tu-dresden.de*

Franz, Jürgen Department of Mathematical Stochastics, Technical University Dresden, Zellescher Weg 12–14, D-01069 Dresden, Germany

Gasmi, Sofiane Fakultät für Mathematik, Otto–von–Guericke–Universität Magdeburg, Box 4120, 39016 Magdeburg, Germany
e-mail: *Sofiane.Gasmi@Mathematik.Uni-Magdeburg.DE*

Göb, Rainer Institut für Angewandte Mathematik und Statistik, Universität Würzburg, Sanderring 2, D–97070 Würzburg, Germany
e-Mail: *goeb@mathematik.uni-wuerzburg.de*

Gorlov, Valeri HTW Mittweida, Technikumplatz 17, D 09648 Mittweida, Germany

Heiligers, Berthold Fakultät für Mathematik, Institut für Mathematische Stochastik, Universität Magdeburg, D–39016 Magdeburg, Germany
e-mail: *Berthold.Heiligers@Mathematik.Uni-Magdeburg.DE*

Herberts, Tina Department of Stochastics, University of Ulm, 89069 Ulm, Germany
e-mail: *therbert@rz.uni-ulm.de*

Jensen, Uwe Department of Stochastics, University of Ulm, 89069 Ulm, Germany
e-mail: *uwe.jensen@mathematik.uni-ulm.de*

Jentsch, Roland HTW Mittweida, Technikumplatz 17, D 09648 Mittweida, Germany
e-mail: *rjentsch@htwm.de*

Kahle, Waltraud Fakultät für Mathematik, Otto–von–Guericke–Universität Magdeburg, Box 4120, 39016 Magdeburg, Germany
e-mail: *Waltraud.Kahle@Mathematik.Uni-Magdeburg.DE*

Kamps, Udo Department of Mathematics, University of Oldenburg, D-26111 Oldenburg, Germany,
e-mail: *kamps@mathematik.uni-oldenburg.de*

Kiesmueller, Gudrun Institut für Angewandte Mathematik und Statistik, Universität Würzburg, Sanderring 2, 97070 Würzburg, Germany
e-mail: *gudrun@mathematik.uni-wuerzburg.de*

Knoth, Sven Department of Statistics, Europe University Viadrina, P.B. 776, D-15207 Frankfurt(Oder), Germany
e-mail: *knoth@euv-frankfurt-o.de*

Kramer, Holger Department of Statistics, Europe University Viadrina, P.B. 776, D-15207 Frankfurt (Oder), Germany
e-mail: *schmid@euv-frankfurt-o.de*

Lehmann, Axel Fakultät für Mathematik, Otto–von–Guericke–Universität Magdeburg, D-39016 Magdeburg, Germany
e-mail: *axel.lehmann@mathematik.uni-magdeburg.de*

Liebner, Simone Fakultät für Mathematik, Otto–von–Guericke–Universität Magdeburg, Institut für Mathematische Stochastik, PF 4120, D–39016 Magdeburg, Germany
e-mail: *Simone.Liebner@Mathematik.Uni-Magdeburg.DE*

Magiera, Ryszard Technical University of Wrocław, Institute of Mathematics, Wybrzeże Wyspiańskiego 27, 50-370 Wrocław, Poland
e-mail: *magiera@graf.im.pwr.wroc.pl*

Offinger, Robert Fakultät für Mathematik, Otto–von–Guericke–Universität Magdeburg, Institut für Mathematische Stochastik, PF 4120, D–39016 Magdeburg, Germany
e-mail: *rooff@imst.math.uni-magdeburg.de*

Ruf, Jürgen Institut für Mathematische Stochastik, Fakultät für Mathematik, Universität Magdeburg, D–39016 Magdeburg, Germany
e-mail: *Juergen.Ruf@Mathematik.Uni-Magdeburg.DE*

Schmid, Wolfgang Department of Statistics, Europe University Viadrina, P.B. 776, D-15207 Frankfurt (Oder), Germany
e-mail: *schmid@euv-frankfurt-o.de*

Schreiber, Karina Fakultät für Mathematik, Otto–von–Guericke–Universität Magdeburg, Institut für Mathematische Stochastik, PF 4120, D–39016 Magdeburg, Germany
e-mail: *Karina.Schreiber@Mathematik.Uni-Magdeburg.DE*

Scholz, Fritz Boeing Shared Services Group, Research and Technology, P.O. Box 3707, MS 7L-22, Seattle WA 98124-2207,
e-mail: *fritz.scholz@boeing.com*

Sobiechowski, Carsten Institute of Mechanics, Otto–von–Guericke–Universität Magdeburg, Universitätsplatz 2, D–39106 Magdeburg, Germany,
e-mail: *carsten.sobiechowski@mb.uni-magdeburg.de*

Tittmann, Peter HTW Mittweida, Technikumplatz 17, D 09648 Mittweida, Germany
e-mail: *peter@htwm.de*

Vangel, Mark National Institute of Standards and Technology, Statistical Engineering Division, Building 820/Room 353, Gaithersburg, MD-20899-0001
e-mail: *vangel@cam.nist.gov*

von Collani, Elart Volkswirtschaftliches Institut, Universität Würzburg, Sanderring 2, 97070 Würzburg, Germany
e-mail: *collani@mathematik.uni-wuerzburg.de*

Wendt, Heide Fakultät für Mathematik, Otto–von–Guericke–Universität Magdeburg, Institut für Mathematische Stochastik, Postfach 4120, 39016 Magdeburg, Germany
e-mail: *heide.wendt @ mathematik.uni-magdeburg.de*

Zierke, Erik Fakultät für Mathematik, Otto–von–Guericke–Universität Magdeburg, Box 4120, 39016 Magdeburg, Germany
e-mail: *Erik.Zierke@Mathematik.Uni-Magdeburg.de*

Tables

Figures

PART I
Lifetime Analysis

The Generalized Discrete Linnik Distributions

Gerd Christoph and Karina Schreiber

Otto–von–Guericke–Universität Magdeburg, Magdeburg, Germany

Abstract: Discrete self-decomposable random variables play an important role in branching processes. Here the generalized discrete Linnik distributions, which are discrete self-decomposable and include the discrete stable as well as the discrete Mittag–Leffler distributions, are introduced and some properties of their probabilities are investigated. Further, the family of survival distributions of a generalized discrete Linnik distributed original observation subjected to a destructive process is determined.

Keywords and phrases: Discrete Linnik distributions, discrete stable and discrete Mittag–Leffler distributions, discrete self-decomposable, stable distributions, survival distributions

1.1 Introductory Example and Preliminaries

Consider a random branching process with discrete time and with a single type of particle. At the end of a unit of time each particle disappears with probability $0 < p \leq \frac{1}{2}$, is transformed into the same particle with probability $1 - 2p$, or is transformed into two particles with probability p. Denote by $Z_0(k)$ the number of particles in the $(k - 1)$-th generation that do not have descendants and disappear at the appearance of the k-th generation. Moreover, let $Z = \sum_{k=1}^{\infty} Z_0(k)$ be the total number of such particles that do not have descendants during the whole evolution process. This problem is a classic one and it was first given in Sevastyanov (1958) as an example for the occurrence of non-normal stable laws in limit theorems. Following Zolotarev (1986, p. 26), and using $\Gamma(x + \frac{1}{2}) = \sqrt{\pi}\, 2^{-2x+1}\, \Gamma(2x)/\Gamma(x)$ [see Abramowitz and Stegun

(1972, no. 6.1.18)] we find for $m = 1, 2, 3, \ldots$

$$P(Z = m) = \frac{\Gamma(m - 1/2)}{2\sqrt{\pi}\,\Gamma(m + 1)} = \frac{1}{2m - 1}\binom{2m}{m}2^{-2m} = (-1)^{m+1}\binom{1/2}{m}. \quad (1.1)$$

Hence the probability generating function (p.g.f.) $g_Z(z)$ of the random variable (r.v.) Z is given by

$$g_Z(z) = 1 - (1 - z)^{1/2} = z\,(1 + (1 - z)^{1/2})^{-1} = g_{M+1}(z), \quad |z| \leq 1, \quad (1.2)$$

where the latter defined r.v. M has a discrete Mittag–Leffler distribution, which is discrete self-decomposable, see Jayakumar and Pillai (1995). Steutel and van Harn (1979) introduced the discrete self-decomposability via p.g.f.: A non-negative integer valued r.v. X or a distribution function (d.f.) on $\mathbf{N}_0$ with p.g.f. $g(z)$ is called *discrete self-decomposable* if for every $\alpha \in (0, 1)$ there exists a r.v. X_α independent of X with p.g.f. $g_\alpha(z)$ such that

$$g(z) = g(1 - \alpha + \alpha z)\,g_\alpha(z), \quad |z| \leq 1, \quad (1.3)$$

holds. In terms of r.v.'s (1.3) has the following form: $X \stackrel{\mathrm{d}}{=} \alpha \odot X + X_\alpha$, where the p.g.f. of $\alpha \odot X$ is given by $g(1 - \alpha + \alpha z)$ and $\stackrel{\mathrm{d}}{=}$ denotes the equality in distribution. This means that $\alpha \odot X \stackrel{\mathrm{d}}{=} N_1 + \cdots + N_X$, where $X, N_1, N_2, \ldots$ are independent and $P(N_j = 0) = 1 - P(N_j = 1) = 1 - \alpha$.

Steutel and van Harn (1979) proved that a p.g.f. g is discrete self-decomposable iff it has the form

$$g(z) = \exp\left\{-\tau \int_z^1 \frac{1 - G(u)}{1 - u}\,\mathrm{d}u\right\}, \quad (1.4)$$

where $\tau > 0$ and G is a p.g.f. with $G(0) = 0$. The pair (τ, G) is unique.

Discrete self-decomposable r.v.'s also occur as limit distributions in branching processes with immigration, see van Harn, Steutel and Vervaat (1982), or Pakes (1995). Consider a pure death branching process with immigration starting with one particle (the ancestor). A particle dies with probability $1 - \mathrm{e}^{-t}$ and it survives with probability e^{-t}. If the p.g.f. of the immigration process is denoted by G and τ describes the intensity of immigration, then the p.g.f. of the limiting law as $t \to \infty$ is determined by equation (1.4). On the other hand this limiting law with p.g.f. (1.4) can be interpreted as a pure death branching process without immigration starting with a random number $X + Q$ of individuals and it is stopped at a random time T. Here T is exponential with intensity τ and the p.g.f.'s of X and Q are given by g and G, respectively. [See Steutel, Vervaat and Wolfe (1983) and Steutel and van Harn (1993).]

The purpose of this paper is to give some more properties of an important class of discrete self-decomposable r.v.'s, the generalized discrete Linnik distributions, which includes discrete Linnik, discrete Mittag–Leffler as well as the discrete stable distributions.

1.2 Assembling Discrete Linnik and Discrete Stable Distributions

A r.v. L is called *generalized discrete Linnik distributed with characteristic exponent $\gamma \in (0, 1]$, scale parameter $\lambda > 0$ and form parameter $\beta > 0$*, written $L \sim \mathrm{DL}(\gamma, \lambda, \beta)$, if it has the p.g.f.

$$
g_L(z) = \begin{cases} (1 + \lambda\,(1 - z)^\gamma/\beta)^{-\beta}, & \text{for } 0 < \beta < \infty, \\ \exp\{-\lambda\,(1 - z)^\gamma\}, & \text{for } \beta = \infty, \end{cases} \qquad |z| \le 1. \qquad (1.5)
$$

We use the notation "scale parameter λ" because of the fact that $\lambda^{-1/\gamma} \odot L \sim \mathrm{DL}(\gamma, 1, \beta)$ provided $L \sim \mathrm{DL}(\gamma, \lambda, \beta)$. Remember that $g_{\alpha \odot L}(z) = g_L(1 - \alpha + \alpha\,z)$.

The *discrete stable distributions* introduced by Steutel and van Harn (1979) occur in our definition (1.5) if $\beta = \infty$ as a natural generalization of the *discrete Linnik distribution* with p.g.f. $(1 + c\,(1 - z)^\gamma)^{-\beta}$ if $\beta < \infty$ investigated by Devroye (1993) for $c = 1$ and by Pakes (1995) for $c > 0$. We denote $\mathrm{DS}(\gamma, \lambda) = \mathrm{DL}(\gamma, \lambda, \infty)$ and write $X \sim \mathrm{DS}(\gamma, \lambda)$ if the r.v. X is discrete stable with the given parameters.

In the case $\beta = 1$ we get the p.g.f. of the *discrete Mittag–Leffler distribution*, see Pillai (1990) and Jayakumar and Pillai (1995). We denote $\mathrm{DML}(\gamma, \lambda) = \mathrm{DL}(\gamma, \lambda, 1)$ and write $M \sim \mathrm{DML}(\gamma, \lambda)$ if the r.v. M is discrete Mittag–Leffler distributed with the given parameters.

In the special case of $\gamma = 1$ some well-known distributions are recovered: the Poisson(λ) distribution if $\beta = \infty$, the negative binomial distribution with parameters $p = \beta/(\lambda + \beta)$ and β and p.g.f. $g(z) = (p/(1 - (1 - p)z))^\beta$ if $\beta < \infty$, and the geometric distribution with $p = (1 + \lambda)^{-1}$ and p.g.f. $g(z) = p/(1 - (1 - p)z)$ if $\beta = 1$. [See Figure 1.1.]

The so called Sibuya distribution plays a crucial role in investigating $\mathrm{DL}(\gamma, \lambda, \beta)$ distributions. Consider a sequence of independent Bernoulli trials, where the k-th trial successfully ends with probability γ/k. Then the random number Z of trials required to achieve the first success possesses the following probabilities:

$$
P(Z = k) = (-1)^{k+1} \binom{\gamma}{k}, \quad k = 1, 2, 3, \ldots; \quad \gamma \in (0, 1]. \qquad (1.6)
$$

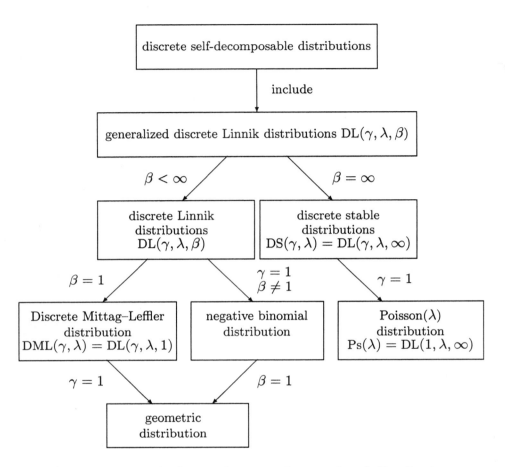

Figure 1.1: Relations between the mentioned distributions

A r.v. with these probabilities is called *Sibuya(γ)* distributed [cf. Devroye (1993)]. It follows from (1.1) and (1.2) that in the example given at the beginning we obtained $Z \sim$ Sibuya($^1\!/_2$) and $M \sim$ DML($^1\!/_2, 1$). Note that the Sibuya(1) distribution is degenerate with $P(Z = 1) = 1$. For the p.g.f. of a Sibuya(γ) distributed r.v. Z we get

$$g_Z(z) = 1 - (1 - z)^\gamma, \qquad \gamma \in (0, 1], \quad |z| \le 1.$$

In the following we need some representations of the introduced generalized discrete Linnik distribution. For that purpose let $Z, Z_1, Z_2, \ldots$ be a sequence of independent and identically Sibuya(γ) distributed r.v.'s and let Y be a Poisson(λ) distributed r.v. which is independent of Z_i.

Devroye (1993) established that for a discrete stable r.v. X the relation

$$X \stackrel{\mathrm{d}}{=} Z_1 + \cdots + Z_Y \tag{1.7}$$

holds, where $Z_1 + \cdots + Z_Y = 0$ if $Y = 0$.

A similar result is valid for the generalized discrete Linnik distribution:

Theorem 1.2.1 *Suppose $0 < \beta < \infty$. Let the r.v. N be negative binomial distributed with parameters β and p, $p \in (0,1)$, and let $Z, Z_1, Z_2, \ldots$ be independent and identically Sibuya(γ) distributed r.v.'s independent of N. Then the r.v. $L \sim DL(\gamma, \beta(1-p)/p, \beta)$ can be obtained as*

$$L \overset{\mathrm{d}}{=} Z_1 + \cdots + Z_N, \tag{1.8}$$

where $Z_1 + \cdots + Z_N = 0$ if $N = 0$.

PROOF. For the p.g.f. of the random sum we obtain:

$$
\begin{aligned}
g_{Z_1 + \cdots + Z_N}(z) &= g_N(g_Z(z)) \\
&= \left(\frac{1}{1 + \lambda(1-z)^\gamma/\beta} \right)^\beta = g_L(z)
\end{aligned}
$$

with $\lambda = \beta(1-p)/p$. ∎

Taking $p = \beta/(\lambda + \beta)$, we have for the p.g.f. of the negative binomial distribution

$$g_N(z) = \left(1 + \frac{\lambda}{\beta}(1-z) \right)^{-\beta} \longrightarrow e^{-\lambda(1-z)} = g_Y(z) \quad \text{as } \beta \to \infty.$$

Hence, the representation of the discrete stable r.v. (1.7) also follows from (1.8) with $p = \beta/(\lambda + \beta)$ as $\beta \to \infty$.

In the special case $\beta = 1$ relation (1.8) characterizes the DML(λ, γ) distributed r.v. as a random sum with geometric distributed summation index.

Another representation of a DL(γ, λ, β) distributed r.v. makes use of a positive r.v. S_γ^λ with Laplace transform

$$\psi_\gamma(z; \lambda) = e^{-\lambda z^\gamma}, \ 0 < \gamma \leq 1, \ \lambda > 0.$$

Note that S_γ^λ with $0 < \gamma < 1$ is (strictly) stable with characteristic exponent γ, skewness parameter 1 and scale parameter $\lambda > 0$, whereas S_1^λ has a degenerate distribution with $P(S_1^\lambda = \lambda) = 1$. Devroye (1993) detected that a discrete stable r.v. with exponent γ and parameter λ is distributed as a Poisson r.v. with random parameter S_γ^λ.

By modifying the Devroye (1993) representation of DL(γ, β, β) distributions we get (see also Pakes (1995)):

Lemma 1.2.1 *Suppose $L \sim DL(\gamma, \lambda, \beta)$. Then L is Poisson distributed with random parameter $(G_\beta)^{1/\gamma} S_\gamma^\lambda$, where G_β is Gamma(β, β) distributed with Laplace transform $\psi_{G_\beta}(z) = (1 + z/\beta)^{-\beta}$ if $\beta < \infty$ and independent of S_γ^λ or $P(G_\infty = 1) = 1$.*

PROOF. It only remains to prove the case $\beta < \infty$. We obtain for the p.g.f.

$$
\begin{aligned}
E(z^L) &= E\left(e^{-G_\beta \lambda (1-z)^\gamma}\right) \\
&= \psi_{G_\beta}(\lambda (1-z)^\gamma) = (1 + \lambda (1-z)^\gamma / \beta)^{-\beta}.
\end{aligned}
$$

∎

Note that $\psi_{G_\beta}(z) \to e^{-z} = \psi_{G_\infty}(z)$ as $\beta \to \infty$.

In the case $\beta = 1$ we find that $M \sim \mathrm{DML}(\gamma, \lambda)$ has a Poisson distribution with random parameter $(G_1)^{1/\gamma} S_\gamma^\lambda$, where G_1 has a standard exponential distribution.

Finally, let us mention that the generalized discrete Linnik distribution is discrete self-decomposable because g_L may be given in the form (1.4) and the pair (τ, G) is determined by $(\gamma \lambda \beta / (\lambda + \beta), g_Z g_M)$ if $\beta < \infty$, where $Z \sim$ Sibuya(γ) and $M \sim \mathrm{DML}(\gamma, \lambda/\beta)$, which leads to $(\lambda \gamma, g_Z)$ if $\beta = \infty$.

1.3   Calculation of Probabilities

In Jayakumar and Pillai (1995) and Christoph and Schreiber (1997) recursion formulae for the probabilities of discrete Mittag–Leffler distributed and discrete stable r.v.'s are shown. Here we want to present explicit formulae for probabilities of the generalized discrete Linnik distributions.

First let us consider a r.v. $L \sim \mathrm{DL}(\gamma, \lambda, \beta)$, $0 < \beta < \infty$. Expanding the p.g.f. in a power series we obtain for $k = 0, 1, 2, \ldots$

$$
P(L = k) = (-1)^k \sum_{j=0}^{\infty} \binom{\gamma j}{k} \binom{j + \beta - 1}{j} \left(-\frac{\lambda}{\beta}\right)^j. \tag{1.9}
$$

Doing the same with the p.g.f. of a $\mathrm{DS}(\gamma, \lambda)$ distributed r.v. X (here $\beta = \infty$) we get for the probabilities for $k = 0, 1, 2, \ldots$

$$
P(X = k) = (-1)^k \sum_{j=0}^{\infty} \binom{\gamma j}{k} \frac{(-\lambda)^j}{j!}.
$$

The following theorem shows a representation of these probabilities with finite sums.

Theorem 1.3.1 *Suppose $L \sim DL(\gamma, \lambda, \beta)$, $0 < \beta < \infty$, and $X \sim DS(\gamma, \lambda)$. Put $\gamma \in (0, 1]$ and $p = \beta/(\lambda + \beta)$. Then for $k = 0, 1, 2, \ldots$ the probabilities are given by*

$$
P(L = k) = (-1)^k p^\beta \sum_{m=0}^{k} \sum_{j=0}^{m} \binom{m}{j} \binom{\gamma j}{k} \binom{-\beta}{m} (-1)^j (p-1)^m \tag{1.10}
$$

and

$$P(X = k) = (-1)^k \, e^{-\lambda} \sum_{m=0}^{k} \sum_{j=0}^{m} \binom{m}{j} \binom{\gamma j}{k} (-1)^j \frac{\lambda^m}{m!}. \qquad (1.11)$$

PROOF. From the definition of DL(γ, λ, β) and DS(γ, λ) distributed r.v.'s by a random sum of Sibuya(γ) distributed r.v.'s we obtain by conditioning on N and Y for $k = 0, 1, 2, \ldots$

$$P(L = k) = \sum_{m=0}^{k} P(Z_1 + \cdots + Z_m = k) \binom{-\beta}{m} (p-1)^m p^{\beta}$$

and

$$P(X = k) = \sum_{m=0}^{k} P(Z_1 + \cdots + Z_m = k) \frac{\lambda^m}{m!} e^{-\lambda},$$

where $P(Z_1 + \cdots + Z_m = k) = 0$ if $m > k$ since $Z \geq 1$. For the calculation of $P(Z_1 + \cdots + Z_m = k)$ for $m \leq k$ we expand the p.g.f. in a power series

$$g_{Z_1 + \cdots + Z_m}(z) = (1 - (1-z)^{\gamma})^m = \sum_{k=0}^{\infty} \sum_{j=0}^{m} \binom{m}{j} \binom{\gamma j}{k} (-1)^{k+j} z^k.$$

Hence

$$P(Z_1 + \cdots + Z_m = k) = (-1)^k \sum_{j=0}^{m} \binom{m}{j} \binom{\gamma j}{k} (-1)^j, \quad m \leq k,$$

which completes the proof of (1.10) and (1.11). ∎

Remark 1.3.1 The probabilities given in (1.10) and (1.11) can be transformed in

$$P(L = k) = (-1)^k \sum_{j=0}^{k} \binom{\gamma j}{k} \binom{-\beta}{j} (1-p)^j \, p^{-j} \, P(N \leq k - j),$$

where N denotes a negative binomial r.v. with parameters $\beta + j$ and p, and

$$P(X = k) = (-1)^k \sum_{j=0}^{k} \binom{\gamma j}{k} \frac{(-\lambda)^j}{j!} \, P(Y \leq k - j),$$

where $Y \sim \mathrm{Ps}(\lambda)$, $k = 0, 1, 2, \ldots$.

Another representation for the probabilities of the discrete Linnik distribution shows the following:

Theorem 1.3.2 *If* $L \sim DL(\gamma, \lambda, \beta)$, $0 < \beta < \infty$, *then* $P(L = 0) = (1 + \lambda/\beta)^{-\beta}$ *and*

$$P(L = k) = (-1)^k \left(1 + \frac{\lambda}{\beta}\right)^{-\beta} \sum \binom{-\beta}{s} s! \left(\frac{\lambda}{\beta + \lambda}\right)^s \prod_{m=1}^{k} \frac{1}{v_m!} \binom{\gamma}{m}^{v_m} \qquad (1.12)$$

for $k = 1, 2, 3, \ldots$, where the summation is carried out over all non-negative integer solutions $(v_1, v_2, \ldots, v_k)$ of the equation $v_1 + 2\,v_2 + \cdots + k\,v_k = k$ and $s = v_1 + v_2 + \cdots + v_k$.

PROOF. We make use of

$$P(L = k) = \frac{1}{k!} \frac{\mathrm{d}^k}{\mathrm{d}z^k}\, g_L(z)\big|_{z=0} = \frac{1}{k!} \frac{\mathrm{d}^k}{\mathrm{d}z^k}\, (1 + \lambda\,(1 - z)^\gamma/\beta)^{-\beta}\big|_{z=0},$$

where the derivatives are calculated by Lemma 5.6 of Petrov (1995, p. 170). ∎

Letting $\beta \to \infty$ we get the same formula for the probabilities of discrete stable r.v.'s as in Christoph and Schreiber (1997).

Using any of the equations (1.9), (1.10) or (1.12) we find for the first probabilities of a DL(γ, λ, β) r.v. L

$$P(L = 0) = \left(1 + \frac{\lambda}{\beta}\right)^{-\beta}, \quad P(L = 1) = \lambda\,\gamma \left(1 + \frac{\lambda}{\beta}\right)^{-\beta-1} \quad \text{and}$$

$$P(L = 2) = \frac{1}{2} \left(1 + \frac{\lambda}{\beta}\right)^{-\beta-2} \left[\frac{(\beta + 1)\,\lambda^2\,\gamma^2}{\beta} - \lambda\,\gamma\,(\gamma - 1)\left(1 + \frac{\lambda}{\beta}\right)\right].$$

In the special case of $\beta = \infty$ we obtain

$$P(X = 0) = \mathrm{e}^{-\lambda}, \quad P(X = 1) = \gamma\,\lambda\,\mathrm{e}^{-\lambda} \quad \text{and}$$

$$P(X = 2) = \frac{1}{2}\,\mathrm{e}^{-\lambda}\,(\lambda^2\,\gamma^2 - \lambda\,\gamma\,(\gamma - 1)).$$

Note that all formulae for the probabilities of the discrete stable r.v. X can also be obtained from the formulae for the probabilities of the generalized discrete Linnik r.v. L by taking the limits $\beta \to \infty$.

In Figures 1.2–1.4 probabilities and distribution functions of discrete stable r.v.'s are shown for various γ and $\lambda = 5$. For the d.f. of a DML one obtains similar figures as in the discrete stable case.

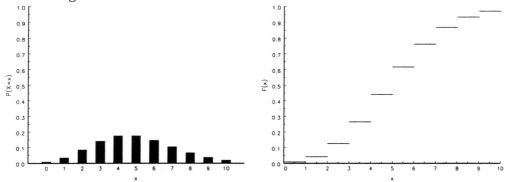

Figure 1.2: Probabilities and d.f. of DS$(1, 5)$ r.v.

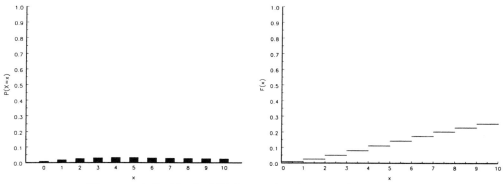

Figure 1.3: Probabilities and d.f. of DS(0.5, 5) r.v.

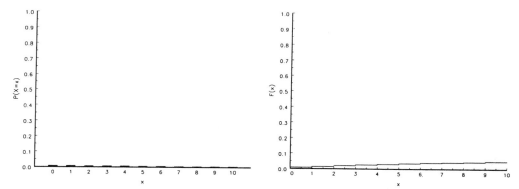

Figure 1.4: Probabilities and d.f. of DS(0.2, 5) r.v.

1.4 Characterization via Survival Distributions

In Fosam (1995) characterizations of discrete self-decomposable r.v.'s are given by survival distributions. For this purpose a random vector (V, U) of non-negative integer valued r.v.'s is considered, such that the original observation V produced by nature is subjected to a destructive process and we observe the undamaged part U of the original observation. The probability of reduction from the original observation $\{V = n\}$ to k undamaged items due to the destructive model is given by the survival distribution $P(U = k | V = n)$. In the following we suppose that the original observation is distributed according to a discrete self-decomposable law with p.g.f. g having the property (1.3). Then Fosam (1995) established the following lemma:

Lemma 1.4.1 *Let V be a non-negative integer valued r.v. with probabilities $\{p_k\}$ and $p_0 > 0$. It is discrete self-decomposable iff for every $\alpha \in (0, 1)$ there* *exists a family of survival distributions $P(U_\alpha = k|V = n)$ such that for every $n \in \mathbf{N}_0$ for which $p_n > 0$*

$$P(U_\alpha = k|V = n) = \frac{a_k^{(\alpha)} b_{n-k}^{(\alpha)}}{p_n} \qquad \text{for } k = 0, \dots, n; \; n \in \mathbf{N}_0, \qquad (1.13)$$

where $\{b_k^{(\alpha)}\}$ is some non-negative sequence and

$$a_k^{(\alpha)} = \alpha^k \sum_{j=k}^{\infty} (1-\alpha)^{j-k} \binom{j}{k} p_j. \qquad (1.14)$$

In view of this Lemma we refer that U_α denotes the observation of the undamaged part of the original observation V for a special $\alpha \in (0,1)$. Let us note that the sequences $\{a_k^{(\alpha)}\}$ and $\{b_k^{(\alpha)}\}$ are the probabilities corresponding to the p.g.f.s $g(1 - \alpha + \alpha s)$ and $g_\alpha(s)$ in equation (1.3), respectively.

Using (1.13) it is easy to see that

$$P(U_\alpha = k) = \sum_{j=k}^{\infty} P(U_\alpha = k|V = j) \, P(V = j) = \sum_{j=k}^{\infty} a_k^{(\alpha)} b_{j-k}^{(\alpha)} = a_k^{(\alpha)},$$

i.e., $\{a_k^{(\alpha)}\}$ determines the marginal distribution of U_α.

Now we want to determine the families of survival distributions if the original observation is discrete stable and discrete Linnik distributed, respectively.

First, let the original observation X be discrete stable distributed with parameters $\gamma \in (0,1]$ and $\lambda > 0$, which means that

$$p_k = P(X = k) = (-1)^k \sum_{j=0}^{\infty} \binom{\gamma j}{k} \frac{(-\lambda)^j}{j!}.$$

With some computations we get from (1.14)

$$a_k^{(\alpha)} = (-1)^k \sum_{j=0}^{\infty} \binom{\gamma j}{k} \frac{(-\lambda \alpha^\gamma)^j}{j!}. \qquad (1.15)$$

Hence, the sequence $\{a_k^{(\alpha)}\}$ determines the probabilities of a discrete stable distributed r.v. X^* with parameters γ and $\lambda \alpha^\gamma$. Since the sequence $\{b_k^{(\alpha)}\}$ gives the probabilities corresponding to the p.g.f. g_α in equation (1.3) we get

$$b_k^{(\alpha)} = (-1)^k \sum_{j=0}^{\infty} \binom{\gamma j}{k} \frac{(-\lambda(1-\alpha^\gamma))^j}{j!}, \qquad (1.16)$$

i.e., the sequence $\{b_k^{(\alpha)}\}$ defines the probabilities of a discrete stable r.v. X^{**} with parameters γ and $\lambda(1 - \alpha^\gamma)$. By the construction of the sequences $\{a_k^{(\alpha)}\}$ and $\{b_k^{(\alpha)}\}$ it follows that $p_n = \sum_{k=0}^{n} a_k^{(\alpha)} b_{n-k}^{(\alpha)}$. Now we get a characterization of discrete stable laws as a consequence of Theorem 1 of Shanbhag (1977):

Corollary 1.4.1 *If X and U_α are non-negative integer-valued r.v.'s such that*

$$P(U_\alpha = k, X = n) = a_k^{(\alpha)}\, b_{n-k}^{(\alpha)}, \quad k = 0, \ldots, n;\ \ n \in \mathbf{N}_0,$$

where $a_k^{(\alpha)}$ and $b_k^{(\alpha)}$ are determined by equations (1.15) and (1.16), respectively, and let $p_n = \sum\limits_{k=0}^{n} a_k^{(\alpha)} b_{n-k}^{(\alpha)}$, $0 < p_0 < 1$, be a discrete probability distribution and α be some fixed number in $(0,1)$ then

$$P(U_\alpha = k) = P(U_\alpha = k \,|\, X = U_\alpha), \quad k = 0, 1, 2, \ldots,$$

iff X is discrete stable distributed.

In the case of $\gamma = 1$ we get the result of Srivastava and Srivastava (1970): The destructive model is a binomial(n, α) distribution, i.e., $P(U_\alpha = k \,|\, X = n) = \binom{n}{k} \alpha^k (1 - \alpha)^{n-k}$, iff the original observation X is distributed according to a Poisson(λ) law. Then the observation of the undamaged part U_α is Poisson$(\lambda\,\alpha)$ distributed.

Now we consider an original observation L produced by nature which is discrete Linnik distributed with parameters $\gamma \in (0,1]$, $\lambda > 0$ and $0 < \beta < \infty$, i.e., $L \sim \mathrm{DL}(\gamma, \lambda, \beta)$, and let this original observation be subjected to a destructive process. Because of Lemma 1.4.1 and the discrete self-decomposability of discrete Linnik distributions there exists a family of survival distributions $P(U_\alpha = k \,|\, L = n)$ for every $\alpha \in (0,1)$. The marginal distribution of the undamaged part U_α is $\mathrm{DL}(\gamma, \alpha^\gamma \lambda, \beta)$ distributed, which follows from (1.14) and (1.9) with some calculus:

$$
\begin{aligned}
a_k^{(\alpha)} &= \alpha^k \sum_{j=k}^{\infty} (1 - \alpha)^{j-k} \binom{j}{k} (-1)^j \sum_{l=0}^{\infty} \binom{\gamma\,l}{j} \binom{l + \beta - 1}{l} \left(-\frac{\lambda}{\beta}\right)^l \\
&= (-1)^k \sum_{l=0}^{\infty} \binom{\gamma\,l}{k} \binom{l + \beta - 1}{l} \left(-\frac{\alpha^\gamma \lambda}{\beta}\right)^l.
\end{aligned}
$$

Since $\{b_k^{(\alpha)}\}$ is the sequence of probabilities due to g_α in (1.3) and expanding the p.g.f.ś of $g_L(z)$ and $g_L(1 - \alpha + \alpha\,z)$ in series we find:

$$
\begin{aligned}
\sum_{k=0}^{\infty} b_k^{(\alpha)} z^k &= \left(\frac{1 + \alpha^\gamma \lambda\,(1 - z)^\gamma/\beta}{1 + \lambda\,(1 - z)^\gamma/\beta}\right)^\beta \\
&= \sum_{j=0}^{\infty} \binom{\beta}{j} \left(\frac{\lambda}{\beta}\right)^j \alpha^{\gamma j} (1 - z)^{\gamma j} \sum_{l=0}^{\infty} \binom{-\beta}{l} \left(\frac{\lambda}{\beta}\right)^l (1 - z)^{\gamma l} \\
&= \sum_{k=0}^{\infty} (-z)^k \sum_{j=0}^{\infty} \binom{\gamma\,j}{k} \left(\frac{\lambda}{\beta}\right)^j \sum_{l=0}^{j} \binom{\beta}{l} \binom{-\beta}{j - l} \alpha^{\gamma l}.
\end{aligned}
$$

Comparing the coefficients we obtain:

$$b_0^{(\alpha)} = \left(\frac{1 + \alpha^\gamma \lambda/\beta}{1 + \lambda/\beta}\right)^\beta$$

and for $k = 1, 2, 3, \ldots$

$$b_k^{(\alpha)} = (-1)^k \sum_{j=0}^{\infty} \binom{\gamma j}{k} \left(\frac{\lambda}{\beta}\right)^j \sum_{l=0}^{j} \binom{\beta}{l} \binom{-\beta}{j-l} \alpha^{\gamma l}.$$

By taking the limit $\beta \to \infty$ we get formula (1.16).

In the case of $\gamma = 1$ and $\beta = 1$, i.e., the original observation L is geometric distributed, we obtain with $p = 1/(1 + \lambda)$

$$a_k^{(\alpha)} = \left(\frac{\alpha (1 - p)}{p + \alpha (1 - p)}\right)^k \frac{p}{p + \alpha (1 - p)}$$

and

$$b_k^{(\alpha)} = \begin{cases} p + \alpha (1 - p) & \text{for } k = 0, \\ p (1 - \alpha) (1 - p)^k & \text{for } k = 1, 2, \ldots \end{cases}$$

Now we get a characterization of the geometric distribution as a consequence of Theorem 1 by Shanbhag (1977):

Corollary 1.4.2 *Let (L, U_α) be a random vector of non-negative integer valued components such that*

$$P(U_\alpha = k, L = n) = p_n \, P(U_\alpha = k|L = n), \quad k = 0, \ldots, n; \; n \in \mathbf{N}_0,$$

where $\{p_n\}$ is a discrete probability distribution. If $0 < p_0 < 1$ and

$$P(U_\alpha = k|L = n) = \begin{cases} (1 - p^*)^k \, p^* & \text{for } k = 0, \ldots, n - 1, \\ (1 - p^*)^n & \text{for } k = n, \end{cases}$$

with $p^* = p(1 - \alpha)/(p + \alpha (1 - p))$, p any fixed number in $(0, 1)$, then

$$P(U_\alpha = k) = P(U_\alpha = k|L = U_\alpha), \quad k = 0, 1, \ldots,$$

iff the distribution $\{p_n\}$ is geometric with parameter p, i. e., $p_n = (1-p)^n \, p$, $n \geq 0$.

1.5 Asymptotic Behaviour

In Section 1.3 we have seen that the probabilities $P(L = n)$ of the generalized discrete Linnik distribution (with the special cases of discrete stable and discrete Mittag–Leffler) cannot be given in a short form. Moreover, often the asymptotic behaviour of $P(L = n)$ for large n is of interest.

Now let us consider a non-negative integer valued r.v. T, which is Poisson distributed with random parameter $U^{1/\gamma} S_\gamma^\lambda$, $\gamma \in (0,1)$, where U denotes a positive r.v., S_γ^λ is a strictly stable r.v. with characteristic exponent γ, skewness parameter 1 and scale parameter $\lambda > 0$ and the r.v.'s S_γ^λ and U are independent. We write $T \sim \mathrm{Ps}(U^{1/\gamma} S_\gamma^\lambda)$. As special cases we find

- $T \sim \mathrm{DL}(\gamma, \lambda, \beta)$ if $U \sim \mathrm{Gamma}(\beta, \beta)$,
- $T \sim \mathrm{DML}(\gamma, \lambda)$ if $U \sim \mathrm{Exp}(1)$ and
- $T \sim \mathrm{DS}(\gamma, \lambda)$ if $P(U = 1) = 1$.

Let $T \sim \mathrm{Ps}(U^{1/\gamma} S_\gamma^\lambda)$ and U be an arbitrary non-negative r.v. having all moments. Denote the moments of U by $m_k^U = E(U^k)$, $k = 1, 2, \ldots$, and $F_U(x) = P(U \leq x)$. Then it is possible to present the p.g.f. of T in a power series with respect to powers of $(1 - z)^\gamma$:

$$g_T(z) = \int_{\mathbf{R}} e^{-\lambda u (1-z)^\gamma} \, dF_U(u) = 1 - \sum_{k=1}^\infty d_k (1 - z)^{\gamma k},$$

where $d_k = \dfrac{(-1)^{k+1}}{k!} \lambda^k m_k^U$.

Theorem 1.5.1 *Let $T \sim Ps(U^{1/\gamma} S_\gamma^\lambda)$ with $\gamma \in (0,1)$ and $\lambda > 0$, where U is an arbitrary positive r.v. with $E(U^m) < \infty$ for some fixed $m \in \mathbf{N}$. Then we have*

$$P(T = n) = \frac{1}{\pi} \sum_{j=1}^m d_j \sin(\gamma j \pi) B(\gamma j + 1, n - \gamma j) + O(n^{-\gamma(m+1)-1}) \quad (1.17)$$

for $n \to \infty$, where $B(x,y) = \Gamma(x) \Gamma(y)/\Gamma(x+y)$ is the beta function.

PROOF. Define p_γ^λ as the density of S_γ^λ. Since T is Poisson with the given random parameter we obtain

$$P(T = n) = \int_{\mathbf{R}} \int_0^\infty \frac{(u^{1/\gamma} s)^n}{n!} e^{-u^{1/\gamma} s} p_\gamma^\lambda(s) \, ds \, dF_U(u) . \quad (1.18)$$

Let n be large and m be fixed with $m + 1 < n$. Using the representation [see e.g. Zolotarev (1986, p. 90) or Christoph and Wolf (1993, p. 13)]

$$p_\gamma^\lambda(s) = \frac{1}{\pi} \sum_{j=1}^m \frac{(-1)^{j+1}}{j!} \lambda^j \, \Gamma(\gamma\, j + 1) \, \sin(\gamma\, j\, \pi) \, s^{-\gamma j - 1} + A_m(s), \qquad (1.19)$$

for $\gamma \in (0, 1)$ and any fixed $m \geq 1$, where $A_m(s) = O(s^{-\gamma(m+1)-1})$ as $s \to \infty$, and putting (1.19) in (1.18) all integrals involved are gamma functions. Stirling's formula leads to

$$\frac{\Gamma(n - \gamma\, j)}{\Gamma(n + 1)} = n^{-\gamma j - 1} \left(1 + \frac{(\gamma\, j)^2 + \gamma\, j}{2\, n} + O(n^{-2}) \right) \qquad (1.20)$$

for any fixed j as $n \to \infty$. Using (1.20) with $j = m + 1$ for the remaining term, we get (1.17). ∎

Remark 1.5.1 Using (1.20) to replace the beta function in (1.17) we find

$$P(T = n) = \frac{1}{\pi} \sum_{j=1}^{[(\gamma+1)/\gamma]} d_j \, \Gamma(\gamma\, j + 1) \sin(\gamma\, j\, \pi) \, n^{-\gamma j - 1} + O(n^{-\gamma - 2}).$$

For the coefficients d_j we obtain in our special cases:

- $T \sim \mathrm{DL}(\gamma, \lambda, \beta):\quad d_j = (-1)^{j+1} \left(\frac{\lambda}{\beta} \right)^j \binom{j + \beta - 1}{j},$

- $T \sim \mathrm{DML}(\gamma, \lambda):\quad d_j = (-1)^{j+1} \lambda^j,$

- $T \sim \mathrm{DS}(\gamma, \lambda):\qquad d_j = (-1)^{j+1} \dfrac{\lambda^j}{j!}.$

We see that $d_1 = \lambda$ in all cases. This shows that all these distributions belong to the domain of attraction of a strictly stable r.v. S_γ^λ with characteristic exponent γ, scale parameter λ and skewness parameter 1. With (1.6) and some calculus we find for a Sibuya(γ) distributed r.v. Z

$$
\begin{aligned}
P(Z = n) &= \frac{1}{\pi} \sin(\gamma\, \pi) \, B(\gamma + 1, n - \gamma) \\
&= \frac{1}{\pi} \Gamma(\gamma + 1) \sin(\gamma\, \pi) \, n^{-\gamma - 1} + O(n^{-\gamma - 2}) \quad \text{as } n \to \infty,
\end{aligned}
$$

where the equation in the first line is exact for all $n \geq 1$. Hence, the Sibuya(γ) distribution belongs to the domain of attraction of S_γ^1, too.

At the beginning of the paper we have seen that the probabilities of a DML($1/2$, 1) distributed r.v. M are available in a closed form, that is

$$P(M = k) = (-1)^k \binom{1/2}{k + 1}, \qquad k = 0, 1, \ldots .$$

In Pillai (1990) and Jayakumar and Pillai (1995) the discrete Mittag–Leffler distribution was defined as a discrete analogue of the (continuous) Mittag–Leffler distribution having the d.f.

$$F_\gamma(x) = \sum_{k=1}^{\infty} \frac{(-1)^{k-1} x^{k\gamma}}{\Gamma(1 + k\gamma)} I_{[0,\infty)}(x)$$

with Laplace transform $\psi_\gamma(z) = (1 + z^\gamma)^{-1}$ for $\mathrm{Re}(z) \geq 0$, where $I_{[a,b]}$ denotes the indicator function. It is easy to verify that in the case of $\gamma = 1$ we get the standard exponential distribution. In the case of $\gamma = \frac{1}{2}$ the continuous Mittag–Leffler distribution may be given also in a closed form. Splitting $F_{1/2}$ into two sums of even and odd k and using $\sum_{k=0}^{\infty} (\Gamma(k + 3/2))^{-1} u^{2k+1} = \exp(u^2)\,\mathrm{erf}(u)$ [see Prudnikov, Brychkov and Marichev (1986, vol. 1, no. 5.2.7.18)] we obtain

$$F_{1/2}(x) = 1 - \mathrm{e}^x\,(1 - \mathrm{erf}(\sqrt{x})), \quad x \in [0, \infty),$$

where $\mathrm{erf}(.)$ denotes the error function.

Acknowledgement. The authors take this opportunity to thank Prof. R. Theodorescu for his helpful comments and discussions which led to several improvements.

References

1. Abramowitz, M. and Stegun, I. (1972). *Handbook of Mathematical Functions with Formulas, Graphs and Mathematical Tables*, National Bureau of Standards Applied Mathematical Series, 55.

2. Christoph, G. and Schreiber, K. (1997). Discrete stable random variables, Statistics & Probability Letters (to appear).

3. Christoph, G. and Wolf, W. (1993). *Convergence Theorems with a Stable Limit Law*, Mathematical Research, Volume 70, Berlin: Akademie–Verlag.

4. Devroye, L. (1993). A triptych of discrete distributions related to the stable law, *Statistics & Probability Letters*, **18**, 349–351.

5. Fosam, E. B. (1995). A characterization of discrete self-decomposable distributions in terms of survival distributions and the self-decomposability of the discrete logarithmic distribution, *Sankhyā*, **57**, 337–341.

6. van Harn, K., Steutel, F. W. and Vervaat, W. (1982). Self-Decomposable discrete distributions and branching processes, *Z. Wahrsch. verw. Gebiete*, **61**, 97–118.

7. Jayakumar, K. and Pillai, R. N. (1995). Discrete Mittag–Leffler distribution, *Statistics & Probability Letters*, **23**, 271–274.

8. Pakes, A. G. (1995). Characterization of discrete laws via mixed sums and Markov branching processes, *Stochastic Processes and their Applications*, **55**, 285–300.

9. Petrov, V. V. (1995). *Limit Theorems of Probability Theory: Sequences of independent random variables*, Oxford: Clarendon Press.

10. Pillai, R. N. (1990). On Mittag–Leffler functions and related distributions, *Annals of the Institute of Statistical Mathematics*, **42**, 157–161.

11. Prudnikov, A. P., Brychkov, Yu. A. and Marichev, O. I. (1986). *Integrals and Series*, Volume 1, Newark, NJ: Gordon and Breach.

12. Sevastyanov, B. A. (1958). Branching stochastic processes for particles diffusing in a restricted domain with absorbing boundaries, *Theory of Probability and its Applications*, **3**, 121–136.

13. Shanbhag, D. N. (1977). An extension of the Rao–Rubin characterization of the Poisson distribution, *Journal of Applied Probability*, **14**, 640–646.

14. Srivastava, R. C. and Srivastava, A. B. L. (1970). On a characterization of Poisson distribution, *Journal of Applied Probability*, **7**, 497–501.

15. Steutel, F. W. and van Harn, K. (1979). Discrete analogues of self-decomposability and stability, *Annals of Probability*, **7**, 893–899.

16. Steutel, F. W., Vervaat, W. and Wolfe, S. J. (1983). Integer–valued branching processes with immigration, *Advances in Applied Probability*, **15**, 713–725.

17. Steutel, F. W. and van Harn, K. (1993). Stability equations for processes with stationary independent increments using branching processes and Poisson mixtures, *Stochatic Processes and their Applications*, **45**, 209–230.

18. Zolotarev, V. M. (1986). *One–dimensional Stable Distributions*, Translations of Mathematical Monographs, Volume 65.

Acceptance Regions and Their Application in Lifetime Estimation

Klaus Dräger

Institut für Angewandte Mathematik & Statistik, Würzburg, Germany

Abstract: In the first part of this paper, a general method is proposed for determining confidence bounds based on so-called acceptance regions. This method can be applied, if the observed random variables are discrete and may adopt at most a finite number of realizations.

This concept has the following advantages:

- Contrary to asymptotic confidence bounds, the inclusion probability of the confidence bounds based on acceptance regions is never less than the given confidence level.

- The confidence bounds based on acceptance regions may be determined in a way that they are optimal with respect to any quality indicator which may be chosen out of a large class of quality indicators including, e.g., all convex combinations of the realizations of the confidence bounds.

The second part of this paper is an application of this concept to a problem of lifetime estimation, namely the determination of a lower confidence bound for the expectation of a Weibull distribution based on a left- and right-censored sample.

Keywords and phrases: Acceptance regions, optimal confidence bound, left- and right-censored sample

2.1 Confidence Bounds Based on Acceptance Regions

In this section, the determination and investigation of lower confidence bounds for (a function of) distribution parameters based on so-called acceptance regions is performed in a very general model.

2.1.1 Basic notations

Let D denote a discrete random vector depending on an unknown vector

$$\kappa \in K \subseteq \overline{\mathbf{R}}^m \ (m \in \mathbf{N}) \tag{2.1}$$

of distributional parameters. K denotes the parameter space and is assumed to be a connected set.

The problem is to determine a lower confidence bound for the unknown value of a continuous and surjective function

$$h : K \to \left[b^{low}, b^{up}\right] \subseteq \overline{\mathbf{R}} \tag{2.2}$$

as a function of D for a given confidence level $\beta \in (0, 1)$.

The set

$$E := \{d : \exists_{\kappa \in K} : \Pr_\kappa \{D = d\} > 0\} \tag{2.3}$$

is assumed to have finite cardinality:

$$|E| < \infty \tag{2.4}$$

Moreover, the distribution of the vector D is assumed to be known for any $\kappa \in K$:

$$r_d : K \to [0, 1], \ r_d(\kappa) := \Pr_\kappa \{D = d\} \ \text{for} \ d \in E \tag{2.5}$$

$r_d(\kappa)$ is assumed to be continuous with respect to κ for any fixed $d \in E$.

Finally, let $\beta \in [0, 1]$ denote a given confidence level.

2.1.2 Confidence bound and system of acceptance regions

This section shows the close relation between a confidence bound and a system of acceptance regions. Here, the assumption (2.4) may be dropped, i.e. in the case $|E| = \infty$, everything remains true.

For all $c \in \left[b^{low}, b^{up}\right]$, let $AR(c)$ denote subsets of E with

$$\Pr_\kappa \{D \in AR(h(\kappa))\} \geq \beta \ \text{for all} \ \kappa \in K \ . \tag{2.6}$$

Then we call

$$\left\{ AR\left(c\right) : c \in \left[b^{low}, b^{up}\right] \right\} \tag{2.7}$$

a *system of β-acceptance regions* for $h\left(\kappa\right)$.

Let (2.7) be a system of β-acceptance regions, and let the sets $m\left(d\right)$ be defined by

$$m\left(d\right) := \left\{ c : d \in AR\left(c\right) \right\}, \quad d \in E, \tag{2.8}$$

then it follows obviously that the random set

$$M\left(D\right) := m\left(d\right) \text{ for } D = d \tag{2.9}$$

is a β confidence region for $h\left(\kappa\right)$, i.e.

$$\Pr_{\kappa} \left\{ h\left(\kappa\right) \in M\left(D\right) \right\} \geq \beta \text{ for all } \kappa \in K \ . \tag{2.10}$$

This is the relation between a system of acceptance regions and a confidence region.

Let (2.7) have the additional properties (besides (2.6))

$$AR\left(c_1\right) \supseteq AR\left(c_2\right) \text{ for all } b^{up} \geq c_1 \geq c_2 \geq b^{low} \tag{2.11}$$

and

$$\bigcap_{x:x>c} AR\left(x\right) = AR\left(c\right) \text{ for all } c \in \left[b^{low}, b^{up}\right), \tag{2.12}$$

then we call (2.7) a *system of lower β-acceptance regions* for $h\left(\kappa\right)$. In this case, the sets $m\left(d\right)$ are intervals:

$$m\left(d\right) = \left[c\left(d\right), b^{up}\right] \text{ for } d \in E \tag{2.13}$$

with

$$c\left(d\right) := \min \left\{ c : c \in \left[b^{low}, b^{up}\right], d \in AR\left(c\right) \right\} \text{ for } d \in E. \tag{2.14}$$

Thus, $\left[C\left(d\right), b^{up}\right]$ with

$$C\left(D\right) := c\left(d\right) \text{ for } D = d \tag{2.15}$$

is a β confidence interval for $h\left(\kappa\right)$, or equivalently, $C\left(D\right)$ is a lower β confidence bound for $h\left(\kappa\right)$, i.e.

$$\Pr_{\kappa} \left\{ h\left(\kappa\right) \geq C\left(D\right) \right\} \geq \beta \text{ for all } \kappa \in K \ . \tag{2.16}$$

This is the relation between a system of lower acceptance regions and a lower confidence bound. The algorithm proposed in the next section is based on this relationship.

2.1.3 The algorithm "System of lower β-acceptance regions"

The set containing all permutations of the elements of E is denoted by

$$S_{|E|} := \{\pi : \pi \text{ is a one-to-one function from } \{1, \dots, |E|\} \text{ to } E\}. \qquad (2.17)$$

For each permutation $\pi \in S_{|E|}$, the algorithm described in this section determines recursively a system of lower β-acceptance regions for $h(\kappa)$ and a lower β confidence bound for $h(\kappa)$.

The realizations of the lower confidence bound $C(D)$ are successively determined in a way that

$$b^{up} \geq c(\pi(1)) \geq \dots \geq c(\pi(|E|)) \geq b^{low}. \qquad (2.18)$$

First, we define

$$AR(b^{up}) := E \qquad (2.19)$$

and

$$c(\pi(0)) := b^{up} \qquad (2.20)$$

(just for initializing; it is not necessary to interpret $\pi(0)$).

Now, the algorithm successively determines $c(\pi(i))$, $i = 1, \dots, |E|$: With the sets

$$M_i := \left\{ c : c \in \left[b^{low}, b^{up}\right], \min_{\kappa : \kappa \in K,\ h(\kappa) \in \left[b^{low}, c\right]} \Pr_{\kappa}\{D \in AR(c(\pi(i-1))) - \{\pi(i)\}\} \geq \beta \right\} \qquad (2.21)$$

we define $c(\pi(i))$ by

$$c(\pi(i)) := \left\{ \begin{array}{cc} \max M_i & \\ b^{low} & \end{array} \text{ for } \begin{array}{c} M_i \neq \emptyset \\ M_i = \emptyset \end{array} \right. \qquad (2.22)$$

The corresponding acceptance regions are

$$AR(c) := \left\{ \begin{array}{c} E \\ AR(c(\pi(i-1))) - \{(\pi(i-1))\} \end{array} \right. \qquad (2.23)$$

$$\text{for } \left\{ \begin{array}{c} i = 1 \\ c \in [c(\pi(i)), c(\pi(i-1))),\ i \geq 2 \end{array} \right.$$

We now prove that this procedure yields indeed a system of lower β-acceptance regions for $h(\kappa)$:

Let $\kappa \in K$ with $h(\kappa) < b^{up}$.

As always $c(\pi(|E|)) = b^{low}$ (for $\beta > 0$; the case $\beta = 0$ is trivial), it is $h(\kappa) \in [c(\pi(i)), c(\pi(i-1)))$ for an $i \in \{1, \dots, |E|\}$. By induction, it follows from the definitions (2.21), (2.22), and (2.23) that

$$\Pr_{\kappa}\{D \in AR(h(\kappa))\} \geq \beta. \qquad (2.24)$$

Clearly, (2.24) holds also in the case $\kappa \in K$ with $h(\kappa) = b^{up}$ because of (2.19). Thus, the system $\left\{ AR\left(c\right) : c \in \left[b^{low}, b^{up}\right] \right\}$ has the property (2.6). It is trivial that (2.11) and (2.12) hold, too. Thus, $\left\{ AR\left(c\right) : c \in \left[b^{low}, b^{up}\right] \right\}$ is a system of lower β-acceptance regions.

Besides the system of acceptance regions, the algorithm determines the random variable defined by (2.15). As (2.14) holds obviously, this random variable is a lower β confidence bound for $h\left(\kappa\right)$.

Example 2.1.1 Lower 0.9 confidence bound for an unknown probability.

We consider the problem to determine a lower 0.9 confidence bound for the unknown probability p as a function of a random variable

$$D \sim Bi\left(3, p\right). \tag{2.25}$$

Then D may adopt values out of

$$E = \{0, 1, 2, 3\}, \tag{2.26}$$

and the parameter space is given by

$$K = [0, 1]. \tag{2.27}$$

For $d \in E$, the probability function r_d defined by (2.5) becomes

$$r_d : [0, 1] \rightarrow [0, 1], \; r_d\left(p\right) = \Pr_p \{D = d\} = \binom{3}{d} p^d \left(1 - p\right)^{3-d}. \tag{2.28}$$

h introduced in (2.2) is the identity on $[0, 1]$:

$$h : [0, 1] \rightarrow \left[b^{low}, b^{up}\right] = [0, 1], \; h\left(p\right) = p \tag{2.29}$$

Thus, for a given confidence level $\beta \in (0, 1)$ the algorithm determines a random variable $C\left(D\right)$ with the property

$$\Pr_p \{p \geq C\left(D\right)\} \geq \beta \text{ for all } p \in [0, 1], \tag{2.30}$$

i.e., a lower β confidence bound for p.

It depends on the permutation $\pi \in S_4$ (and, certainly, on the confidence level β) which confidence bound is obtained. For

$$\pi(i) := 4 - i, \quad i = 1, 2, 3, 4 \tag{2.31}$$

and $\beta = 0.9$, we obtain the confidence bound

$$C(D) = \begin{cases} 0.4641 & D = 3 \\ 0.1958 & D = 2 \\ 0.0345 & D = 1 \\ 0 & D = 0 \end{cases} \text{ for } \tag{2.32}$$

and the corresponding acceptance regions

$$AR(c) = \begin{cases} \{0,1,2,3\} & c \in [0.4641,1] \\ \{0,1,2\} & c \in [0.1958,0.4641) \\ \{0,1\} & c \in [0.0345,0.1958) \\ \{0\} & c \in [0,0.0345) \end{cases} \quad \text{for} \quad . \tag{2.33}$$

Note that the lower 0.9 confidence bound $C(D)$ given by (2.32) is exactly the same as that one which is generally suggested as lower confidence bound for the probability parameter of a binomial distribution. It is tabled in many books, e.g. in Odeh & Owen (1983). However, determining two-sided confidence intervals for p by means of the concept of acceptance regions yields shorter confidence intervals (in the sense of the average length of their realizations) than those which are widely tabled. For details, see von Collani et al. (1996).

2.1.4   Quality of lower confidence bounds

A very important aspect is the influence of the permutation $\pi \in S_{|E|}$ on the quality of the corresponding confidence bound. Suppose that the same situation as in Example 2.1.1 is considered where the permutation is given by

$$\pi(i) := i - 1, \quad i \in \{1,2,3,4\} \tag{2.34}$$

instead of (2.31). Then the trivial lower 0.9 confidence bound

$$C(D) = 0 \text{ for } D = 0,1,2,3 \tag{2.35}$$

is obtained which is obviously very "bad".

To measure the quality of any lower β confidence bound being a function of D, say $Q(D)$, we define a *class Z of quality indicators* as the set containing all functions

$$z : \{Q(D) : Q(D) \text{ is a lower } \beta \text{ confidence bound for } h(\kappa)\} \to [-\infty,\infty] \tag{2.36}$$

with necessarily

$$q_1(d) \geq q_2(d) \text{ for all } d \in E \Rightarrow z(Q_1(D)) \geq z(Q_2(D)) \tag{2.37}$$

where $Q_1(D)$ and $Q_2(D)$ are lower β confidence bounds for $h(\kappa)$ whose realizations are denoted by $q_1(d)$ and $q_2(d)$ $(d \in E)$, respectively.

For a given quality indicator $z \in Z$, we measure the quality of a lower confidence bound $Q(D)$ by $z(Q(D))$ (the larger, the better).

An example for a quality indicator is the average of all realizations:

$$z(Q(D)) := \frac{1}{|E|} \sum_{d \in E} q(d) \tag{2.38}$$

This quality indicator yields $z\left(C\left(D\right)\right) = 0.1736$ for the confidence bound given by (2.32) and $z\left(C\left(D\right)\right) = 0$ for the confidence bound given by (2.35).

This example illustrates the definitions:

Let $Q_1\left(D\right)$ and $Q_2\left(D\right)$ denote lower β confidence bounds for $h\left(\kappa\right)$. For a quality indicator $z \in Z$, $Q_1\left(D\right)$ is called *z-better* or *z-larger* than $Q_2\left(D\right)$, if

$$z\left(Q_1\left(D\right)\right) \geq z\left(Q_2\left(D\right)\right). \tag{2.39}$$

$Q_1\left(D\right)$ is called *z-optimal* or *z-maximal*, if

$$z\left(Q_1\left(D\right)\right) \geq z\left(Q\left(D\right)\right) \text{ for each lower } \beta \text{ confidence bound } Q\left(D\right). \tag{2.40}$$

2.1.5 Optimality of the algorithm

The optimality of the algorithm can be described from different points of view: First, suppose any lower confidence bound $Q(D)$ (which is not necessarily determined by the algorithm) for $h(\kappa)$ is given. Then the quality of the confidence bound determined by the algorithm (where a special permutation $\pi \in S_{|E|}$ is used which follows from $Q(D)$) is at least as good as the quality of $Q(D)$. This property does not depend on the quality indicator! More precisely:

Theorem 2.1.1 *Let $\beta \in [0,1]$, and let $Q(D)$ be any lower β confidence bound for $h(\kappa)$ whose realizations are*

$$q\left(1\right), \dots, q\left(|E|\right) \in \left[b^{low}, b^{up}\right]. \tag{2.41}$$

If $C(D)$ is a lower β confidence bound for $h(\kappa)$ determined by the algorithm where a permutation $\pi \in S_{|E|}$ with

$$q\left(\pi\left(1\right)\right) \geq \dots \geq q\left(\pi\left(|E|\right)\right). \tag{2.42}$$

is used, then $C(D)$ is z-larger than $Q(D)$ for all quality indicators $z \in Z$, or equivalently, $C(D) \geq Q(D)$.

PROOF. Let $\left\{AR\left(c\right) : c \in \left[b^{low}, b^{up}\right]\right\}$ denote the system of lower β-acceptance regions for $h(\kappa)$ corresponding to $C(D)$.

Let $i \in 1, \dots, |E|$. We show that the assumption

$$c\left(\pi\left(i\right)\right) < q\left(\pi\left(i\right)\right) \tag{2.43}$$

yields a contradiction. As $h(\kappa)$ is assumed to be continuous and surjective, there is, according to (2.21) and (2.22), a $\kappa_0 \in K$ with

$$c\left(\pi\left(i\right)\right) < h\left(\kappa_0\right) < q\left(\pi\left(i\right)\right) \tag{2.44}$$

and

$$\Pr_{\kappa_0}\left\{D \in AR\left(h\left(\kappa_0\right)\right) - \left\{\pi\left(i\right)\right\}\right\} < \beta . \tag{2.45}$$

As $c(\pi(i)) < h(\kappa_0)$, it follows from (2.23) that

$$E - \bigcup_{j=1}^{i} \{\pi(j)\} \subseteq AR(h(\kappa_0)) - \{\pi(i)\}. \tag{2.46}$$

Thus,

$$\mathrm{Pr}_{\kappa_0} \left\{ D \in E - \bigcup_{j=1}^{i} \{\pi(j)\} \right\} < \beta. \tag{2.47}$$

As $h(\kappa_0) < q(\pi(i))$, we obtain with (2.42)

$$\{d : d \in E, h(\kappa_0) \geq q(d)\} \subseteq E - \bigcup_{j=1}^{i} \{\pi(j)\}. \tag{2.48}$$

Thus,

$$\mathrm{Pr}_{\kappa_0} \{D \in \{d : d \in E, h(\kappa_0) \geq q(d)\}\} < \beta. \tag{2.49}$$

Now,

$$\mathrm{Pr}_{\kappa_0} \{h(\kappa_0) \geq Q(D)\} = \mathrm{Pr}_{\kappa_0} \{D \in \{d : d \in E, h(\kappa_0) \geq q(d)\}\}. \tag{2.50}$$

Thus,

$$\mathrm{Pr}_{\kappa_0} \{h(\kappa_0) \geq Q(D)\} < \beta. \tag{2.51}$$

This is a contradiction to the assumption that $Q(D)$ is a lower β confidence bound for $h(\kappa)$. Thus, the assumption (2.43) is wrong, i.e. it is shown that

$$c(\pi(i)) \geq q(\pi(i)) \text{ for } i \in 1, \ldots, |E|. \tag{2.52}$$

Thus, $C(D) \geq Q(D)$, and it follows from (2.37) that

$$z(C(D)) \geq z(Q(D)), \tag{2.53}$$

where $z \in Z$ may be any quality indicator. ∎

However, another point of view is even more important, as normally, the problem is not to determine a confidence bound which is at least as good as a given one, but to determine a confidence bound which is optimal with respect to a given quality indicator $z \in Z$. This problem may also be solved by the algorithm:

Theorem 2.1.2 *Let $\beta \in [0,1]$ and $z \in Z$. Let $C_\pi(D)$ ($\pi \in S_{|E|}$) denote the lower β confidence bound determined by the algorithm where the permutation π is used. Let $\pi^* \in S_{|E|}$ be a permutation with*

$$z(C_{\pi^*}(D)) \geq z(C_\pi(D)) \text{ for all } \pi \in S_{|E|}. \tag{2.54}$$

Then the $C_{\pi^*}(D)$ is a z-maximal lower β confidence bound for $h(\kappa)$.

PROOF. Let $Q(D)$ be any lower β confidence bound for $h(\kappa)$, and let π be a permutation given by (2.42). According to (2.53) and (2.54),

$$z(Q(D)) \leq z(C_\pi(D)) \leq z(C_{\pi^*}(D)), \tag{2.55}$$

i.e. $z(C_{\pi^*}(D))$ is z-maximal. ∎

z-maximal strategies

Theorem 2.1.2 suggests how to determine a z-maximal lower confidence bound for $h(\kappa)$. The corresponding strategy is called a *z-maximal strategy*:

For any given quality indicator $z \in Z$, determine lower confidence bounds

$$C_\pi(D), \ \pi \in S_{|E|} \tag{2.56}$$

by the algorithm for all $\left|S_{|E|}\right| = |E|!$ different permutations, calculate the corresponding quality indicators $z(C_\pi(D))$ and take a lower confidence bound $C_{\pi^*}(D)$ with (2.54); it is z-maximal.

Note that this procedure does not necessarily yield a unique confidence bound.

Remark. In the case that a lower β confidence bound $C(D)$ for p is determined by the algorithm as a function of a random variable

$$D \sim Bi(n, p), \tag{2.57}$$

and that the quality indicator (2.38) is taken, it can be shown that the permutation

$$\pi^*(i) := n - i + 1, \quad i = 1, \ldots, n+1 \tag{2.58}$$

yields a z-maximal confidence bound $C_{\pi^*}(D)$, i.e. in this case, it is not necessary to determine all $(n+1)!$ confidence bounds $C_\pi(D)$, $\pi \in S_{n+1}$.

However, generally (i.e., without assuming (2.57)) for determining a z-maximal lower confidence bound for a certain quality indicator $z \in Z$, there seems to be no alternative to the z-maximal strategy which requires the determination of $|E|!$ confidence bounds.

2.1.6 Quick determination vs. good quality

From a practical point of view, the computing time should be taken into account when confidence bounds are to be determined. In this aspect, the z-maximal strategy has a disadvantage: the algorithm has to be iterated $|E|!$ times which may take a long time. Thus, we look for alternative strategies which, on the one hand, yield confidence bounds with a relative large value of the given quality indicator and, on the other hand, with considerably less computing time compared with a z-maximal strategy.

We propose two types of such strategies here.

$\hat{H}(D)$-oriented strategies

The idea of $\hat{H}(D)$-oriented strategies is to select (not any, but) a proper permutation $\pi \in S_{|E|}$ before the algorithm starts using π.

We assume that there is an estimator $\hat{H}(D)$ for the unknown value $h(\kappa)$. The estimates are denoted by $\hat{h}(d)$ $(d \in E)$. Let $\pi \in S_{|E|}$ be a (not necessarily unique) permutation with

$$\hat{h}(\pi(1)) \geq \ldots \geq \hat{h}(\pi(|E|)), \tag{2.59}$$

then we call a lower confidence bound $C(D)$ for $h(\kappa)$ determined by the algorithm where π is used a $\hat{H}(D)$-*oriented* confidence bound. The determination of a $\hat{H}(D)$-oriented confidence bound is called a $\hat{H}(D)$-oriented strategy.

In the binomial model,

$$D \sim Bi(n, p), \tag{2.60}$$

the maximum likelihood estimator

$$\hat{H}(D) := \frac{D}{n} \tag{2.61}$$

of p yields the permutation

$$\pi(i) := n - i + 1, \quad i = 1, \ldots, n + 1. \tag{2.62}$$

This is the same permutation as in (2.58). Thus, in the binomial model, the $\frac{D}{n}$-oriented and the z-maximal lower confidence bound coincide (if z is given by (2.38)).

The quality indicator of a $\hat{H}(D)$-oriented confidence bound is generally much larger than that one for a confidence bound for an arbitrarily selected permutation $\pi \in S_{|E|}$, while the computing times are almost equal. Thus, the $\hat{H}(D)$-oriented strategy seems to be quite reasonable.

Recursive strategies

A recursive strategy means the determination of a lower confidence bound for $h(\kappa)$ according to the following modification of the algorithm described in Section 2.1.3. The permutation $\pi \in S_{|E|}$ is here determined itself recursively for $i = 1, \ldots, |E|$. The idea is that $\pi(i)$ is selected in a way that the corresponding realization $c(\pi(i))$ is as large as possible (remember that for each quality indicator $z \in Z$, large realizations of the confidence bound yield good quality).

Following that idea, we now describe more precisely the modification of the algorithm:

Suppose that $\pi(1), \ldots, \pi(i-1)$ and the realizations $c(\pi(0))), \ldots, c(\pi(i-1))$ are known. Then the lower β-acceptance region $AR(c(\pi(i-1)))$ is also known. Each element of the set $E - \cup_{j=1}^{i-1}\{\pi(j)\}$ is a "candidate" to be chosen as $\pi(i)$. Thus, for $d \in E - \cup_{j=1}^{i-1}\{\pi(j)\}$, we define the sets

$$M_i(d) :=$$
$$\left\{ c : c \in \left[b^{low}, b^{up} \right], \min_{\kappa:\kappa \in K,\, h(\kappa) \in \left[b^{low}, c \right]} \Pr_\kappa \{ D \in AR(c(\pi(i-1))) - \{d\} \} \geq \beta \right\}. \tag{2.63}$$

The "candidates" for the next realization to be determined, namely $c\left(\pi\left(i\right)\right)$, are

$$\begin{cases} \max\limits M_i(d) \\ b^{low} \end{cases} \text{ for } \begin{array}{l} M_i(d) \neq \emptyset \\ M_i(d) = \emptyset \end{array}, \quad d \in E - \bigcup_{j=1}^{i-1}\left\{\pi\left(j\right)\right\}. \quad (2.64)$$

We choose the "best" candidate: Let $\pi\left(i\right) \in E - \cup_{j=1}^{i-1}\left\{\pi\left(j\right)\right\}$ with

$$c\left(\pi\left(i\right)\right) := \max M_i\left(\pi\left(i\right)\right)\cup\left\{b^{low}\right\} = \max\limits_{d\in E-\cup_{j=1}^{i-1}\left\{\pi\left(j\right)\right\}} \max M_i(d)\cup\left\{b^{low}\right\}. \quad (2.65)$$

Note that $\pi\left(i\right)$ is not necessarily unique; this is why the corresponding strategy is called *a* instead of *the* recursive strategy.

We call a confidence bound $C\left(D\right)$ determined by this modified algorithm a *recursive* confidence bound.

Example 2.1.2 Recursive lower 0.9 confidence bound for an unknown probability.

A lower 0.9 confidence bound for the unknown parameter p is to be determined as a function of the random variable

$$D \sim Bi\left(2,p\right). \quad (2.66)$$

We consider this simple example in order to illustrate how a recursive strategy works.

For $i = 1$, we obtain

$$M_1(d) = \left\{c : c \in [0,1], \min\limits_{p:p\in[0,c]} \Pr_p\left\{D \in AR(c(\pi\left(0\right))) - \{d\}\right\} \geq 0.9\right\}$$

$$= \begin{cases} \left\{c : c \in [0,1], \min\limits_{p:p\in[0,c]} \Pr_p\left\{D \in \{1,2\}\right\} \geq 0.9\right\} & d = 0 \\ \left\{c : c \in [0,1], \min\limits_{p:p\in[0,c]} \Pr_p\left\{D \in \{0,2\}\right\} \geq 0.9\right\} & \text{for } d = 1 \\ \left\{c : c \in [0,1], \min\limits_{p:p\in[0,c]} \Pr_p\left\{D \in \{0,1\}\right\} \geq 0.9\right\} & d = 2 \end{cases}$$

$$= \begin{cases} \left\{c : c \in [0,1], \min\limits_{p:p\in[0,c]} \left(1 - (1-p)^2\right) \geq 0.9\right\} & d = 0 \\ \left\{c : c \in [0,1], \min\limits_{p:p\in[0,c]} \left(1 - 2p\left(1-p\right)\right) \geq 0.9\right\} & \text{for } d = 1 \\ \left\{c : c \in [0,1], \min\limits_{p:p\in[0,c]} \left(1 - p^2\right) \geq 0.9\right\} & d = 2 \end{cases}$$

$$= \begin{cases} \emptyset & d = 0 \\ \left\{c : c \in [0,1], \left(1 - 2c\left(1-c\right)\right) \geq 0.9\right\} & \text{for } d = 1 \\ \left\{c : c \in [0,1], \left(1 - c^2\right) \geq 0.9\right\} & d = 2 \end{cases} \quad (2.67)$$

Now, the "best" candidate is chosen to be $c(\pi(1))$:

$$c(\pi(1)) \;:=\; \max_{d\in\{0,1,2\}} \max M_1(d) \cup \{0\}$$

$$= \max_{d\in\{0,1,2\}} \left\{ \begin{array}{ll} 0 & d=0 \\ 0.0528 \quad \text{for} & d=1 \\ 0.3162 & d=2 \end{array} \right.$$

$$= 0.3162 \tag{2.68}$$

The corresponding vector $\pi(1)$ turns out to be unique: $\pi(1) = 2$. This terminates the case $i = 1$.

Analogously, we obtain

$$c(\pi(2)) = 0.0513, \tag{2.69}$$

$$\pi(2) = 1, \tag{2.70}$$

$$c(\pi(3)) = 0, \tag{2.71}$$

and

$$\pi(3) = 0. \tag{2.72}$$

Conclusion. The recursive confidence bound (which is here unique) is given by

$$C(D) = \left\{ \begin{array}{ll} 0.3162 & D=2 \\ 0.0513 \quad \text{for} & D=1 \\ 0 & D=0 \end{array} \right. \tag{2.73}$$

and the corresponding permutation by

$$\pi(i) = 3 - i, \quad i = 1, 2, 3. \tag{2.74}$$

This is the same permutation as in (2.58). Thus, in the binomial case, the $\frac{D}{2}$-oriented, the recursive, and the z-maximal lower 0.9 confidence bound for the parameter p coincide (if z is given by (2.38)).

A recursive strategy takes rather more computing time than a $\hat{H}(D)$-oriented strategy, but takes far less computing time than a z-maximal strategy. However, the quality indicator of a recursive confidence bound turns out to be considerably larger than the corresponding quality indicator of a $\hat{H}(D)$-oriented confidence bound in many examples. Thus, we propose to choose a recursive strategy, as it seems to be a good compromise between the requirements "short computing time" and "large quality indicator".

2.2 Confidence Bound for the Expectation of a Weibull Distribution

In this section, the aim is to determine a lower confidence bound for the expectation of a Weibull distribution based on a left- and right-censored sample which is a special censoring pattern described below. Thus, we apply the concept of determining confidence bounds based on acceptance regions to this special case.

2.2.1 The model

Let $n \in \mathbf{N}$ be the size of a random sample $A_1, ..., A_n$ of a Weibull distribution:

$$\Pr \{A_1 \leq t\} = \begin{cases} 0 & \\ 1 - e^{-(\frac{t}{\theta})^\sigma} & \end{cases} \quad \text{for} \quad \begin{array}{l} t \leq 0 \\ t > 0 \end{array} \qquad (2.75)$$

$\theta, \sigma \in \mathbf{R}^+$ are called the *scale* and the *shape parameter*, respectively. Both are assumed to be unknown. The aim is to determine a lower confidence bound for the expectation

$$\mathrm{E}\,[A_1] = \theta \Gamma \left(1 + \frac{1}{\sigma} \right). \qquad (2.76)$$

However, often the sample elements A_i $(i = 1, \ldots, n)$ are not observed directly, but only indirectly due to censoring. Here, we assume the following censoring pattern: Let

$$\begin{pmatrix} t_1 & \cdots & t_k \\ n_1 & \cdots & n_k \end{pmatrix} \qquad (2.77)$$

denote an experimental design where $0 < t_i < t_j < \infty$ for $i < j$ and $n_1, \ldots, n_k \in \mathbf{N}$ with $\sum_{j=1}^k n_j = n$. This design determines the following rule:

1. Divide the n sample items into k groups $I_1, \ldots, I_k$ of sizes $n_1, \ldots, n_k$ and put them on test.

2. Observe after t_j $(j = 1, \ldots, k)$ time units of test the (random) number L_j of items of group no. j which have failed.

It follows that instead of observing A_j, we observe the random variables L_j with

$$L_j := \sum_{i \in I_j} 1_{\{A_i \leq t_j\}} \text{ for } j = 1, \ldots, k \qquad (2.78)$$

where 1_M denotes the indicator function of M. Thus, the estimation techniques are (only) to be based on the design vectors $(t_1, \ldots, t_k)$ and $(n_1, \ldots, n_k)$ and on the random vector

$$L := (L_1, \ldots, L_k). \qquad (2.79)$$

L is called a *left- and right-censored sample*. This name is due to the fact that for each item, only the information $1_{\{A_i \le t_j\}}$ is available where A_i is the lifetime of the item and t_j is a fixed time point. If $1_{\{A_i \le t_j\}} = 0$, then the item no. i has not failed up to t_j (right-censored case); if $1_{\{A_i \le t_j\}} = 1$, then the item no. i has failed up to t_j (left-censored case). This censoring pattern plays an important rule in many applications; e.g., it is often cheaper to collect censored data than to collect non-censored data. For some details, see Dubey (1965).

A trivial inference of (2.78) is that $L_1, \ldots, L_k$ are independent and

$$L_j \sim Bi\left(n_j, F\left(t_j\right)\right) = Bi\left(n_j, 1 - e^{-\left(\frac{t_j}{\theta}\right)^\sigma}\right) \quad \text{for } j = 1, \ldots, k. \tag{2.80}$$

2.2.2 Applying the algorithm "System of lower β-acceptance regions"

In the terms introduced in (2.1) to (2.5),

$$D = L \tag{2.81}$$

is the discrete random vector whose distribution depends on (θ, σ),

$$K = \left(\mathbf{R}^+\right)^2 \tag{2.82}$$

is the parameter space,

$$E = \{0, \ldots, n_1\} \times \ldots \times \{0, \ldots, n_k\} \tag{2.83}$$

is the set containing the realizations of L, and

$$
\begin{aligned}
r_l : \left(\mathbf{R}^+\right)^2 &\to [0, 1], \\
r_l(\theta, \sigma) &= \mathrm{Pr}_{\theta, \sigma}\{L = l\} \\
&= \prod_{j=1}^{k} \binom{n_j}{l_j} \left(1 - e^{-\left(\frac{t_j}{\theta}\right)^\sigma}\right)^{l_j} \left(e^{-\left(\frac{t_j}{\theta}\right)^\sigma}\right)^{n_j - l_j}
\end{aligned}
\tag{2.84}
$$

for $l \in E$ is the probability function. Moreover,

$$h : \left(\mathbf{R}^+\right)^2 \to \mathbf{R}^+, \ h(\theta, \sigma) = \theta \Gamma\left(1 + \frac{1}{\sigma}\right) \tag{2.85}$$

is the function giving the distributional parameter of interest, namely the expectation of the Weibull distribution.

The problem to determine a lower β confidence bound for $h(\theta, \sigma)$ as a function of L may be solved by the algorithm.

Example 2.2.1 Lower 0.9 confidence bound for $h(\theta, \sigma)$.

Let

$$
\begin{pmatrix} t_1 & t_2 \\ n_1 & n_2 \end{pmatrix} = \begin{pmatrix} 1 & 2 \\ 3 & 3 \end{pmatrix} \tag{2.86}
$$

be the experimental design and let

$$
\beta = 0.9 \tag{2.87}
$$

be the confidence level.

Let $\hat{H}(L)$ denote the estimator defined by Table 2.1 where each realization is given. This estimator has a maximum likelihood interpretation: For each realization $\hat{h}(l)$, $l \in \{0, \ldots, 3\} \times \{0, \ldots, 3\}$, it is

$$
\hat{h}(l) = \arg \max_{\theta, \sigma \in \mathbf{R}^+} \mathrm{Pr}_{\theta, \sigma}\{L = l\} , \tag{2.88}
$$

if the corresponding maximum exists uniquely. If not, then

$$
\hat{h}(l) = \lim_{i \to \infty} h(\theta_i, \sigma_i) \tag{2.89}
$$

holds where $(\theta_i, \sigma_i)_{i \in \mathbf{N}}$ is a sequence in $(\mathbf{R}^+)^2$ with

$$
\lim_{i \to \infty} \mathrm{Pr}_{\theta_i, \sigma_i}\{L = l\} \geq \mathrm{Pr}_{\theta, \sigma}\{L = l\} \text{ for all } (\theta, \sigma) \in \mathbf{R}^+ . \tag{2.90}
$$

A permutation (which is obviously not unique) $\pi \in S_{16}$ with

$$
\hat{h}(\pi(1)) \geq \ldots \geq \hat{h}(\pi(16)) \tag{2.91}
$$

is given by

$$
\begin{aligned}
\pi(1) &= (3,1), & \pi(2) &= (2,2), & \pi(3) &= (3,0), & \pi(4) &= (2,1), \\
\pi(5) &= (2,0), & \pi(6) &= (3,2), & \pi(7) &= (1,1), & \pi(8) &= (1,0), \\
\pi(9) &= (0,0), & \pi(10) &= (0,1), & \pi(11) &= (0,2), & \pi(12) &= (1,2), \\
\pi(13) &= (0,3), & \pi(14) &= (1,3), & \pi(15) &= (2,3), & \pi(16) &= (3,3).
\end{aligned} \tag{2.92}
$$

In Table 2.1, the $\hat{H}(L)$-oriented lower 0.9 confidence bound $B(L)$ with permutation (2.92) and a recursive lower 0.9 confidence bound $C(L)$, respectively, are given. The corresponding realizations are denoted by $b(l)$ and $c(l)$ ($l \in E$), respectively.

Table 2.1: $\hat{H}(L)$-oriented and online lower 0.9 confidence bound for $h(\theta, \sigma)$

l	$\hat{h}(l)$	$b(l)$	$c(l)$	l	$\hat{h}(l)$	$b(l)$	$c(l)$
$(0,0)$	4.0000	0.3176	1.9356	$(2,0)$	∞	0.6691	∞
$(0,1)$	2.0000	0.3176	1.5907	$(2,1)$	∞	0.6700	1.1232
$(0,2)$	2.0000	0.3175	1.2180	$(2,2)$	∞	0.7046	0.4795
$(0,3)$	1.5000	0.3126	0.9678	$(2,3)$	1.0000	0.1166	0.1166
$(1,0)$	∞	0.3176	3.0066	$(3,0)$	∞	0.7029	∞
$(1,1)$	∞	0.3176	1.3964	$(3,1)$	∞	∞	∞
$(1,2)$	1.7006	0.3135	0.8420	$(3,2)$	∞	0.3177	0.2836
$(1,3)$	1.0000	0.2836	0.5948	$(3,3)$	0.5000	0.0000	0.0000

2.2.3   Quality of lower confidence bounds for the expectation

To measure the quality of lower confidence bounds, we take two quality indicators out of the class Z. For a confidence bound $Q(L)$ with realizations $q(l)$ ($l \in E$), the first quality indicator is given by

$$z_1(Q(L)) := \frac{1}{|E|} \sum_{l \in E} \left(1 - e^{-q(l)}\right), \tag{2.93}$$

i.e., the average of the transformed realizations of the confidence bounds where the transformation function is $1 - e^{-x}$ which we define as 1 for $x = \infty$. The transformation function makes sense here, because the simple average of the realizations is infinity, if at least one realization of the corresponding confidence bound is infinity, which makes a comparison of such confidence bounds impossible.

For the second quality indicator, we identify (θ, σ) as a realization of two independent random variables (Θ, Σ) each distributed according to $Unif(0.5, 2.5)$ and define

$$
\begin{aligned}
&z_2(Q(L))\\
&:= E\left[E_{\Theta,\Sigma}\left[1 - e^{-Q(L)}\right]\right]\\
&= E\left[\sum_{l \in E}\left(1 - e^{-q(l)}\right)\Pr_{\Theta,\Sigma}\{L = l\}\right]\\
&= E\left[\sum_{l \in E}\left(1 - e^{-q(l)}\right)\prod_{j=1}^{k}\binom{n_j}{l_j}\left(1 - e^{-\left(\frac{t_j}{\Theta}\right)^{\Sigma}}\right)^{l_j}\left(e^{-\left(\frac{t_j}{\Theta}\right)^{\Sigma}}\right)^{n_j - l_j}\right]\\
&= \sum_{l \in E}\left(1 - e^{-q(l)}\right)\frac{1}{4}\int_{0.5}^{2.5}\int_{0.5}^{2.5}\binom{n_j}{l_j}\left(1 - e^{-\left(\frac{t_j}{\theta}\right)^{\sigma}}\right)^{l_j}\left(e^{-\left(\frac{t_j}{\theta}\right)^{\sigma}}\right)^{n_j - l_j}d\theta d\sigma
\end{aligned}
\tag{2.94}
$$

i.e., the second quality indicator is a convex combination of the transformed realizations of the confidence bounds where the factors are given by

$$w_2(l) := \frac{1}{4}\int_{0.5}^{2.5}\int_{0.5}^{2.5}\binom{n_j}{l_j}\left(1 - e^{-\left(\frac{t_j}{\theta}\right)^{\sigma}}\right)^{l_j}\left(e^{-\left(\frac{t_j}{\theta}\right)^{\sigma}}\right)^{n_j - l_j}d\theta d\sigma, \quad l \in E. \tag{2.95}$$

We consider the lower 0.9 confidence bounds determined in Example 2.2.1. For the $\hat{H}(L)$-oriented confidence bound $B(L)$, we obtain

$$z_1(B(L)) = 0.3446 \tag{2.96}$$

and

$$z_2(B(L)) = 0.2354. \tag{2.97}$$

For the recursive confidence bound $C(L)$, we obtain

$$z_1(C(L)) = 0.6318 \tag{2.98}$$

and

$$z_2 \left(C \left(L \right) \right) = 0.4207. \tag{2.99}$$

Thus, $C \left(L \right)$ is z_1-better and z_2-better than $B \left(L \right)$.

We do not apply a z-maximal strategy, because already in this simple example this would mean that the algorithm has to be iterated $16! > 10^{13}$ times. This illustrates our recommendation given in Section 2.1.6: From a practical point of view, a recursive strategy seems to be most reasonable.

References

1. von Collani, E., Dräger, K. and Hottendorf, J. (1996). *Tables for Optimal Two-Sided Confidence Intervals and Tests for an Unknown Probability*, Universität Würzburg, Monograph Series in Stochastics 1.

2. Dubey, S. D. (1965). Asymptotic properties of several estimators of Weibull parameters, *Technometrics*, **7**, 423–434.

3. Odeh, R. E. and Owen, D. B. (1983). *Attribute Sampling Plans, Tables of Tests, and Confidence Limits for Proportions*, New York, Basel: Marcel Dekker.

3

On Statistics in Failure-Repair Models Under Censoring

Jürgen Franz

Technical University Dresden, Dresden, Germany

Abstract: The paper deals with estimation models using censored life time data. First, a short survey on parametric and nonparametric estimation is given for i.i.d. life times under censorship without repair. Special investigations are made for Koziol-Green models. In the second part, a general failure-repair model described by counting processes is considered. Under the Koziol-Green condition some results concerning Bayes estimation and the nuisance parameter case are obtained.

Keywords and phrases: Estimation in parametric and nonparametric life time models, censoring time, Koziol-Green model, failure-repair process, counting process, random observation time, Bayes estimator, optimum estimating equation, nuisance parameter

3.1 Introduction

The purpose of the paper is to present statistical methods and results for censored data. We consider models of random censorship where life time observations are censored from the right. This situation is typical for many investigations in practice, for instance in clinical studies and in life testing problems of technical systems. The life time (survival time) of any unit is connected with a random censoring time.

The paper gives a certain survey on models. Some parametric and nonparametric statistical results are discussed. First, we introduce the model of i.i.d. life times and censoring times [cf. Fleming and Harrington (1991), Breslow (1992), Hurt (1992)]. The likelihood function is the starting point for parametric studies. Using the Koziol-Green condition we specialize the model of random censorship, and we give a characterization of this model. For nonparametric estimation of the survival function the Kaplan-Meier and the Abdushukurov-Cheng-Lin estimators are considered [cf. Csörgö (1988), Pawlitschko (1996)].

A second part of the paper contains results concerning failure-repair models [cf. Stadje and Zuckerman (1991), Franz (1994)]. We introduce a general model preferring the description by counting processes. The failure-repair process is observed up to a stopping time τ. Special results are given under the Koziol-Green condition. We consider the Bayes estimation and derive a conjugate prior distribution used for Bayes estimators for a class of parameter functions. Optimum estimating equations in the situation when a distribution density function contains a nuisance parameter were characterized by Godambe and Thompson (1974) and Godambe (1976, 1984). Ferreira (1982) considered sequential estimation through estimating equations in the nuisance parameter case for a sequence of random variables. Ignoring the nuisance parameter, one can search for estimators of the parameter of interest which are based on partial likelihood. In the paper, a factorization of the likelihood function is given needed for optimum estimating functions under nuisance parameters.

3.2   Survival Data Analysis Under Censoring

First we introduce some basic notions and statistical concepts needed in life testing problems for technical systems or in clinical survival studies.

Assume that T is a random life time with the distribution function $F_T(t)$ and possessing the density $f_T(t)$. The function $R_T(t) = 1 - F_T(t)$ is said to be the survival function, and $r_T(t) = \frac{f_T(t)}{R_T(t)}$ $(t \geq 0, \quad R_T(t) > 0)$ is the hazard rate. Obviously, we have

$$R_T(t) = \exp\left\{-\int_0^t r_T(x)dx\right\}.$$

In most practical cases, we can only observe censored samples. We shall use the model of random censorship where the data are censored from the right. We introduce the random censoring time C, independent of T, having the distribution function $F_C(t)$ and the density $f_C(t)$. Now it is usual to suppose that only the pair

$$M = \min(T, C), \qquad \Delta = \mathbf{1}\{T \leq C\}$$

can be observed. The following properties are known [see, for instance, Hurt (1992)]:

1. $R_M(t) := P(M > t) = P(T > t)P(C > t) = R_T(t)R_C(t)$,

2. The joint (generalized) density of (M, Δ) is given by

$$f(t,i) = [f_T(t)R_C(t)]^i [f_C(t)R_T(t)]^{1-i}$$
$$(t \geq 0; \ i = 1 \text{ if } T \leq C, \ i = 0 \text{ if } T > C).$$

In the special case of deterministic censorship by $C = c_0$:

$$f(t,i) = [f_T(t)]^i [R_T(c_0)]^{1-i}.$$

3. The hazard rate of M: $\lambda_M(t) = \lambda_T(t) + \lambda_C(t)$.

4. Let (T_j, C_j), $j = 1, 2, ..., n$, be pairs of i.i.d. random variables distributed like (T, C). For the sample $(M_1, \Delta_1), ..., (M_n, \Delta_n)$ the likelihood function is given by

$$L_n = \prod_{j:T_j \leq C_j} f_T(T_j)R_C(T_j) \times \prod_{j:T_j > C_j} f_C(C_j)R_T(C_j). \tag{3.1}$$

Example 3.2.1 Assume that T and C are 2-parametric-Weibull distributed with

$$f_T(t) = f_T(t;\vartheta) = \frac{b}{a^b}t^{b-1}\exp(-(\frac{t}{a})^b) =: e^\alpha \beta t^{\beta-1}\exp(-e^\alpha t^\beta),$$

$$f_C(t) = f_C(t;\vartheta) = \frac{d}{c^d}t^{d-1}\exp(-(\frac{t}{c})^d) =: e^\gamma \delta t^{\delta-1}\exp(-e^\gamma t^\delta).$$

Then formula (3.1) changes into

$$L_n = \exp[(\alpha + \ln\beta)N_1 + (\beta - 1)W_1 - e^\alpha S_1(\beta)]$$
$$\times \exp[(\gamma + \ln\delta)N_0 + (\delta - 1)W_0 - e^\gamma S_0(\delta)], \tag{3.2}$$

where

$$S_1(\beta) = \sum_{j=1}^n M_j^\beta, \quad S_0(\delta) = \sum_{j=1}^n M_j^\delta, \quad N_1 + N_0 = n,$$

$$N_1 = \sum_{j=1}^n \Delta_j, \quad W_1 = \sum_{j=1}^n \Delta_j \ln M_j, \quad W_0 = \sum_{j=1}^n (1 - \Delta_j) \ln M_j.$$

Remark 3.2.1 The likelihood function (3.2) and also the function (3.1) are factorized in two terms depending on failure time parameters and on censoring time parameters, respectively. The parameter estimation can be separately done based on the corresponding partial likelihood function.

For the maximum likelihood estimators (m.l.e.) of the parameters contained in the vector ϑ, under certain regularity conditions, it holds

$$n^{1/2}(\hat\vartheta - \vartheta) \xrightarrow[n \to \infty]{\mathcal{L}} N(0, I^{-1}(\vartheta))$$

($\hat{\vartheta}$ m.l.e. of ϑ, $I(\vartheta)$ Fisher information matrix).

Now, let us look at the so-called informative censorship.

In parametric studies a natural question arises about the distribution of the censoring time. Often there are good reasons to apply the model of random censorship introduced by Koziol and Green (1976):

Assumption (KG). $R_C(t) = [R_T(t)]^\gamma$ for all $t \geq 0$ where $\gamma \geq 0$ (γ is called censoring exponent).

Remark 3.2.2 Considering the values of γ, $\gamma = 0$ corresponds to the non-censored case, $0 < \gamma < 1$ means a greater portion of uncensored observations and $\gamma > 1$ a greater portion of censored observations. The model of random censorship fulfilling Assumption (KG) is shortly called KG-model.

The following theorem characterizes the KG-model.

Theorem 3.2.1 *Let* $p := P(T < C)$ *the expected portion of uncensored observations and let* $0 < p < 1$. *Then*

(i) $p = \frac{1}{1+\gamma}$, $R_M(t) = [R_T(t)]^{1/p}$,

(ii) $f_M(t) = (1/p)f_T(t)[R_T(t)]^\gamma$,

(iii) $\lambda_C(t) = \gamma\lambda_T(t)$,

(iv) *Assumption (KG) holds if and only if* M_j *and* Δ_j ($j = 1, 2, ..., n$) *are independent,*

(v) $L_n(\theta) = \gamma^{N_0} \prod_{j=1}^n f_T(M_j)[R_T(M_j)]^\gamma$.

PROOF. See Hurt (1992, Theorems I.1 and I.2, and Remark I.1).   ■

The assertion (i) in Theorem (3.2.1) is of great importance in view of the estimation of $R_T(t)$. The Koziol-Green model is studied in more details in the next sections.

3.3   Nonparametric Estimators for $R_T(t)$

We emphasize to investigate censored life time data. In the noncensored case the empirical distribution function is used to get a good estimator for $R_T(t)$. Now the estimator first studied by Kaplan and Meier (1958) is introduced:

$$\widehat{R_{KM}}(t) = \begin{cases} \prod_{j=1}^n \left(1 - \frac{\Delta_{(j)}}{n-j+1}\right)^{1\{M_{(j)} \leq t\}}, & t < M_{(n)} \\ 0, & t \geq M_{(n)} \end{cases} \tag{3.3}$$

where $M_{(1)}, ..., M_{(n)}$ form the ordered sample of observed failure times $M = \min(T, C)$ and $\Delta_{(j)}$ $(j = 1, ..., n)$ are the corresponding life time indicators. The estimator (3.3) was examined in many papers; a certain survey is given by Breslow (1992). The version of the estimator given in (3.3) was introduced by Stute and Wang (1993). For fixed $t > 0$ the Kaplan-Meier estimator is asymptotically normal:

Theorem 3.3.1 *Let $R_T(t)$ and $\widehat{R_{KM}}(t)$ be the survival function and the corresponding Kaplan-Meier estimator (3.3), respectively. Then*

$$\{n^{1/2}(\widehat{R_{KM}}(t) - R_T(t)), 0 \le t \le c\} \xrightarrow[n \to \infty]{\mathcal{L}} G \qquad (3.4)$$

in Skorohod space D[0,c] where G is a Gaussian process with mean zero and with the covariance function

$$\sigma_{KM}(s, t) := -R_T(s)R_T(t) \int_0^s \frac{dR_T(u)}{R_T^2(u) R_C(u)} \qquad (0 \le s \le t \le c)$$

with $c = \sup\{t > 0 : R_M(t) > 0\}$.

For the proof, see Csörgő (1988) and Hurt (1992).

From Theorem 3.3.1 the following asymptotic confidence region for $R_T(t)$ is obtained ($z_{1-\alpha/2}$ denotes the $(1 - \alpha/2)$-quantile of the standard-normal distribution):

$$[\widehat{R_{KM}}(t) - z_{1-\alpha/2}(Var(\hat{R}_{KM}(t)))^{1/2}; \widehat{R_{KM}}(t) + z_{1-\alpha/2}(Var(\hat{R}_{KM}(t)))^{1/2}],$$

where an estimator of the variance of the Kaplan-Meier estimator is needed:

$$Var(\widehat{\hat{R}}_{KM}(t)) = [\hat{R}_{KM}(t)]^2 \sum_{j:M_{(j)} \le t} \frac{\Delta_{(j)}}{(n - j)(n - j + 1)}.$$

We now turn to the KG-model, and we are interested in estimators of $R_T(t)$ with best (asymptotic) properties.

Since $R_T(t) = [R_M(t)]^p$, Abdushukurov (1984) and Cheng and Lin (1984) introduced the estimator

$$\widehat{R_{ACL}}(t) = [\widehat{R_M}(t)]^{\hat{p}} \qquad (3.5)$$

where $\widehat{R_M}(t) = 1/n \sum_{j=1}^n \mathbf{1}\{M_j > t\}$ and $\hat{p} = 1/n \sum_{j=1}^n \Delta_j$. The denotation $\widehat{R_{ACL}}(t)$ is given by Csörgő (1988); he also studied properties of this estimator. A detailed representation of properties of the ACL-estimator is contained in the Ph.D. thesis of Pawlitschko (1996). One of these results is

Theorem 3.3.2 *The estimator $\widehat{R_{ACL}}(t)$ given by (3.5) is uniformly strongly consistent, and the process*

$$\{n^{1/2}(\widehat{R_{ACL}}(t) - R_T(t)), 0 \le t \le c\} \tag{3.6}$$

with $c = sup\{t > 0 : R_M(t) > 0\}$ converges in law to a Gaussian process.

PROOF. See Csörgő (1988), Pawlitschko (1996). ∎

Theorem 3.3.2 shows that the estimator of $R_T(t)$ is asymptotically normal, too. If the Assumption (KG) is fulfilled, Pawlitschko (1996) proved that $\widehat{R_{ACL}}(t)$ is uniformly asymptotically more efficient than $\widehat{R_{KM}}(t)$. However, most data in practice do not satisfy the Koziol-Green assumption. In order to keep the useful properties and the structure of the ACL-estimator and to apply the estimation for general data structures, the KG-model has been generalized. For instance, Pawlitschko (1996) investigates models with mixed censorship.

We consider triplets (T_j, C_j, Z_j), $j = 1, ..., n$, of random variables where T_j and C_j are described above ($R_C(t) = [R_T(t)]^\gamma$) and Z_j is a certain noninformative censoring time.

Now,

$$M_j = \min(T_j, C_j, Z_j), \quad \Delta_j = \begin{cases} 1, & T_j \le \min(C_j, Z_j), \\ 0, & C_j \le \min(T_j, Z_j), \\ -1 & otherwise. \end{cases} \tag{3.7}$$

We have $R_M(t) = P(\min(T, C, Z) > t) = R_T(t)R_C(t)R_Z(t) = [R_T(t)]^{1+\gamma}R_Z(t)$ and $p = \frac{1}{1+\gamma}$. Using $R_U(t) := [R_T(t)]^{1/p}$ and $\epsilon_j = 1$ if $\Delta_j \ne -1$, $\epsilon_j = 0$ if $\Delta_j = -1$ $(j = 1, ..., n)$, then $R_U(t)$ is estimated by the Kaplan-Meier estimator

$$\widehat{R}_{U.KM}(t) = \begin{cases} \prod_{j=1}^n (1 - \frac{\epsilon_{(j)}}{n-j+1})^{\mathbf{1}\{M_{(j)} \le t\}}, & 0 \le t \le M_{(n)}, \\ 0, & t > M_{(n)}. \end{cases} \tag{3.8}$$

The estimator of p is given by

$$\hat{p} = \frac{\hat{p}_1}{\hat{p}_1 + \hat{p}_0}, \quad \hat{p}_r := 1/n \sum_{j=1}^n \mathbf{1}\{\Delta_j = r\} \quad (r = 0, 1).$$

Therefore, the mixed ACL-estimator (ACLm) of $R_T(t)$ has the form

$$\widehat{R_{ACLm}}(t) = [\widehat{R_{U.KM}}(t)]^{\hat{p}}.$$

3.4 The Failure-Repair Model Under Censoring

3.4.1 The general model

A technical appliance working continuously is considered, for instance an equipment for certain control functions. The appliance works continuously up to a shut-down caused by a failure or by an inspection. We assume that a shut-down immediately leads to a certain repair action. Working periods and repair periods alternate and are independent, and the consideration starts with a new appliance.

Let us introduce the failure-repair process $\{X(t), t \geq 0\}$ on some probability space $(\Omega, \mathcal{F}, P_\vartheta)$ having values only zero and one.

The trajectories of $\{X(t), t \geq 0\}$ have jump time points being end points of working or of repair periods, respectively. Consider the pairs (t_n, ξ_n), $n = 1, 2, \ldots$ Connected with the shut-down points t_{2k-1}, labels ξ_{2k-1} $(k = 1, 2, \ldots)$ are given according to a certain label distribution or to a strategy fixed at the beginning. The label values determine the degree (type) of the following repairs; the values could be chosen as follows:

$$\xi_n = \begin{cases} -1, & n = 2k, & \text{(end of repair)}, \\ 0, & n = 2k-1, & \text{(censoring case)}, \\ 1, & n = 2k-1, & \text{(failure, minimal repair)}, \\ 2, & n = 2k-1, & \text{(failure, perfect repair)}, \\ 3, & n = 2k-1, & \text{(failure, partial repair)}. \end{cases} \tag{3.9}$$

In the cases of shut-downs the labels define degrees of repair $d_k = d(\xi_{2k-1}) \in [0, 1]$; in what follows we choose $d_k = 1$ for a minimal repair, $d_k = 0$ for a perfect repair and $d_k = c$, $0 < c < 1$, in the case of partial repair. Moreover, for the remainder it is assumed that the repair after censoring is perfect, i.e. $\xi_{2k-1} = 0$ implies $d_k = d(0) = 0$.

Equivalent to the marked point process (t_n, ξ_n), $n = 1, 2, \ldots$, a description with counting processes is usual. We shall prefer this one. We introduce $\{N(t), t \geq 0\}$ defined on $(\Omega, \mathcal{F}, P_\vartheta)$ with $N(t) = (N_0(t), N_1(t), N_2(t))$ and with values in $(R^3, \mathcal{B}_{R^3})$, where ϑ is an unknown parameter with values in an open set $\Theta \subset R^m$. $N^+(t) := N_0(t) + N_1(t)$ is the number of shut-down cases and $N_2(t)$ counts the finished repairs. The counting components are given by

$$N_0(t) = \sum_{k=1}^{N^+(t)} \mathbf{1}\{t_{2k-1} \leq t, \xi_{2k-1} = 0\} \tag{3.10}$$

$$N_1(t) = \sum_{k=1}^{N^+(t)} \mathbf{1}\{t_{2k-1} \leq t, \xi_{2k-1} > 0\} \tag{3.11}$$

$$N_2(t) \; = \; \sum_{k=1}^{N^+(t)} \mathbf{1}\{t_{2k} \le t\} \tag{3.12}$$

Let $P_{\vartheta,t}$ denote the restriction of P_ϑ to the σ-algebra $\mathcal{F}_t = \sigma\{N(s) : s \le t\} \subset \mathcal{F}$. Suppose that for each $t \ge 0$ the family $P_{\vartheta,t}, \vartheta \in \Theta$, is dominated by the corresponding restricted Poisson measure ν_t with the mean parameter $\lambda = 1$. Assume that there exist intensity functions $\lambda_i(t,\vartheta)$ associated with $N_i(t), i = 0,1,2$. Then, under certain regularity conditions, the density function of $N(t)$ (likelihood function) is of the form

$$L(t,\vartheta) = \frac{dP_{\theta,t}}{d\nu_t} = \exp\left\{\sum_{i=0}^{2}\left[\int_0^t \log \lambda_i(u,\vartheta)dN_i(u) - \int_0^t \lambda_i(u,\vartheta)du\right]\right\} \tag{3.13}$$

[see Liptser and Shiryaev (1978) or Pruscha (1985)]. The process $\Lambda_i(t,\vartheta) = \int_0^t \lambda_i(u,\vartheta)\,du, \; t \ge 0 \; (i = 0,1,2)$, is the compensator of $N_i(t), t \ge 0$, and $N_i(t) - \Lambda_i(t,\vartheta)$ is a zero-mean martingale with respect to $(\mathcal{F}_t, P_\vartheta)$.

Before turning to the detailed structure of intensities we introduce the virtual-age process $\{V(t), t \ge 0\}$ characterizing the wear and tear of an appliance, with

$$V(t) = \begin{cases} 0, & t = 0, \\ t - t_{2k-2} + V(t_{2k-2}), & t_{2k-2} < t < t_{2k-1} \quad (k = 1,2,...), \\ 0, & t_{2k-1} \le t < t_{2k}, \\ d(\xi_{2k-1})V(t_{2k-1} - 0), & t = t_{2k} \end{cases} \tag{3.14}$$

where $V(t_{2k-1} - 0) = \lim_{t\uparrow t_{2k-1}} V(t)$.

Assumption (I). The intensity functions corresponding to the counting process components $N_i(t)$ ($i = 0,1,2$) are of the form

$$\lambda_i(t,\vartheta) = \exp(\theta_{1i})\mu_i(t,\theta_{2i}),$$

where $\vartheta = (\theta_1, \theta_2)$, $\theta_1 = (\theta_{10}, \theta_{11}, \theta_{12})$, $\theta_2 = (\theta_{20}, \theta_{21}, \theta_{22})$, and $\mu_i(t,\theta_{2i}) \ge 0$ is a deterministic or random function twice continuously differentiable with respect to θ_{2i}.

The class of counting processes satisfying Assumption (I) covers many processes which are of importance in view of their large applicability in reliability systems theory. Some special models are:

1. Weibull-type processes: $\mu.(s,\theta_{2.}) = \theta_{2.}(s)^{\theta_{2.}-1}$,

2. log-linear-type processes: $\mu.(s,\theta_{2.}) = \theta_{2.}\exp(\theta_{2.}s)$,

3. Pareto-type processes: $\mu.(s,\theta_{2.}) = \theta_{2.}/(1 + \theta_{2.}s)$,

where $s = s(t)$ is the actual time; the intensity of the considered process depends on $s(t)$, e.g. $s(t) = t$ for a nonhomogeneous Poisson process, $s(t) = t - t_{j-1}$, $(t_{j-1} \leq t < t_j, t_j \ (j = 0, 1, \ldots; t_0 = 0)$ are renewal time points) for a renewal process and $s(t) = V(t)$ in the case of the failure counting process introduced above.

Let τ be a stopping time relative to $\mathcal{F}_t, t \geq 0$, such that $P_\vartheta(\tau < \infty) = 1$ for each $\vartheta \in \Theta$. Suppose that the process can be observed permanently during the random time interval $[0, \tau]$. Under Assumption (I) the likelihood function is [see, for instance, Franz (1994)]

$$L(\tau, \vartheta) = \exp\left\{ \sum_{i=0}^{2} [\theta_{1i} N_i(\tau) - \exp(\theta_{1i}) S_i(\tau, \theta_{2i}) + M_i(\tau, \theta_{2i})] \right\}, \quad (3.15)$$

where

$$S_i(\tau, \theta_{2i}) = \int_0^\tau \mu_i(s_i(u), \theta_{2i}) du,$$

$$M_i(\tau, \theta_{2i}) = \sum_{j=1}^{N_i(\tau)} \log \mu_i(s_i(t_{ij} - 0), \theta_{2i})$$

(the t_{ij}'s are the jump points of $N_i(t)$, $s_0(u) = u - t_{0j}$ $(t_{0j} \leq u < t_{0,j+1})$, $s_1(u) = V(u)$ and $s_2(u) = u - t_{2j}$ $(t_{2j} \leq u < t_{2,j+1})$).

Remark 3.4.1

1. For estimating the parameter components of the vector ϑ one can propose the maximum likelihood estimators $\hat{\theta}_{ri}$ of θ_{ri} $(r = 1, 2; i = 1, \ldots, k)$. These are a solution of the equation $\mathcal{U}(\tau, \vartheta) = (\partial/\partial\vartheta)(\log L(\tau, \vartheta)) = 0$. The likelihood function (3.15) takes a convenient form with the components linear in $(N_i(\tau), M_i(\tau, \theta_{2i}), S_i(\tau, \theta_{2i}))$, and the score vector is

$$\mathcal{U}(\tau, \vartheta) = \begin{pmatrix} -\exp(\theta_{20})\dot{S}_0(\tau, \theta_{10}) + \dot{M}_0(\tau, \theta_{10}) \\ N_0(\tau) - \exp(\theta_{20})S_0(\tau, \theta_{10}) \\ \vdots \\ -\exp(\theta_{22})\dot{S}_2(\tau, \theta_{12}) + \dot{M}_2(\tau, \theta_{12}) \\ N_2(\tau) - \exp(\theta_{22})S_2(\tau, \theta_{12}) \end{pmatrix}.$$

(Here an upper dot denotes the derivative with respect to the unknown parameter.) In general, the estimators for θ_{ri} can only be obtained by means of numerical procedures.

2. $(N_i(\tau), S_i(\tau, \theta_{2i}))$ is a semi-sufficient statistic with respect to θ_{1i} in the sense of Willing (1987).

3.4.2 Model under Koziol-Green assumption

In what follows we focus on relations between failures and censoring cases; we shall neglect the repair periods (for instance, because of known repair time distributions). In certain practical problems, it is reasonable to use the Koziol-Green model for investigating the censorship. Equivalently to Assumption (KG) we use

$$\lambda_0(t,\vartheta) = \gamma\lambda_1(t,\vartheta) \tag{3.16}$$

with $\gamma > 0$. We put $\theta_{10} = \theta_{11} = \alpha$ and $\theta_{20} = \theta_{21} = \beta$. Then we obtain

$$\begin{aligned}
\lambda_1(t,\vartheta) &= \exp(\alpha)\mu_1(V(t),\beta), \\
\lambda_0(t,\vartheta) &= \gamma\exp(\alpha)\mu_1(t-t_{2k-1},\beta) \quad (t \in [t_{2k-1},t_{2k}]).
\end{aligned} \tag{3.17}$$

As a special case of (3.15) the likelihood function under Assumption (KG) is

$$L_{KG}(\tau,\vartheta) = \exp\left\{\alpha N^+(\tau) - \exp(\alpha)S(\tau,\beta,\gamma) + M(\tau,\beta) + (\log\gamma)N_0(\tau)\right\} \tag{3.18}$$

where

$$S(\tau,\beta,\gamma) = \gamma S_0(\tau,\beta) + S_1(\tau,\beta), \quad M(\tau,\beta) = M_0(\tau,\beta) + M_1(\tau,\beta)$$

and $S_0(\tau,\beta), S_1(\tau,\beta), M_0(\tau,\beta)$ and $M_1(\tau,\beta)$ are explained by (3.17) and (3.15).

In order to characterize the counting processes, the distributions of operating times and jump probabilities are useful too.

Theorem 3.4.1 *Let $T_k = t_{2k-1} - t_{2k-2}$ $(k = 1, 2, ...)$ be the k-th active period and $\{N_0(t),\ t \geq 0\}$ and $\{N_1(t),\ t \geq 0\}$ belong to a Weibull type. Then, the distribution of T_k, given $t_{2k-2} = s$, is obtained by*

$$P_\vartheta(T_k > t|t_{2k-2} = s) = \exp\left\{-e^\alpha[(t+V(t))^\beta - (V(t))^\beta + \gamma t^\beta]\right\}. \tag{3.19}$$

For the censoring probability we have

$$P_\vartheta(\xi_{2k-1} = 0|T_k = u, t_{2k-2} = s) = \frac{\gamma u^\beta}{[u+V(s)]^\beta + \gamma u^\beta} \cdot \tag{3.20}$$

PROOF. Is directly obtained using Theorem 1 in Franz (1994). ∎

Immediately we see that results by analogy with (3.19) and (3.20) are obtained in log-linear and Pareto types of intensities.

The following theorem gives a useful factorization of the likelihood function (3.18).

Theorem 3.4.2 *Let $\tau = \inf\{t > 0 : N^+(t) = n\}$ where n is a given integer. Then*

(a) the likelihood function $L_{KG}(\tau, \vartheta)$ has the form

$$L_{KG}(\tau, \vartheta) = g_S(\tau; \beta, \gamma)h(S(\tau), \vartheta), \qquad (3.21)$$

with

$$g_S(\tau; \beta, \gamma) = \frac{(n-1)!\gamma^{N_0}}{[S(\tau)]^{n-1}} \exp[M(\tau, \beta)] \qquad (3.22)$$

and

$$h(S(\tau); \vartheta) = \frac{e^{\alpha n}}{(n-1)!}[S(\tau)]^{n-1} \exp[-e^{\alpha}S(\tau)], \qquad (3.23)$$

where $S(\tau) = S(\tau; \beta, \gamma)$;

(b) for β and γ fixed, the variable $S(\tau; \beta, \gamma)$ is gamma distributed with parameters n and e^{α}.

The proof is similar to that of Franz and Magiera (1997b).

Note that for β and γ fixed, the random variable $S(\tau; \beta, \gamma)$ is a sufficient statistic for α relative to $\mathcal{F}_\tau$.

Now, let us turn to Bayes parameter estimation in the case of the Koziol-Green model. In order to simplify, the consideration is restricted to counting processes of Weibull types. For the Bayes set-up, let ϑ be the true process parameter, and let Q denote a prior probability distribution on Θ corresponding to the random variable θ. The failure-repair process can be observed up to τ; the likelihood function is given by (3.18). Using the prior distribution and quadratic loss, the Bayes estimator is obtained as the posterior mean value of θ. The question arises what prior distribution is suitable. In the specialized model we find the following semi-conjugate prior density.

Theorem 3.4.3 *Let Q possess the density $q(\vartheta)$ where $\vartheta = (\alpha, \beta, \gamma) \in R \times R^+ \times R^+$. Then, the posterior density $q_\tau(\vartheta)$ exists, and*

$$q(\vartheta) = q(\vartheta, z) = \frac{b^d \gamma^{d-1}}{a^{d-c}\Gamma(d)\Gamma(c-d)} \exp\left\{c\alpha - e^{\alpha}(a+b\gamma)\right\} p(\beta) \qquad (3.24)$$

is semi-conjugate relative to (α, γ), where $z = (a, b, c, d)$ is a vector of positive hyper-parameters with $d < c$ and $p(\beta)$ is any marginal density w.r.t. β.

PROOF. The assertion is immediately obtained from (3.18) and (3.24). ∎

We remark that the exponential part of (3.24) belongs to a log-gamma-beta density; in the case of posterior density the hyper-parameters change to $z_\tau = (a + S_0, b + S_1, c + N^+, d + N_0)$, the marginal density becomes $p_\tau(\beta) = w(\beta)[\int w(u)du]^{-1}$, $w(u) = \exp(M(u, \tau))p(u)\frac{(a+S_0(u,\tau))^{d-c-N_0}}{(b+S_1(u,\tau))^{d+N_1}}$.

Using the prior density (3.24) we study the structure of the Bayes estimator for functions

$$h(\vartheta) = \gamma^v \exp(u\alpha - e^\alpha(w_0 + \gamma w_1))$$
$$(u > -c, v > -d, w_0 < -a, w_1 > -b, u - v > d - c). \quad (3.25)$$

Theorem 3.4.4 *Under a quadratic loss function and by use of the prior density (3.24) the Bayes estimator for (3.25) is given by*

$$\widehat{h(\tau)} = \frac{\Gamma(v + d + N_0)\Gamma(u + c - v - d + N_1)}{\Gamma(d + N_0)\Gamma(c - d + N_1)} \quad (3.26)$$

$$\times \int_0^\infty \frac{(b + S_1(x))^{d+N_1}(a + S_0(x))^{c-d+N_0}p_\tau(x)}{(b + S_1(x) + w_1)^{v+d+N_1}(a + S_0(x) + w_0)^{u-v+c-d+N_0}} \, dx.$$

$$(3.27)$$

The assertion of Theorem 3.4.4 is a generalization of Theorem 2 in Franz (1994). The proof is analogous to that one.

Example 3.4.1

1. $h(\vartheta) = \gamma \longleftrightarrow \widehat{h(\tau)} = \frac{d+N_0}{c-d+N_1-1} \int_0^\infty \frac{(a+S_0(x))p_\tau(x)}{b+S_1(x)} \, dx.$

2. $h(\vartheta) = \gamma \exp(\alpha) \longleftrightarrow \widehat{h(\tau)} = (d + N_0) \int_0^\infty \frac{p_\tau(x)}{b+S_1(x)} \, dx.$

Finally, we are concerned with parameter estimation in the case of a nuisance parameter using estimating functions in sense of Godambe (1960). Consider $\vartheta = (\vartheta_1, \vartheta_2)$ where $\vartheta_1 \in \Theta_1$ is the parameter of interest, and $\vartheta_2 \in \Theta_2$ is a nuisance parameter. In the Godambe's estimation approach which in many cases turns out to be the more preferable way in point estimation theory, not estimators but estimating functions $D(\tau, \vartheta_1)$ are considered, where $D(\tau, \vartheta_1)$ is $\mathcal{F}_\tau$-measurable for each $\vartheta \in \Theta$. $D(\tau, \vartheta_1)$ is a function of both data and unknown parameter ϑ_1. It is supposed that the plan $\delta := (\tau, D(\tau, \vartheta_1))$ satisfies the conditions:

(i) $E_\vartheta D(\tau, \vartheta_1) = 0$ ($\vartheta \in \Theta$),

(ii) $\partial D(\tau, \vartheta_1)/\partial \vartheta_1$ exists ν_τ-a.s. ($\vartheta \in \Theta$),

(iii) $0 < E_\vartheta[\partial D(\tau, \vartheta_1)/\partial \vartheta_1]^2 < \infty$ for all $\vartheta \in \Theta$,

and some regularity conditions concerning the interchanging of integration and differentiation w.r.t. ϑ_1.

The parameter ϑ_1 will be estimated by $\widehat{\vartheta}_1(\tau)$ as a solution of the *estimating equation* $D(\tau, \vartheta_1) = 0$, and the plan $\delta = (\tau, D(\tau, \vartheta_1))$ will then be called an *estimating equation plan*. A plan δ satisfying the regularity conditions given above will be called a *regular estimating plan*. The class of all regular estimating plans for ϑ_1 will be denoted by $\mathcal{D}_1$.

Definition 3.4.1 A plan $\delta^* = (\tau, D^*(\tau, \vartheta_1)) \in \mathcal{D}_1$ is said to be optimum if

$$E_\vartheta \left\{ \frac{D^*(\tau, \vartheta_1)}{E_\vartheta \left[\frac{\partial D^*(\tau, \vartheta_1)}{\partial \vartheta_1} \right]} \right\}^2 \leq E_\vartheta \left\{ \frac{D(\tau, \vartheta_1)}{E_\vartheta \left[\frac{\partial D(\tau, \vartheta_1)}{\partial \vartheta_1} \right]} \right\}^2$$

for all $\delta = (\tau, D(\tau, \vartheta_1)) \in \mathcal{D}_1$ and all $\vartheta \in \Theta$.

Now, one can prove that the maximum likelihood estimating function $\frac{\partial \log L(\tau, \vartheta)}{\partial \vartheta}$ belongs to an optimum plan in the case of the full likelihood function problem (see Franz and Magiera (1997b)). Using Theorem 3.4.2 and $\vartheta_1 = (\beta, \gamma)$, ($\alpha$ is taken as a nuisance parameter) the estimating function $\frac{\partial \log g_S(\tau, \vartheta_1)}{\partial \vartheta_1}$ where $g_S(\tau, \vartheta_1)$ is given by (3.22) is optimal relative to the conditional likelihood function.

Theorem 3.4.5 *For the complete plans* $\delta = (\tau, D(\tau, \vartheta_1)) \in \mathcal{D}_1$, *the inequality*

$$E_\vartheta \left\{ \frac{D(\tau, \vartheta_1)}{E_\vartheta \left[\frac{\partial D(\tau, \vartheta_1)}{\partial \vartheta_1} \right]} \right\}^2 \geq \frac{1}{E_\vartheta \left[\frac{\partial \log g_S(\tau, \vartheta_1)}{\partial \vartheta_1} \right]^2}$$

is valid for all $\vartheta \in \Theta$, *the equality being attained when* $D(\tau, \vartheta_1)$ *is given by*

$$D^*(\tau, \vartheta_1) = \frac{\partial \log g_S(\tau, \vartheta_1)}{\partial \vartheta_1} \tag{3.28}$$

up to a non-zero multiplicative constant $c(\vartheta_1)$ *and up to sets of* μ_τ-*measure zero.*

The proof proceeds as in Godambe (1976) (compare to Franz and Magiera (1997b)).

Under the stated conditions Theorem 3.4.5 justifies replacing the maximum likelihood equation

$$\frac{\partial \log L(\tau, \vartheta)}{\partial \vartheta_1} = 0$$

by the maximum conditional likelihood equation

$$\frac{\partial \log g_S(\tau, \vartheta_1)}{\partial \vartheta_1} = 0,$$

in order to eliminate the influence of the nuisance parameter ϑ_2.

Example 3.4.2 Let $\{N_0(t), t \geq 0\}$ and $\{N_1(t), t \geq 0\}$ be Weibull-type processes. Then $M(\tau, \beta) = (\log \beta)n + (\beta + 1)M(\tau)$ and $M(\tau)$ is a special sum obtained from (3.17) and (3.15). The optimum estimating function $D^*(\tau, \vartheta_1)$ for (β, γ) is given by

$$D^*(\tau, \vartheta_1) = \begin{pmatrix} n\beta^{-1} + M(\tau) - (n-1)[S(\tau, \beta, \gamma)]^{-1}\frac{\partial}{\partial \beta}S(\tau, \beta, \gamma) \\ \gamma^{-1}N_0(\tau) - (n-1)[S(\tau, \beta, \gamma)]^{-1}S_0(\tau, \beta) \end{pmatrix}.$$

The estimation of β and γ is carried out without knowledge of the estimator of α.

References

1. Abdushukurov, A. A. (1984). On some estimates of the distribution function under random censorship, In *Conference of Young Scient. Math. Inst. Acad. Sci. Uzbek.*, Taschkent, No. 8756-V (in Russian).

2. Breslow, N. E. (1992). Introduction to Kaplan and Meier (1958), Nonparametric estimation from incomplete observations, In *Breakthroughs in Statistics*, Vol. II (Eds., S. Kotz and N. L. Johnson), pp. 311–318, Methodology and Distribution, New York: Springer-Verlag.

3. Cheng, P. E. and Lin, G. D. (1984). Maximum likelihood estimation of survival function under the Koziol-Green proportional hazard model, *Technical Report B-84-5*, Institute of Statistics, Academica Sinica, Taipei, Taiwan

4. Csörgö, S. (1988). Estimation in the proportional hazards model of random censorship, *Statistics*, **19**, 437–463.

5. Ferreira, P. E. (1982). Sequential estimation through estimating equations in the nuisance parameter case, *Annals of Statistics*, **10**, 167–173.

6. Fleming, T. R. and Harrington, D. P. (1991). *Counting Processes and Survival Analysis*, New York: John Wiley & Sons.

7. Franz, J. (1994). On estimation problems in random censored repair models, *EQC*, **9**, 125–142.

8. Franz, J. and Magiera, R. (1997a). On information inequalities in sequential estimation for stochastic processes, *Mathematical Methods of Operatations Research (ZOR)*, **46**, 1–27.

9. Franz, J. and Magiera, R. (1997b). Sequential estimation for a family of counting processes in the nuisance parameter case, *Statistical Papers* (to appear).

10. Godambe, V. P. (1960). An optimum property of regular maximum likelihood estimation, *Annals of Mathematical Statistics*, **31**, 1208–1212.

11. Godambe, V. P. (1976). Conditional likelihood and unconditional optimum estimating functions, *Biometrika*, **63**, 277–284.

12. Godambe, V. P. (1984). On ancillarity and Fisher information in the presence of a nuisance parameter, *Biometrika*, **71**, 626–629.

13. Godambe, V. P. and Thompson, M. E. (1974). Estimating equations in the presence of a nuisance parameter, *Annals of Statistics*, **2**, 568–574.

14. Hurt, J. (1992). On statistical methods for survival data analysis, *Proceedings of the Summer School JČMF (ROBUST 1992)*, Prague 1992, 54–74.

15. Kaplan, E. L. and Meier, P. (1958). Nonparametric estimation from incomplete observations, *Journal of the American Statistical Association*, **53**, 457–481.

16. Koziol, J. A. and Green, S. B. (1976). A Cramér-von-Mises statistic for randomly censored data, *Biometrika*, **63**, 456–474.

17. Koziol, J. A. (1980). Goodness of fit tests for randomly censored data, *Biometrika*, **67**, 693–696.

18. Liptser, R. S. and Shiryaev, A. N. (1978). *Statistics of Random Processes*, Vol. II, New York: Springer-Verlag.

19. Pawlitschko, J. (1996). Die Schätzung einer Überlebensfunktion in Verallgemeinerung des Koziol-Green-Modells, *Dissertation*, Universität Dortmund.

20. Pruscha, H. (1985). Parametric inference in Markov branching processes with time-dependent random immigration rate, *Journal of Applied Probability*, **22**, 503–517.

21. Stadje, W. and Zuckerman, D. (1991). Optimal maintenance strategies for repairable systems with general degrees of repair, *Journal of Applied Probability*, **28**, 384–396.

22. Stute, W. and Wang, J.-L. (1993). The strong law under random censorship, *Annals of Statistics*, **21**, 1591–1607.

23. Willing, R. (1987). Semi-sufficiency in accelerated life testing, In *Probability and Bayesian Statistics* (Ed., R. Viertl), New York: Plenum Press.

4

Parameter Estimation in Renewal Processes with Imperfect Repair

---

Sofiane Gasmi and Waltraud Kahle

Otto–von–Guericke–Universität Magdeburg, Magdeburg, Germany

Abstract: In this paper we develop statistical methods for a general repair model from Last and Szekli (1995).

For determining the model parameters the maximum likelihood estimator is considered. Special results are obtained by the use of Pareto, Log-linear and Weibull-type intensities. Estimations for the degree of repair in a simple model are developed.

Keywords and phrases: Renewal processes, imperfect repair, parameter estimation

---

4.1   Introduction

Most of the papers concerning the stochastic behavior of repairable systems assume two types of repair, perfect repair and minimal repair. It is well known in practice that repair may not yield a functioning item which is "as good as new". On the other hand, the minimal repair assumption seems to be to pessimistic in realistic repair strategies. From this it is seen that the imperfect repair is of great signification in practice.

In this paper we assume that all repair times are small and can be neglected.

The idea of virtual age process has been introduced by Kijima (1989) and Stadje and Zuckerman (1991).

Kijima (1989) considered general repairs and introduced the concept of virtual age in general repair models for a repairable system. Brown and Proschan (1983) assumed that repair is perfect with probability p and minimal with probability $1 - p$. Block et al. (1993) generalized this model to the case where p is time-dependent. Stadje and Zuckerman (1991) discussed optimal maintenance policies under general repair.

Each of these models includes the renewal process and the nonhomogeneous Poisson process (NHPP) as special cases.

For the renewal process, the inter-failure times are independent and identically distributed so that we receive an as well as new item after substitution of a failed item through a new, or its repair. Such repairs are said to be *perfect*.

For the NHPP, the survivor function of the time to the next failure given that there was a failure at time x is $\bar{F}(\cdot + x)/\bar{F}(x)$, where $\bar{F} = 1 - F$ and F denotes the distribution function of the time to the first failure. Thus, we may view the repair makes a system again able to work, but after the minimal repair the failure rate of the system is the same as just before the failure. Such repairs are called *minimal*.

Most of the models concerning the stochastic behavior of repairable systems are insufficiently flexible to permit for an *imperfect repair*, this can make the state of the system immediately after repair intermediate between that of a new system and that obtained by minimal repair.

The Kijima's models I and II are modified by Baxter, Kijima, and Tortorella (1996). The new developed model is more flexible than its predecessors and it is defined in terms of two distribution functions that admit natural realistic interpretations.

Interesting statistical studies of parameter estimation in systems with various degrees of repair are given by Bathe and Franz (1996).

4.2   A General Model

We consider a general model of repairable systems introduced by Last and Szekli (1995). This model includes the special cases of classical perfect repair, minimal repair, imperfect repair and general repairable systems of Kijima (1989), Brown and Proschan (1983), Stadje and Zuckerman (1991). In this paper we will examine processes counting unplanned repairs only.

The model of Last and Szekli is based on the following assumptions:

The system starts working with an initial item having a prescribed failure rate $\lambda_1(t) = \lambda(t)$. Let T_1 denote the random time, when the item fails. At this time T_1 the item will be repaired with the random degree Z_1. The degree of repair is between 0 and 1, where the case of 0 corresponds to the minimal repair and the case of 1 to the perfect repair. The age of the item is decreased to $(1 - Z_1)T_1$ which is called the virtual age of the item at time T_1 and is denoted by V_1. The distribution of the time until the next failure has then the failure rate $\lambda_2(t) := \lambda(t - T_1 + V_1)$. Assume now that T_n is the time of the n-th $(n \geq 1)$ failure and that Z_n is the degree of repair at that time. After repair the failure rate of the $(n+1)$–th waiting time until the next failure is determined by

$$\lambda_{n+1}(t) := \lambda(t - T_n + V_n), \quad T_n \leq t < T_{n+1}, \; n \geq 0,$$

where $V_n := (1 - Z_n)(V_{n-1} + T_n - T_{n-1}), \quad V_0 := 0, \; T_0 := 0.$

The process defined by $V(t) := t - T_n + V_n, \; T_n \leq t < T_{n+1}, \; n \geq 1$ is called the *virtual age process*. A realization of this virtual age process is shown in Figure 4.1. Here t_i, $i = 1, \ldots, 5$ are the realizations of T_i, at times t_1 and t_3 we have a perfect repair, at times t_4 and t_2 we have a minimal repair and an imperfect repair, respectively. The corresponding failure rate is shown in Figure 4.2.

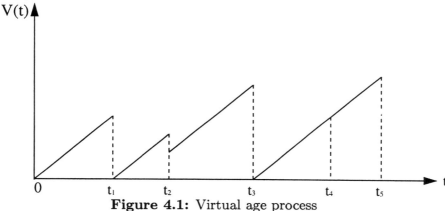

Figure 4.1: Virtual age process

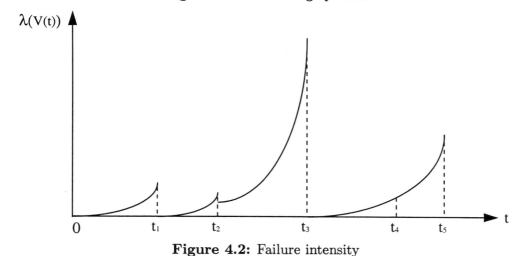

Figure 4.2: Failure intensity

4.3 Specifications

Now we want to deduce special cases from the general model.

(i) Perfect repair:

If we choose the consecutive repair degrees deterministically equal to one, then $V_n = 0$ for all $n \geq 1$ and the consecutive failure rates are given by $\lambda_{n+1}(t) = \lambda(t - T_n)$, $T_n \leq t < T_{n+1}$. Then the failure counting process is a renewal process with the failure rate $\lambda(t)$.

(ii) Minimal repair:

Choosing the repair degrees constant and equal to zero, then the virtual age V_n is equal to T_n for all $n \geq 1$. The consecutive failure rates are given by $\lambda_{n+1}(t) = \lambda(t)$, $T_n \leq t < T_{n+1}$. Then the failure counting process is a nonhomogeneous Poisson process with the intensity function $\lambda(t)$.

(iii) The model of Brown and Proschan (1983):

Let (Z_k) $k = 1, \ldots, n$ denote a sequence of i.i.d. random variables independent of $(T_1, \ldots, T_k)$ with $1 - P(Z_k = 0) = P(Z_k = 1) = p$, $p \in (0, 1)$. The failure rate process is given by $\lambda_{n+1}(t) = \lambda(t - T_n + (1 - Z_n)M_n)$, where M_n is the time from T_n to the last perfect repair. The idea of this model is: After a failure the item is replaced by a new one with probability p, that means, we realize a perfect repair, or the item is minimally repaired with probability $(1 - p)$.

(iv) The model I of Kijima (1989):

Let us take a sequence of i.i.d. random variables $(A_k)_{k \geq 1}$ uniformly distributed on $[0, 1]$. It is assumed that (A_k) is independent of (T_k) for $k = 1, \ldots, n$. Then $V(T_n) = V(T_{n-1}) + A_n X_n$, with $X_n = T_n - T_{n-1}$, $T_0 = 0$ and $A_0 = 0$. The consecutive failure rate is determined by

$$\lambda_{n+1}(t) = \lambda(t - T_n + A_n X_n + \ldots + A_1 X_1), \quad T_n \leq t < T_{n+1}.$$

If we put in the general model

$$Z_n := \frac{(1 - A_n)X_n}{A_1 X_1 + \ldots + A_{n-1} X_{n-1} + X_n} \quad n \geq 1,$$

then we obtain model I of Kijima.

(v) The model II of Kijima (1989):

This model is based on the virtual age notion too. We use the same condition as in model I of Kijima. But here we assume that $V(T_n) = A_n(V(T_{n-1}) + X_n)$, $(T_0 = 0$ and $A_0 = 0)$, and therefore $\lambda_{n+1}(t) = \lambda(t - T_n + A_n X_n + A_n A_{n-1} X_{n-1} + \ldots + A_n A_{n-1} \ldots A_1 X_1)$, $T_n \leq t < T_{n+1}$.
If we put in the general model $Z_n := 1 - A_n$, $n \geq 1$ then we obtain model II of Kijima.

In both models the consecutive failure rates depend on the whole history of the repair degree processes and the failure counting process.

(vi) The model of Stadje and Zuckerman (1991):

This model is similar to Kijima's model II, but the Z_n's are dependent on the underlying virtual age process $(V(t))$.

If we assume $Z_n := l(V(T_n-))/V(T_n-)$, where $V(T_n-)$ denotes the virtual age of the working item just before a failure and l is a deterministic function which fulfills $l(x) \leq x$, then we obtain the model of Stadje and Zuckerman.

4.4 Parameter Estimation in the General Model

Let $(T_k)_{k=1,2,\ldots}$ be the sequence of failure times and let (t_k) denote the sequence of a realization of (T_k). We introduce the sequence $(Z_k)_{k=1,2,\ldots}$ of degree of repair with realizations (z_k).

Let $N(t) = \sum_{k=1}^{\infty} \mathbf{1}(T_k \leq t)$ be the number of failures until time t.

We consider now a marked point process $\Phi = ((T_k, Z_k))$. The virtual age process $V(t)$ is defined as before. Φ is described by the counting process $\{N(t)\,,\ t \geq 0\}$ with intensity $\lambda_t = \lambda(V(t))$, where $V(t) = t - t_k + V_k$, $t_k \leq t < t_{k+1}$ and $k = 0, 1, 2, \ldots$, and by marks Z_k describing the degree of repair at time point T_k.

We denote by $D_1(\cdot \mid t_1)$ the conditional distribution function of Z_1 given $T_1 = t_1$. For $k \geq 1$, we define

$$D_{k+1}(z \mid t_1, z_1, \ldots, t_k, z_k, t_{k+1})$$
$$:= P(Z_{k+1} \leq z \mid T_1 = t_1, Z_1 = z_1, \ldots, T_k = t_k, Z_k = z_k, T_{k+1} = t_{k+1}),$$
$$z \in [0, 1].$$

$D_{k+1}(z \mid t_1, z_1, \ldots, t_k, z_k, t_{k+1})$ is the conditional distribution of the corresponding degree of repair conditionally on the past evolution of the process.

Calculating the intensity function of the marked point process Φ and using successive conditioning it follows from Last and Brandt (1995), formula (10.1.14), that the likelihood function of one observation has the form

$$L(t; \theta) = \left(\prod_{i=1}^{N(t)} \lambda(V(t_i-)) \cdot d_i(z \mid t_1, z_1, \ldots, t_{i-1}, z_{i-1}, t_i) \right)$$
$$\cdot \exp\left[\int_0^t (1 - \lambda(V(s)))ds \right],$$

where $d_i(z \mid t_1, z_1, \ldots, t_{i-1}, z_{i-1}, t_i)$ is the density function of Z_i conditionally on the past evolution of the process, and $V(t_i-) = t_i - t_{i-1} + V_{i-1}$.

If we use parametric models for the failure intensity $\lambda(t)$ and the distribution D_k of degree of repair, then it is necessary to estimate the parameters of the corresponding distributions by the observation of the repair process up to time t. The loglikelihood function

$$\ln L(t;\theta) = \sum_{i=1}^{N(t)} \ln \lambda(V(t_i-)) + \int_0^t (1 - \lambda(V(s)))ds$$
$$+ \sum_{i=1}^{N(t)} \ln d_i(z \mid t_1, z_1, \ldots, t_{i-1}, z_{i-1}, t_i)$$

consists of two terms: the first depends on the failure intensity λ and the second on the distribution of degree of repair only. If we assume that the degree of repair is independent of the past evolution of the process and that its distribution is known and does not consist of unknown parameters, it is possible to estimate intensity parameters from

$$\ln L(t;\theta) = \sum_{i=1}^{N(t)} \ln \lambda(V(t_i-)) + \int_0^t (1 - \lambda(V(s)))ds \,. \tag{4.1}$$

Of course the virtual ages V_i, $i = 1, \ldots, N(t)$ depend on the degree of repair of the realizations which must be observable.

4.4.1   Estimation of the parameters of failure intensity

In the following it is assumed that the degrees of repair (Z_i) are independent of each other and of (T_i) for $i = 1, \ldots, N(t)$, and have a known distribution function on $[0, 1]$.

   Now let us assume that the failure intensity exists and has the following form

$$\lambda(x) = \frac{\beta}{\alpha}\, \eta(\theta, x)\,, \ \theta = (\alpha, \beta) \in (0, \infty) \times (0, \infty),$$

where $\eta(\cdot)$ is a deterministic function of x. For various selections of η we can distinguish the following classes of intensity functions:

- Weibull type   : $\eta(\theta, x) = \left(\frac{x}{\alpha}\right)^{\beta-1}$,
- Pareto type   : $\eta(\theta, x) = \left(1 + \frac{x}{\alpha}\right)^{-1}$,
- Log–linear type : $\eta(\theta, x) = \exp\left(\frac{x}{\alpha}\right)$.

In the following we want to estimate the parameters of these distribution types. We assume that the virtual age process $V(t)$ is defined as before and that $t_0 = 0$.

Example 4.4.1 Weibull type

The failure intensity is given by

$$\lambda(x) = \frac{\beta}{\alpha} \left(\frac{x}{\alpha}\right)^{\beta-1}, \ \beta > 1, \ \alpha > 0,$$

where $\theta = (\alpha, \beta)$ is the unknown parameter. To calculate $\int_0^t \frac{\beta}{\alpha} \left(\frac{V(x)}{\alpha}\right)^{\beta-1} dx$
we can use the fact that $V(x)$ is linear between two failure times:

$$\int_0^t \frac{\beta}{\alpha} \left(\frac{V(x)}{\alpha}\right)^{\beta-1} dx = \sum_{i=1}^{N(t)} \int_{t_{i-1}}^{t_i} \frac{\beta}{\alpha} \left(\frac{V(x)}{\alpha}\right)^{\beta-1} dx + \int_{t_{N(t)}}^{t} \frac{\beta}{\alpha} \left(\frac{V(x)}{\alpha}\right)^{\beta-1} dx$$

$$= \sum_{i=1}^{N(t)} \left[\left(\frac{x - t_{i-1} + V_{i-1}}{\alpha}\right)^{\beta}\right]_{t_{i-1}}^{t_i} + \left[\left(\frac{x - t_{N(t)} + V_{N(t)}}{\alpha}\right)^{\beta}\right]_{t_{N(t)}}^{t}$$

$$= \sum_{i=1}^{N(t)} \left(\frac{t_i - t_{i-1} + V_{i-1}}{\alpha}\right)^{\beta} + \left(\frac{t - t_{N(t)} + V_{N(t)}}{\alpha}\right)^{\beta} - \sum_{i=1}^{N(t)} \left(\frac{V_i}{\alpha}\right)^{\beta}.$$

Then we get from (4.1)

$$\ln L(t; \theta) = N(t)(\ln \beta - \beta \ln \alpha) + (\beta - 1) \sum_{i=1}^{N(t)} \ln V(t_i-) + t$$

$$- \sum_{i=1}^{N(t)} \left(\frac{t_i - t_{i-1} + V_{i-1}}{\alpha}\right)^{\beta} - \left(\frac{t - t_{N(t)} + V_{N(t)}}{\alpha}\right)^{\beta} + \sum_{i=1}^{N(t)} \left(\frac{V_i}{\alpha}\right)^{\beta}. \quad (4.2)$$

The point estimators can be found in the usual way:

$$\hat{\alpha} = \left(\frac{\sum_{i=1}^{N(t)} (t_i - t_{i-1} + V_{i-1})^{\hat{\beta}} + (t - t_{N(t)} + V_{N(t)})^{\hat{\beta}} - \sum_{i=1}^{N(t)} V_i^{\hat{\beta}}}{N(t)}\right)^{1/\hat{\beta}} \quad (4.3)$$

and $\hat{\beta}$ is the solution of the following equation

$$\frac{1}{\hat{\beta}} - \left[\sum_{i=1}^{N(t)} (t_i - t_{i-1} + V_{i-1})^{\hat{\beta}} \ln(t_i - t_{i-1} + V_{i-1}) + (t - t_{N(t)} + V_{N(t)})^{\hat{\beta}}\right.$$

$$\left. \ln(t - t_{N(t)} + V_{N(t)}) - \sum_{i=1}^{N(t)} V_i^{\hat{\beta}} \ln V_i \right] \left[\sum_{i=1}^{N(t)} (t_i - t_{i-1} + V_{i-1})^{\hat{\beta}}\right.$$

$$\left. + (t - t_{N(t)} + V_{N(t)})^{\hat{\beta}} - \sum_{i=1}^{N(t)} V_i^{\hat{\beta}}\right]^{-1} + \frac{\sum_{i=1}^{N(t)} \ln V(t_i-)}{N(t)} = 0. \quad (4.4)$$

Remark 4.4.1

i) Perfect repair:

In the case of perfect repair we obtain

$$\hat{\alpha} = \left(\frac{\sum\limits_{i=1}^{N(t)} (t_i - t_{i-1})^{\hat{\beta}} + (t - t_{N(t)})^{\hat{\beta}}}{N(t)} \right)^{1/\hat{\beta}} \tag{4.5}$$

and the point estimators for the parameter β can be found by the numerical solution of the following equation

$$\frac{1}{\hat{\beta}} - \frac{\sum\limits_{i=1}^{N(t)} (t_i - t_{i-1})^{\hat{\beta}} \ln(t_i - t_{i-1}) + (t - t_{N(t)})^{\hat{\beta}} \ln(t - t_{N(t)})}{\sum\limits_{i=1}^{N(t)} (t_i - t_{i-1})^{\hat{\beta}} + (t - t_{N(t)})^{\hat{\beta}}}$$

$$+ \frac{\sum\limits_{i=1}^{N(t)} \ln(t_i - t_{i-1})}{N(t)} = 0. \tag{4.6}$$

ii) Minimal repair:

If we consider the case of minimal repair, then we obtain the explicit solutions

$$\hat{\alpha} = \left(\frac{t^{\hat{\beta}}}{N(t)} \right)^{1/\hat{\beta}} \quad \text{and} \tag{4.7}$$

$$\hat{\beta} = \left(\ln t - \frac{1}{N(t)} \sum\limits_{i=1}^{N(t)} \ln t_i \right)^{-1}. \tag{4.8}$$

Example 4.4.2 Pareto type

The failure intensity has the form

$$\lambda(x) = \frac{\beta}{\alpha} \left(1 + \frac{x}{\alpha} \right)^{-1}, \ \beta > 0, \ \alpha > 0,$$

with the unknown parameter $\theta = (\alpha, \beta)$. In the same way as before we find

$$\hat{\beta} = \left[\frac{\sum\limits_{i=1}^{N(t)} \ln(\hat{\alpha} + t_i - t_{i-1} + V_{i-1}) + \ln\left(\frac{\hat{\alpha} + t - t_{N(t)} + V_{N(t)}}{\hat{\alpha}} \right) - \sum\limits_{i=1}^{N(t)} \ln(\hat{\alpha} + V_i)}{N(t)} \right]^{-1} \tag{4.9}$$

and

$$\hat{\beta}\left[\sum_{i=1}^{N(t)}\frac{1}{\hat{\alpha}+t_i-t_{i-1}+V_{i-1}}+\frac{1}{\hat{\alpha}+t-t_{N(t)}+V_{N(t)}}-\frac{1}{\hat{\alpha}}-\sum_{i=1}^{N(t)}\frac{1}{\hat{\alpha}+V_i}\right]$$

$$+\sum_{i=1}^{N(t)}\frac{1}{\hat{\alpha}+V(t_i-)}=0. \tag{4.10}$$

The equation (4.10) has a unique solution $\hat{\alpha}$, but in order to get explicit values numerical methods are necessary.

Remark 4.4.2

i) Perfect repair:
   The point estimators are

$$\hat{\beta}=\left[\frac{\sum\limits_{i=1}^{N(t)}\ln(\hat{\alpha}+t_i-t_{i-1})+\ln(\hat{\alpha}+t-t_{N(t)})-(N(t)+1)\ln\hat{\alpha}}{N(t)}\right]^{-1} \tag{4.11}$$

and $\hat{\alpha}$ is the solution of the following equation

$$(\hat{\beta}+1)\sum_{i=1}^{N(t)}\frac{1}{\hat{\alpha}+t_i-t_{i-1}}+\hat{\beta}\left[\frac{1}{\hat{\alpha}+t-t_{N(t)}}-\frac{1+N(t)}{\hat{\alpha}}\right]=0. \tag{4.12}$$

ii) Minimal repair:
   The point estimators for the parameters can be found from

$$\frac{1}{N(t)}\sum_{i=1}^{N(t)}\frac{1}{\hat{\alpha}+t_i}+(\ln(\hat{\alpha}+t)-\ln\hat{\alpha})^{-1}\left(\frac{1}{\hat{\alpha}+t}-\frac{1}{\hat{\alpha}}\right)=0 \tag{4.13}$$

$$\text{and }\hat{\beta}=\left[\frac{1}{N(t)}\left(\ln(\hat{\alpha}+t)-\ln\hat{\alpha}\right)\right]^{-1}. \tag{4.14}$$

Example 4.4.3 Log-linear type

The failure intensity is of the form

$$\lambda(x)=\frac{\beta}{\alpha}\exp\left(\frac{x}{\alpha}\right),\ \beta>0,\ \alpha>0.$$

The point estimators can be found as before in the usual way:

$$\hat{\beta}=\left[\frac{\sum\limits_{i=1}^{N(t)}\exp\left(\frac{t_i-t_{i-1}+V_{i-1}}{\hat{\alpha}}\right)+\exp\left(\frac{t-t_{N(t)}+V_{N(t)}}{\hat{\alpha}}\right)-1-\sum\limits_{i=1}^{N(t)}\exp\left(\frac{V_i}{\hat{\alpha}}\right)}{N(t)}\right]^{-1} \tag{4.15}$$

and $\hat{\alpha}$ is the solution of the following equation

$$
\hat{\alpha} + \frac{\sum\limits_{i=1}^{N(t)} V(t_i-)}{N(t)} - \frac{\hat{\beta}}{N(t)} \left[\sum_{i=1}^{N(t)} (t_i - t_{i-1} + V_{i-1}) \exp\left(\frac{t_i - t_{i-1} + V_{i-1}}{\hat{\alpha}} \right) \right.
$$

$$
\left. + (t - t_{N(t)} + V_{N(t)}) \exp\left(\frac{t - t_{N(t)} + V_{N(t)}}{\hat{\alpha}} \right) - \sum_{i=1}^{N(t)} V_i \exp\left(\frac{V_i}{\hat{\alpha}} \right) \right] = 0. \quad (4.16)
$$

Remark 4.4.3

i) Perfect repair:
   In this case we obtain

$$
\hat{\beta} = \left[\frac{\sum\limits_{i=1}^{N(t)} \exp\left(\frac{t_i - t_{i-1}}{\hat{\alpha}} \right) + \exp\left(\frac{t - t_{N(t)}}{\hat{\alpha}} \right) - (1 + N(t))}{N(t)} \right]^{-1} \quad (4.17)
$$

   and $\hat{\alpha}$ is the solution of the following equation

$$
\hat{\alpha} + \frac{t_{N(t)}}{N(t)} - \frac{\sum\limits_{i=1}^{N(t)} (t_i - t_{i-1}) \exp\left(\frac{t_i - t_{i-1}}{\hat{\alpha}} \right) + (t - t_{N(t)}) \exp\left(\frac{t - t_{N(t)}}{\hat{\alpha}} \right)}{\sum\limits_{i=1}^{N(t)} \exp\left(\frac{t_i - t_{i-1}}{\hat{\alpha}} \right) + \exp\left(\frac{t - t_{N(t)}}{\hat{\alpha}} \right) - (1 + N(t))} = 0.
$$

$$
(4.18)
$$

ii) Minimal repair:
    The point estimators for the parameters can be found from

$$
\hat{\alpha} + \frac{\sum\limits_{i=1}^{N(t)} t_i}{N(t)} - \frac{t \exp\left(\frac{t}{\hat{\alpha}} \right)}{\exp\left(\frac{t}{\hat{\alpha}} \right) - 1} = 0 \quad (4.19)
$$

$$
\text{and } \hat{\beta} = \left[\frac{\exp\left(\frac{t}{\hat{\alpha}} \right) - 1}{N(t)} \right]^{-1}. \quad (4.20)
$$

4.4.2 A simple model for estimating the degree of repair

Now we want to describe a simple model for estimating the degree of repair. Let us assume that after a failure the item is immediately repaired to a fixed level see Figure 4.3, that means the failure intensity at time t_k $(k = 1, 2, \dots)$ has a jump to $\lambda(a)$.

We put in the general model $Z_n = \frac{T_n - T_{n-1}}{a + T_n - T_{n-1}}$, $n \geq 1$ and parameter estimation can be found from (4.1)

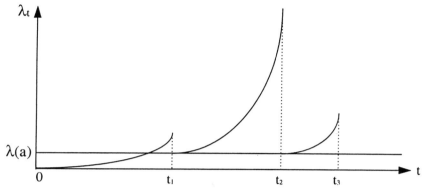

Figure 4.3: Realizations of intensities λ_t

In the case of Weibull type intensity the loglikelihood function of the failure-repair process has the form (4.2), with $V_0 = 0$, $V_i = a$ $(i = 1, \ldots, N(t))$, $V(t_1-) = t_1$ and $V(t_i-) = t_i - t_{i-1} + a$ $(i = 2, \ldots, N(t))$.

Then we obtain:

$$\ln L(t; \theta) = N(t)(\ln \beta - \beta \ln \alpha) + (\beta - 1) \left[\ln t_1 + \sum_{i=2}^{N(t)} \ln(t_i - t_{i-1} + a) \right] + t$$

$$- \sum_{i=2}^{N(t)} \left(\frac{t_i - t_{i-1} + a}{\alpha} \right)^\beta - \left(\frac{t_1}{\alpha} \right)^\beta - \left(\frac{t - t_{N(t)} + a}{\alpha} \right)^\beta + N(t) \left(\frac{a}{\alpha} \right)^\beta, \quad (4.21)$$

with $\theta = (\alpha, \beta, a)$, $\alpha > 0$, $\beta > 1$ and $0 \leq a < t_1$.

The point estimators for the parameters β and a can be found by the numerical solution of the following equations

$$\frac{1}{\hat{\beta}} - \left[t_1^{\hat{\beta}} \ln t_1 + \sum_{i=2}^{N(t)} (t_i - t_{i-1} + \hat{a})^{\hat{\beta}} \ln(t_i - t_{i-1} + \hat{a}) + (t - t_{N(t)} + \hat{a})^{\hat{\beta}} \right.$$

$$\ln(t - t_{N(t)} + \hat{a}) - N(t)\hat{a}^{\hat{\beta}} \ln \hat{a} \left] \left[t_1^{\hat{\beta}} + \sum_{i=2}^{N(t)} (t_i - t_{i-1} + \hat{a})^{\hat{\beta}} + (t - t_{N(t)} + \hat{a})^{\hat{\beta}} \right.$$

$$\left. - N(t)\hat{a}^{\hat{\beta}} \right]^{-1} + \frac{\ln t_1 + \sum\limits_{i=2}^{N(t)} \ln(t_i - t_{i-1} + \hat{a})}{N(t)} = 0. \quad (4.22)$$

$$\frac{(\hat{\beta} - 1) \hat{a}^{\hat{\beta}}}{\hat{\beta}} \sum_{i=2}^{N(t)} \frac{1}{t_i - t_{i-1} + \hat{a}} - \sum_{i=2}^{N(t)} (t_i - t_{i-1} + \hat{a})^{\hat{\beta}-1} - (t - t_{N(t)} + \hat{a})^{\hat{\beta}-1}$$

$$+ N(t) \hat{a}^{\hat{\beta}-1} = 0, \quad (4.23)$$

and the estimator of the parameter α can explicitly be given:

$$\hat{\alpha} = \left(\frac{t_1^{\hat{\beta}} + \sum_{i=2}^{N(t)} (t_i - t_{i-1} + \hat{a})^{\hat{\beta}} + (t - t_{N(t)} + \hat{a})^{\hat{\beta}} - N(t)\hat{a}^{\hat{\beta}}}{N(t)} \right)^{1/\hat{\beta}} \qquad (4.24)$$

So it is possible to estimate the degree of repair which may be unknown in practical problems.

Parameter estimators of α and a with known $\beta = 2$ are illustrated in the following example:

A sample with size 100 was observed until time $t = 10$. Let be $\alpha = 1$, $a = 0.4$ and let $s = 20$ be the number of simulations.

$\hat{\alpha}_j$ and $\hat{a}_j$, $j = 1, \ldots, s$ are parameter estimators of α and a, respectively.

Table 4.1: Parameter estimators of α and a

j	$\hat{\alpha}_j$	$\hat{a}_j$
1	1.0278	0.4137
2	1.0877	0.4224
3	1.0446	0.4210
4	0.9854	0.3981
5	1.0034	0.4012
6	1.1030	0.4342
7	1.0327	0.4139
8	0.9624	0.3942
9	1.0119	0.4090
10	0.9702	0.3954
11	1.0138	0.4129
12	1.0577	0.4213
13	1.0346	0.4201
14	0.9554	0.3817
15	1.0134	0.4127
16	1.1035	0.4348
17	1.0627	0.4219
18	0.9621	0.3914
19	1.0189	0.4135
20	0.9502	0.3802

The mean squared error of $\hat{\alpha}$ is small and equal to 0.0007.

References

1. Bathe, F. and Franz, J. (1996). Modelling of repairable systems with various degrees of repair, *Metrika*, **43**, 149–164.

2. Baxter, L., Kijima, M. and Tortorella, M. (1996). A point process model for the reliability of a maintained system subject to general repair, *Communications in Statistics–Stochastic Models*, **12**, 37–65.

3. Block, H. W., Langberg, N. and Savits, T. H. (1993). Repair replacement policies, *Journal of Applied Probability*, **30**, 194–206.

4. Brown, M. and Proschan, F. (1983). Imperfect repair, *Journal of Applied Probability*, **20**, 851–859.

5. Gasmi, S. (1995). Statistik der Erneuerungstheorie fr verschiedene Erneuerungsarten, *Diplomarbeit*.

6. Kijima, M. (1989). Some results for repairable systems, *Journal of Applied Probability*, **26**, 89–102.

7. Last, G. and Brandt, A. (1995). Marked Point Processes: The Dynamic Approach, New York: Springer-Verlag.

8. Last, G. and Szekli, R. (1995). Stochastic comparison of repairable systems by coupling, *Bericht 95/11*.

9. Stadje, W. and Zuckerman, D. (1991). Optimal maintenance strategies for repairable systems with general degrees of repair, *Journal of Applied Probability*, **28**, 384–396.

5

Investigation of Convergence Rates in Risk Theory in the Presence of Heavy Tails

Simone Liebner

Otto–von–Guericke–Universität Magdeburg, Magdeburg, Germany

Abstract: Asmussen and Klüppelberg (1996) established limit theorems for various random variables occurring in risk theory. In the present paper rates of convergence in certain limit theorems are given, which may be improved in some special cases.

Keywords and phrases: Subexponential distributions, extreme value theory, rates of convergence, integrated tail

5.1 A Model in Risk Theory

In non-life insurance very large claims happen more frequently than in other fields of insurance. So Asmussen and Klüppelberg (1996) considered models where ruin of an insurance company is caused by one large claim.

Exactly speaking they investigated the following claim surplus process S of a risk process R in continuous time:

$$S_t = \sum_{n \in N : 0 < T_n \leq t} U_n - c\,t \quad \text{and} \quad R_t = u - S_t, \quad t \geq 0,$$

where $\{U_n\}_{n \in N}$ are i.i.d. random variables (the claims) with common distribution B (the claim size distribution), the T_n are epochs of a Poisson process with intensity β and c is a positive constant (the premium rate). Without loss of generality we may assume that $c = 1$. The risk reserve is denoted by the positive constant u. As we see the process S is downwards skipfree.

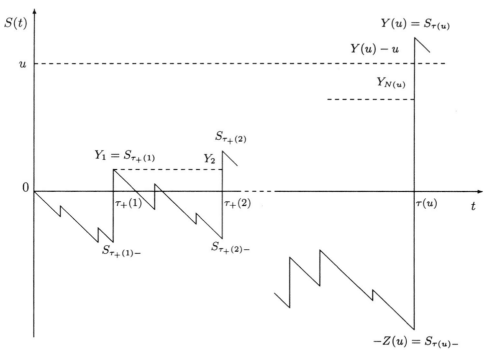

In this path decomposition $\tau(u) = \inf\{t \geq 0 : S_t > u\}$ is the ruin time and we define $-Z(u) = S_{\tau(u)-}$ as the level of the process just before and $Y(u) = S_{\tau(u)}$ as the level of the process just after the crossing of level u. The sequence $\{Y_n\}_{n=1,2,\ldots}$ with $Y_n = S_{\tau_+(n)} - S_{\tau_+(n-1)}$ denotes the decrements of the initial reserve u. The ladder epochs $\tau_+(n)$, $n \geq 0$, are defined by

$$
\begin{aligned}
\tau_+(0) &= 0, \\
\tau_+(1) &= \tau(0) = \inf\{t > 0 : S_t > 0\}, \\
\tau_+(k+1) &= \inf\{t > \tau_+(k) : S_t > S_{\tau_+(k)}\}.
\end{aligned}
$$

Let $P^{(u)}(\cdot) = P(\cdot \mid \tau(u) < \infty)$ denote the conditional distribution of an event given that the ruin time is finite.

Asmussen and Klüppelberg (1996) proved limit theorems for the random variable $Z^{(u)}$, which is defined as follows (see the proof of Theorem 1.3 of Asmussen and Klüppelberg (1996)):

Using the above defined decrements $\{Y_n\}_{n=1,2,\ldots}$ of the process S_t it is proved that

$$
\|P^{(u)}(Z(u) \in \cdot) - P(Z^{(u)} - Y_1 - \cdots - Y_{N-1} \in \cdot)\| \to 0 \tag{5.1}
$$

as $u \to \infty$, where N is geometric with parameter ρ, the Y_i's are independent of N having the distribution B_0, $Z^{(u)}$ has distribution $B_0^{(u)}$, and $\|\cdot\|$ denotes the supremum norm. B_0 is the integrated tail distribution of B with density $\overline{B}/\mu_B$, where μ_B is the mean of B and $\overline{B} = 1 - B$, i.e.,

$$
B_0(u) = \frac{1}{\mu_B} \int_0^u \overline{B}(x)\,\mathrm{d}x .
$$

Now $B_0^{(u)}$ is defined as

$$B_0^{(u)}(x) = 1 - \frac{\overline{B}_0(u+x)}{\overline{B}_0(u)}.$$

The distribution of $Z^{(u)}/a(u)$ follows as

$$
\begin{aligned}
P^{(u)}\left(\frac{Z^{(u)}}{a(u)} > x\right) &= P^{(u)}\left(Z^{(u)} > a(u)\,x\right) \\
&= \overline{B}_0^{(u)}(a(u)\,x) = \frac{\overline{B}_0(u + a(u)\,x)}{\overline{B}_0(u)}.
\end{aligned}
$$

Defining $t(u) := a(u)\,x + u$ we get

$$P^{(u)}\left(\frac{Z^{(u)}}{a(u)} > x\right) = \frac{\overline{B}_0(t(u))}{\overline{B}_0(u)}. \tag{5.2}$$

As mentioned above we want to state risk models where the ruin occurs because of one large claim. It is known that the subexponential distributions are very useful in modeling this fact, see Pitman (1980). Hence, distributions B from the domain of attraction of the Gumbel and Fréchet distributions are used, which give a subexponential distribution B_0.

Remember the following definitions and theorems about the domain of attraction of extreme value distributions, subexponential distributions and the relations between these two types of distribution functions.

Theorem 5.1.1 (see Resnick (1987, Proposition 0.3)) *Consider   i.i.d. random variables $X_1, X_2, \ldots, X_n$ with common distribution function F and $M_n = \max\{X_1, \ldots, X_n\}$. Let $a_n > 0$ and $b_n \in \mathbf{R}$, $n \geq 1$, be sequences of constants such that*

$$P\left(\frac{M_n - b_n}{a_n} \leq x\right) = F^n(a_n\,x + b_n) \xrightarrow{w} G(x)$$

for $n \to \infty$, where " $\xrightarrow{w}$ " means the weak convergence and G is nondegenerated. Then G is one of the following three types:

1. $\Phi_\alpha(x) = \begin{cases} 0, & x \leq 0 \\ \exp\{-x^{-\alpha}\}, & x > 0 \end{cases}$ *for some $\alpha > 0$;*

2. $\Psi_\alpha(x) = \begin{cases} \exp\{-(-x)^\alpha\}, & x < 0 \\ 1, & x \geq 0 \end{cases}$ *for some $\alpha > 0$;*

3. $\Lambda(x) = \exp\{-e^{-x}\}$ $x \in \mathbf{R}$.

We call Φ_α, Ψ_α and Λ the extreme value distributions.

Two limiting distributions $G_1(x)$ and $G_2(x)$ are of the same type, if they differ only by a positive linear transformation of the arguments, i.e., if $G_1(x) = G_2(ax + b)$ for an $a > 0$ and $b \in \mathbf{R}$.

Definition 5.1.1 A measurable function $U : \mathbf{R}_+ \to \mathbf{R}$ is regular varying at ∞ with exponent ϱ, $-\infty < \varrho < \infty$, written $U \in RV_\varrho$, if for $x > 0$

$$\lim_{t\to\infty} \frac{U(t\,x)}{U(t)} = x^\varrho .$$

Definition 5.1.2 A distribution function F is in the domain of attraction of an extreme value distribution G ($F \in D(G)$), if there exist normalizing constants $a_n > 0$ and $b_n \in \mathbf{R}$ such that

$$F^n(a_n x + b_n) = P(M_n \le a_n\,x + b_n) \overset{w}{\to} G(x).$$

Some conditions for $F \in D(\Phi_\alpha)$ and $F \in D(\Lambda)$ are given in Resnick (1987), Feller (1971) and Bingham, Goldie and Teugels (1987):

Proposition 5.1.1 $F \in D(\Lambda)$ *is equivalent to the existence of a positive function g such that*

$$\lim_{x\to x_0} \frac{\overline{F}(x + t\,g(x))}{\overline{F}(x)} = e^{-t} \quad \text{for all} \quad t \in \mathbf{R},$$

where x_0 is the upper endpoint of F.

Proposition 5.1.2 $F \in D(\Phi_\alpha)$ *iff* $\overline{F} \in RV_{-\alpha}$, $\alpha > 0$.

Remark 5.1.1 Consider $\overline{F}_0(u) = 1/\mu_F \int_u^\infty \overline{F}(x)\,\mathrm{d}x$. If $F \in D(\Phi_{\alpha+1})$, $\alpha > 0$, then $F_0 \in D(\Phi_\alpha)$ by Karamata's Theorem (see Resnick (1987)) and $\overline{F}_0 \in RV_{-\alpha}$ by Proposition 5.1.2.

Definition 5.1.3 A distribution function F belongs to the class $\mathcal{S}(\gamma)$ for $\gamma \ge 0$ if $F(x) < 1$ for all x and

1.

$$\lim_{x\to\infty} \frac{\overline{F}(x - y)}{\overline{F}(x)} = e^{\gamma y} \quad \text{for all} \quad y \in \mathbf{R},$$

2.

$$\lim_{x\to\infty} \frac{1 - F * F(x)}{1 - F(x)} = \lim_{x\to\infty} \frac{P(X_1 + X_2 > x)}{P(X_1 > x)} =: d < \infty$$

are satisfied, where X_1 and X_2 are i.i.d. with common distribution function F and $*$ means the convolution sign. We call $\mathcal{S}(0)$ the class of subexponential distributions denoted by $\mathcal{S}$.

Now we will give some conditions that $F_0 \in \mathcal{S}$ if F belongs to a domain of attraction of a Gumbel or Fréchet distribution.

Theorem 5.1.2 (Bingham, Goldie and Teugels (1987)) *If* $\overline{F} \in RV_{-\alpha}$ *($\alpha > 0$) then $F \in \mathcal{S}$.*

Remark 5.1.2 With our last remark it follows that F_0 is subexponential, if $F \in D(\Phi_{\alpha+1})$.

If $F \in D(\Lambda)$, then F_0 is not necessarily subexponential. The following theorem ensures that F_0 is subexponential, if $F \in D(\Lambda)$.

Theorem 5.1.3 (Goldie and Resnick (1988, Corollary 2.5.)) *Let* F *be such a distribution function that* $a(u) = \mu_F \overline{F}_0(u)/\overline{F}(u) \to \infty$, $a'(u) \to 0$ *as* $u \to \infty$ *and $a(u)$ is eventually nondecreasing. If there exist $\lambda > 1$ such that*

$$\liminf_{u \to \infty} \frac{a(\lambda u)}{a(u)} > 1,$$

then $F \in D(\Lambda)$ and $F_0 \in \mathcal{S}$.

Because of the asymptotical conditions on a any function $a(u) \sim \mu_F$ $\overline{F}_0(u)/\overline{F}(u)$ can be chosen as normalizing function, where $a_1(u) \sim a_2(u)$ denotes the asymptotically equivalence, i.e., $\lim\limits_{u \to \infty} \dfrac{a_1(u)}{a_2(u)} = 1$ holds.

For more details about subexponential distribution functions see Goldie and Resnick (1988) and Pitman (1980).

5.2 Limit Theorem

Let V_α, $0 < \alpha \leq \infty$, be a random variable with a generalized Pareto distribution, i.e., V_α is positive and its tail is given by

$$P(V_\alpha > x) = \begin{cases} (1 + \frac{x}{\alpha})^{-\alpha}, & \alpha < \infty, \\ e^{-x}, & \alpha = \infty, \end{cases} \quad \text{where} \quad x > 0.$$

Remember, that B is the distribution function of the claims $\{U_n\}$ in the surplus process S and $Z^{(u)}$ is defined in (5.1).

Theorem 5.2.1 (Asmussen and Klüppelberg (1996, Theorem 3.3))

1. *If* $\overline{B} \in RV_{-\alpha-1}$, $\alpha > 0$, *then*

$$\alpha \frac{Z^{(u)}}{u} \to V_\alpha \quad \text{for} \quad u \to \infty$$

in $P^{(u)}$*-distribution.*

2. *If $B \in D(\Lambda)$ and $B_0 \in \mathcal{S}$, then*

$$\frac{Z^{(u)}}{a(u)} \to V_\infty \quad for \quad u \to \infty$$

in $P^{(u)}$–distribution, where $a(u) \sim \mu_B \, \overline{B}_0(u)/\overline{B}(u) \to \infty$.

5.3   Rates of Convergence

According to Theorem 5.2.1 we will determine the rate of convergence under stronger conditions on the distribution function B of the claims. Therefore we consider at first a distribution function B with $\overline{B} \in RV_{-\alpha-1}$ having the form

$$\overline{B}(u) = c \, u^{-\alpha-1}(1 + r(u)) \ \text{ with } \ r(u) = O(u^{-\beta}), \ \beta > 0, \ \alpha > 0, \text{ as } u \to \infty.$$
(5.3)

Because of Theorem 5.1.2 the distribution function B belongs to $\mathcal{S}$ as well as B_0, defined in section 1. We get for $\overline{B}_0$, $a(u)$ and $P^{(u)}(Z^{(u)}/a(u) > x)$ the following expressions:

$$
\begin{aligned}
\overline{B}_0(u) &= \frac{1}{\mu_B} \int_u^\infty \overline{B}(x) \, \mathrm{d}x \\
&= \frac{c}{\mu_B} \left(\int_u^\infty x^{-\alpha-1} \, \mathrm{d}x + \int_u^\infty x^{-\alpha-1} r(x) \, \mathrm{d}x \right) \\
&= \frac{c}{\alpha \, \mu_B} u^{-\alpha} (1 + \delta(u)),
\end{aligned}
$$

where $\delta(u) = \alpha \, u^\alpha \int_u^\infty x^{-\alpha-1} r(x) \, \mathrm{d}x$ and with (5.2)

$$P^{(u)} \left(\frac{Z^{(u)}}{a(u)} > x \right) = \frac{\overline{B}_0(t(u))}{\overline{B}_0(u)} = \left(\frac{t(u)}{u} \right)^{-\alpha} \frac{1 + \delta(t(u))}{1 + \delta(u)}.$$
(5.4)

Moreover

$$\mu_B \frac{\overline{B}_0(u)}{\overline{B}(u)} = \frac{u}{\alpha} \frac{1 + \delta(u)}{1 + r(u)} \sim \frac{u}{\alpha} = a(u).$$

For the above described distribution function the following theorem holds.

Theorem 5.3.1 *Suppose, that $\overline{B}(u)$ has the form (5.3) and choose $a(u) = u/\alpha$. Let $r(u)$ take such a form that $\delta(u) = O(u^{-\tau})$ with $\tau \geq \beta$ as $u \to \infty$. Then*

$$P^{(u)} \left(\frac{Z^{(u)}}{a(u)} > x \right) = \left(1 + \frac{x}{\alpha} \right)^{-\alpha} + O(u^{-\tau}) \quad for \quad u \to \infty.$$

PROOF. We have

$$t(u) \;=\; a(u)\,x + u = u\left(1 + \frac{x}{\alpha}\right)$$

and with (5.4) it follows that

$$
\begin{aligned}
P^{(u)}\left(\frac{Z^{(u)}}{a(u)} > x\right) &= \left(1 + \frac{x}{\alpha}\right)^{-\alpha} (1 + \delta(t(u)))\,(1 + \delta(u))^{-1} \\
&= \left(1 + \frac{x}{\alpha}\right)^{-\alpha} \left(1 + \delta(t(u)) - \delta(u) + O(u^{-2\tau})\right) \\
&= \left(1 + \frac{x}{\alpha}\right)^{-\alpha} + O(u^{-\tau}).
\end{aligned}
$$

∎

In some cases it is possible to obtain a higher rate of convergence, if we use another normalizing function. As we can see in Theorem 5.2.1 it is clear that any function $a(u) \sim \mu_B\,\overline{B_0}(u)/\overline{B}(u)$ can be chosen as normalizing function.

Theorem 5.3.2 *Under the same conditions as in Theorem 5.3.1 with*

$$\delta(u) = \frac{c_0}{u^\tau} + \frac{c_1}{u^{\tau+s}} + O(u^{-2\tau}),$$

where $0 < s < \tau$, $c_0, c_1 \in \mathbf{R}$, *and by choosing*

$$a(u) = \frac{u}{\alpha}\left(1 + \frac{x+\alpha}{x\,\alpha}\{\delta(\kappa(u)) - \delta(u)\}\right) \quad \text{with} \quad \kappa(u) = u\left(1 + \frac{x}{\alpha}\right)$$

we obtain

$$P^{(u)}\left(\frac{Z^{(u)}}{a(u)} > x\right) = \left(1 + \frac{x}{\alpha}\right)^{-\alpha} + O(u^{-2\tau}) \quad \text{for} \quad u \to \infty.$$

PROOF. We have

$$
\begin{aligned}
t(u) &= a(u)\,x + u = u\frac{x}{\alpha} + \frac{u\,(x+\alpha)}{\alpha^2}\{\delta(\kappa(u)) - \delta(u)\} + u \\
&= \kappa(u)\left(1 + \frac{1}{\alpha}\{\delta(\kappa(u)) - \delta(u)\}\right).
\end{aligned}
$$

Now we find

$$\left(\frac{t(u)}{u}\right)^{-\alpha} = \left(1 + \frac{x}{\alpha}\right)^{-\alpha}\left(1 - \delta(\kappa(u)) + \delta(u) + O(u^{-2\tau})\right)$$

and

$$\delta(t(u)) = c_0 \kappa^{-\tau}(u) \left(1 + \frac{1}{\alpha}\{\delta(\kappa(u)) - \delta(u)\}\right)^{-\tau} +$$
$$+ c_1 \kappa^{-\tau-s}(u) \left(1 + \frac{1}{\alpha}\{\delta(\kappa(u)) - \delta(u)\}\right)^{-\tau-s} + O(u^{-2\tau})$$
$$= c_0 \kappa^{-\tau}(u) + c_1 \kappa^{-\tau-s}(u) + O(u^{-2\tau}).$$

Hence, we get with (5.4)

$$P^{(u)}\left(\frac{Z^{(u)}}{a(u)} > x\right) = \left(1 + \frac{x}{\alpha}\right)^{-\alpha} \left(1 - \delta(\kappa(u)) + \delta(u) + O(u^{-2\tau})\right) \cdot$$
$$\cdot \left(1 + c_0 \kappa^{-\tau}(u) + c_1 \kappa^{-\tau-k}(u)\right) \cdot \left(1 - \delta(u) + O(u^{-2\tau})\right)$$
$$= \left(1 + \frac{x}{\alpha}\right)^{-\alpha} \left(1 - \delta(\kappa(u)) + \delta(u) + O(u^{-2\tau})\right) \cdot$$
$$\cdot \left(1 - \delta(u) + c_0 \kappa^{-\tau}(u) + c_1 \kappa^{-\tau-k}(u) + O(u^{-2\tau})\right)$$
$$= \left(1 + \frac{x}{\alpha}\right)^{-\alpha} + O(u^{-2\tau}).$$

∎

We want to demonstrate this by an example.

Example 5.3.1 Consider

$$\overline{B}(u) = \begin{cases} 1 & u < 1, \\ \frac{1}{2} u^{-\alpha-1}\left(1 + u^{-2}\right) & u \geq 1. \end{cases}$$

At first we take $a(u) = u/\alpha$. We get

$$t(u) = u\left(1 + \frac{x}{\alpha}\right) \quad \text{and} \quad \delta(u) = \alpha u^\alpha \int_u^\infty x^{-\alpha-3}\,dx = \frac{\alpha}{\alpha+2} u^{-2}.$$

Thus by using (5.4)

$$P^{(u)}\left(\frac{Z^{(u)}}{a(u)} > x\right) = \left(1 + \frac{x}{\alpha}\right)^{-\alpha} \left(1 + \frac{\alpha u^{-2}}{\alpha+2}\left(1 + \frac{x}{\alpha}\right)^{-2}\right) \cdot$$
$$\cdot \left(1 - \frac{\alpha}{\alpha+2} u^{-2} + O(u^{-4})\right)$$
$$= \left(1 + \frac{x}{\alpha}\right)^{-\alpha} \left(1 - \frac{\alpha u^{-2}}{\alpha+2}\left\{1 - \left(1 + \frac{x}{\alpha}\right)^{-2}\right\} + O(u^{-4})\right).$$

Only a convergence rate of $O(u^{-2})$ is available, which cannot be improved by choosing $a(u) = u/\alpha$. But $\delta(u)$ has the form described in Theorem 5.3.2, hence we get a higher rate of convergence if we use

$$a(u) = \frac{u}{\alpha}\left(1 + \frac{x+\alpha}{x\,\alpha}\left\{\delta(\kappa(u)) - \delta(u)\right\}\right) \quad \text{with} \quad \kappa(u) = u\left(1 + \frac{x}{\alpha}\right).$$

Hence, we have

$$a(u) = \frac{u}{\alpha}\left(1 - \frac{x+\alpha}{x\,(\alpha+2)}\,u^{-2}\left[1 - \left(1 + \frac{x}{\alpha}\right)^{-2}\right]\right).$$

For $t(u) := a(u)\,x + u$ we obtain

$$t(u) = u\left(1 + \frac{x}{\alpha}\right)\left(1 - \frac{1}{\alpha+2}\,u^{-2}\left[1 - \left(1 + \frac{x}{\alpha}\right)^{-2}\right]\right)$$

and hence with (5.4) it follows

$$
\begin{aligned}
P^{(u)}\left(\frac{Z^{(u)}}{a(u)} > x\right) &= \left(1 + \frac{x}{\alpha}\right)^{-\alpha}\left(1 - \frac{\alpha}{\alpha+2}\,u^{-2} + O(u^{-4})\right) \cdot \\
&\quad \cdot\left(1 + \frac{\alpha\,u^{-2}}{\alpha+2}\left(1 + \frac{x}{\alpha}\right)^{-2} + O(u^{-4})\right) \cdot \\
&\quad \cdot\left(1 + \frac{\alpha}{\alpha+2}\,u^{-2}\left[1 - \left(1 + \frac{x}{\alpha}\right)^{-2}\right] + O(u^{-4})\right) \\
&= \left(1 + \frac{x}{\alpha}\right)^{-\alpha} \cdot \\
&\quad \cdot\left(1 - \frac{\alpha}{\alpha+2}\,u^{-2}\left[1 - \left(1 + \frac{x}{\alpha}\right)^{-2}\right] + O(u^{-4})\right) \cdot \\
&\quad \cdot\left(1 + \frac{\alpha}{\alpha+2}\,u^{-2}\left[1 - \left(1 + \frac{x}{\alpha}\right)^{-2}\right] + O(u^{-4})\right) \\
&= \left(1 + \frac{x}{\alpha}\right)^{-\alpha}\left(1 + O(u^{-4})\right).
\end{aligned}
$$

Consider at next distribution functions from the domain of attraction of the Gumbel distribution. Let $\overline{B}$ be of the form

$$\overline{B}(u) = u^{\tau}\,e^{-c u^{\alpha}}, \quad u > 0, \ \alpha \in (0,1), \ \tau \in \mathbf{R}, \ c > 0. \tag{5.5}$$

We need $\alpha \in (0,1)$ in order to ensure that $B_0(u) \in \mathcal{S}$, see Theorem 5.1.3. We get the following expressions for $\overline{B}_0$.

$$\overline{B}_0(u) = \frac{1}{\mu_B}\int_u^{\infty} x^{\tau}\,e^{-c x^{\alpha}}\,\mathrm{d}x = \frac{1}{\alpha\,\mu_B}\int_{u^{\alpha}}^{\infty} z^{\frac{\tau+1}{\alpha}-1}\,e^{-c z}\,\mathrm{d}z$$

$$= \begin{cases} \dfrac{1}{c\,\alpha\,\mu_B} e^{-c\,u^\alpha} u^{\tau+1-\alpha} \left(1 + \dfrac{\tau+1-\alpha}{c\,\alpha} u^{-\alpha} + O(u^{-2\alpha})\right) \\ \qquad\qquad \text{if } \tau \neq \alpha - 1, \\[2em] \dfrac{1}{c\,\alpha\,\mu_B} e^{-c\,u^\alpha} \\ \qquad\qquad \text{if } \tau = \alpha - 1, \end{cases}$$

so that we can calculate $P^{(u)}(Z^{(u)}/a(u) > x)$ as

$$P^{(u)}\left(\frac{Z^{(u)}}{a(u)} > x\right) = \frac{\overline{B}_0(t(u))}{\overline{B}_0(u)}$$

$$= \begin{cases} \exp\{-c\,t^\alpha(u) + c\,u^\alpha\} \left(\dfrac{t(u)}{u}\right)^{\tau+1-\alpha} \cdot \\ \quad \cdot \left(1 + \dfrac{\tau+1-\alpha}{c\,\alpha} [t^{-\alpha}(u) - u^{-\alpha}] + O(u^{-2\alpha})\right), & \tau \neq \alpha - 1, \qquad (5.6) \\[1.5em] \exp\{-c\,t^\alpha(u) + c\,u^\alpha\}, & \tau = \alpha - 1. \end{cases}$$

At last it follows that $a(u) \sim \dfrac{u^{1-\alpha}}{c\,\alpha}$ since

$$\mu_B \frac{\overline{B}_0(u)}{\overline{B}(u)} = \begin{cases} \dfrac{1}{c\,\alpha} u^{1-\alpha} \left(1 + \dfrac{\tau+1-\alpha}{c\,\alpha} u^{-\alpha} + O(u^{-2\alpha})\right) & \text{if } \tau \neq \alpha - 1, \\[1.5em] \dfrac{1}{c\,\alpha} u^{1-\alpha} & \text{if } \tau = \alpha - 1. \end{cases}$$

Theorem 5.3.3 *Let $\overline{B}(u)$ be of the form (5.5) and $a(u) = u^{1-\alpha}/(c\,\alpha)$. Then*

$$P^{(u)}\left(\frac{Z^{(u)}}{a(u)} > x\right) = e^{-x} + O(u^{-\alpha}), \quad u \to \infty.$$

PROOF. First of all

$$t(u) := a(u)\,x + u = u\left(1 + \frac{x}{c\,\alpha} u^{-\alpha}\right)$$

and thus

$$t^\alpha(u) = u^\alpha \left(1 + \frac{x}{c} u^{-\alpha} + \frac{\alpha-1}{2} \frac{x^2}{c^2\,\alpha} u^{-2\alpha} + O(u^{-3\alpha})\right).$$

So we obtain with (5.6) in the case $\tau \neq \alpha - 1$

$$P^{(u)}\left(\frac{Z^{(u)}}{a(u)} > x\right) = \exp\left\{-c\,u^\alpha \left(1 + \frac{x}{c} u^{-\alpha} + \frac{\alpha-1}{2} \frac{x^2}{c^2\,\alpha} u^{-2\alpha} + O(u^{-3\alpha})\right) + \right.$$

$$+ c u^\alpha \Big\} \left(1 + \frac{(\tau + 1 - \alpha)\, x}{c\alpha} u^{-\alpha} + O(u^{-2\alpha})\right) \left(1 + O(u^{-2\alpha})\right)$$

$$= e^{-x} \left(1 - \frac{\alpha - 1}{2} \frac{x^2}{c\alpha} u^{-\alpha} + O(u^{-2\alpha})\right) \cdot$$

$$\cdot \left(1 + \frac{(\tau + 1 - \alpha)\, x}{c\alpha} u^{-\alpha} + O(u^{-2\alpha})\right)$$

$$= e^{-x} \left(1 + u^{-\alpha} x \left[\frac{(1 - \alpha)\, x}{2\, c\,\alpha} + \frac{\tau + 1 - \alpha}{c\,\alpha}\right] + O(u^{-2\alpha})\right)$$

and for $\tau = \alpha - 1$ it follows in the same way

$$P^{(u)} \left(\frac{Z^{(u)}}{a(u)} > x\right) = e^{-x} \left(1 + \frac{1 - \alpha}{2} \frac{x^2}{c\alpha} u^{-\alpha} + O(u^{-2\alpha})\right).$$

$\blacksquare$

By choosing another normalizing function $a(u)$ we can improve the result as we see in the following statement.

Theorem 5.3.4 *Let $\overline{B}(u)$ be of the form (5.5) and*

$$a(u) = \frac{1}{c\alpha} u^{1-\alpha} \left(1 + \left[\frac{\tau + 1 - \alpha}{c\alpha} + \frac{(1 - \alpha)\, x}{2\, c\,\alpha}\right] u^{-\alpha}\right)$$

then

$$P^{(u)} \left(\frac{Z^{(u)}}{a(u)} > x\right) = e^{-x} + O(u^{-2\alpha}) \quad \text{for} \quad u \to \infty.$$

PROOF. To start with

$$t(u) = u \left(1 + \frac{x}{c\alpha} u^{-\alpha} + \frac{x}{c\alpha} u^{-2\alpha} \left[\frac{\tau + 1 - \alpha}{c\alpha} + \frac{(1 - \alpha)\, x}{2\, c\,\alpha}\right]\right)$$

we obtain

$$t^\alpha(u) = u^\alpha \left(1 + \frac{x}{c} u^{-\alpha} + \frac{x}{c} u^{-2\alpha} \left[\frac{\tau + 1 - \alpha}{c\alpha} + \frac{(1 - \alpha)\, x}{2\, c\,\alpha}\right] + \right.$$

$$\left. + \frac{\alpha - 1}{2} \frac{x^2}{c^2\,\alpha} u^{-2\alpha} + O(u^{-3\alpha})\right)$$

$$= u^\alpha \left(1 + \frac{x}{c} u^{-\alpha} + \frac{(\tau + 1 - \alpha)\, x}{\alpha\, c^2} u^{-2\alpha} + O(u^{-3\alpha})\right).$$

Hence, we get by using (5.6) if $\tau \neq \alpha - 1$

$$P^{(u)}\left(\frac{Z^{(u)}}{a(u)} > x\right) = \exp\left\{-c\,u^\alpha\left(1 + \frac{x}{c}u^{-\alpha} + \frac{(\tau+1-\alpha)\,x}{\alpha\,c^2}u^{-2\alpha} + O(u^{-3\alpha})\right) + \right.$$

$$\left. + c u^\alpha\right\}\cdot\left(1 + \frac{(\tau+1-\alpha)\,x}{c\,\alpha}u^{-\alpha} + O(u^{-2\alpha})\right)\left(1 + O(u^{-2\alpha})\right)$$

$$= e^{-x}\left(1 - \frac{(\tau+1-\alpha)\,x}{c\,\alpha}u^{-\alpha} + O(u^{-2\alpha})\right)\cdot$$

$$\cdot\left(1 + \frac{(\tau+1-\alpha)\,x}{c\,\alpha}u^{-\alpha} + O(u^{-2\alpha})\right)$$

$$= e^{-x} + O(u^{-2\alpha}).$$

For $\tau = \alpha - 1$ it follows in the same way that

$$P^{(u)}\left(\frac{Z^{(u)}}{a(u)} > x\right) = \exp\left\{-c\,u^\alpha\left(1 + \frac{x}{c}u^{-\alpha} + O(u^{-3\alpha})\right) + c\,u^\alpha\right\}$$

$$= e^{-x} + O(u^{-2\alpha}).$$

$\blacksquare$

We want to illustrate this by an

Example 5.3.2 Consider the tail distribution

$$\overline{B}(u) = \begin{cases} 1, & u < 1/2, \\ u^{-1}e^{-c\,u^{1/2}}, & u \geq 1/2. \end{cases}$$

Using (5.5) and $a(u) = u^{1-\alpha}/(c\,\alpha)$ with $\alpha = 1/2$, $\tau = -1$ and $c = \sqrt{2}\ln 2$ we obtain

$$a(u) = \frac{2}{c}u^{1/2} \quad \text{and} \quad t(u) := a(u)\,x + u = u\left(1 + \frac{2\,x}{c}u^{-1/2}\right).$$

So we get

$$t^{1/2}(u) = u^{1/2}\left(1 + \frac{x}{c}u^{-1/2} - \frac{x^2}{2\,c^2}u^{-1} + O(u^{-3/2})\right).$$

Finally it follows with (5.6)

$$P^{(u)}\left(\frac{Z^{(u)}}{a(u)} > x\right) = \exp\left\{-c\,u^{1/2}\left(1 + \frac{x}{c}u^{-1/2} - \frac{x^2}{2\,c^2}u^{-1} + O(u^{-3/2})\right) + \right.$$

$$\left. + c\,u^{1/2}\right\}\left(1 - \frac{x}{c}u^{-1/2} + O(u^{-1})\right)\left(1 + O(u^{-1})\right)$$

$$= e^{-x}\left(1 + \frac{x^2}{2\,c}u^{-1/2} + O(u^{-1})\right)\left(1 - \frac{x}{c}u^{-1/2} + O(u^{-1})\right)$$

$$= e^{-x}\left(1 - \frac{x}{c}\left[1 + \frac{x}{2}\right]u^{-1/2} + O(u^{-1})\right).$$

Thus we obtain a rate of convergence of $O(u^{-1/2})$ only. But by using

$$a(u) = \frac{2}{c} u^{1/2} \left(1 + \left[-\frac{1}{c} + \frac{x}{2c} \right] u^{-1/2} \right)$$

as normalizing function we can increase this rate as we see now. We get

$$t(u) := a(u)\, x + u = u \left(1 + \frac{2x}{c} u^{-1/2} + \frac{2x}{c} \left[-\frac{1}{c} + \frac{x}{2c} \right] u^{-1} \right)$$

and hence

$$t^{1/2}(u) = u^{1/2} \left(1 + \frac{x}{c} u^{-1/2} - \frac{x}{c^2} u^{-1} + O(u^{-3/2}) \right).$$

So we obtain with (5.6)

$$
\begin{aligned}
P^{(u)} \left(\frac{Z^{(u)}}{a(u)} > x \right) &= \exp \left\{ -c\, u^{1/2} \left(1 + \frac{x}{c} u^{-1/2} - \frac{x}{c^2} u^{-1} + O(u^{-3/2}) \right) + c\, u^{1/2} \right\} \cdot \\
&\quad \cdot \left(1 - \frac{x}{c} u^{-1/2} + O(u^{-1}) \right) \left(1 + O(u^{-1}) \right) \\
&= e^{-x} \left(1 + \frac{x}{c} u^{-1/2} + O(u^{-1}) \right) \left(1 - \frac{x}{c} u^{-1/2} + O(u^{-1}) \right) \\
&= e^{-x} \left(1 + O(u^{-1}) \right).
\end{aligned}
$$

Remark 5.3.1 As we see, the normalizing function $a(u)$ giving a higher order of convergence depends on x in both cases.

References

1. Asmussen, S. and Klüppelberg, C. (1996). Large deviations results in the presence of heavy tails, with applications to insurance risk, *Stochastic Processes and Applications*, **64**, 103–125.

2. Bingham, N. H., Goldie, C. M. and Teugels, J. L. (1987). Regular variation, *Encyclopedia of Mathematics and its Applications*, Vol. 27, pp. 21–44, 430–432, Cambridge University Press.

3. Feller, W. (1971). *An Introduction to Probability and Its Applications, Volume II*, Second edition, pp. 275–284, New York: John Wiley & Sons.

4. Goldie, C. M. and Resnick, S. (1988). Distributions that are both subexponential and in the domain of attraction of an extreme-value distribution, *Advances in Applied Probability*, **20**, 706–718.

5. Pitman, E. J. G. (1980). Subexponential distribution functions, *Journal of the Australian Mathematics Society, Series A*, **29**, 337–347.

6. Resnick, S. (1987). Extreme values, regular variation and point processes, *Applied Probability*, Vol. 4, pp. 1–62, New York: Springer-Verlag.

6

Least Squares and Minimum Distance Estimation in the Three-Parameter Weibull and Fréchet Models with Applications to River Drain Data

---

Robert Offinger

Otto-von-Guericke-Universität Magdeburg, Magdeburg, Germany

Abstract: After introducing three-Parameter Weibull and Fréchet models we define various least squares and minimum distance estimation methods in general. We then show how these methods can be applied to the Weibull and Fréchet models and examine the quality of two special estimators in a small simulation study. These studies show that the estimators are a good alternative to already known estimators. Finally we discuss the application of the models to river drain data and some involved problems.

Keywords and phrases: Point estimation, censored samples, Gumbel distribution, maximum likelihood method, univariate optimization by hybrid method, simulation, bias, mean square error, EDF tests, Cramér-von Mises statistic, fitting empirical data, skewness, kernel estimation

---

6.1 Introduction

Let $X_1, \ldots, X_n$ be independent and $\text{Wei}_{\gamma, \lambda, \alpha}$-distributed random variables with unknown parameters $\gamma \in \mathbf{R}, \lambda \in (0, \infty), \alpha \in (0, \infty)$, where $\text{Wei}_{\gamma, \lambda, \alpha}$ denotes the **Weibull distribution** on $(\mathbf{R}, \mathcal{B}^1)$ with Lebesgue density

$$f_{\gamma, \lambda, \alpha}(z) = \begin{cases} 0 & , z \leq \gamma \\ \alpha \lambda^\alpha (z - \gamma)^{\alpha - 1} \exp\left(-(\lambda(z - \gamma))^\alpha\right) & , z > \gamma \end{cases} .$$

The cumulative distribution function of $\text{Wei}_{\gamma, \lambda, \alpha}$ is given by

$$F_{\gamma, \lambda, \alpha}(z) = \begin{cases} 0 & , z \leq \gamma \\ 1 - \exp\left(-(\lambda(z - \gamma))^\alpha\right) & , z > \gamma \end{cases} .$$

Since

$$F_{\gamma,\lambda,\alpha}(z) = F_{0,1,1}\left((\lambda(z-\gamma))^{\alpha}\right)$$

for $z \geq \gamma$, γ is a **location parameter**, λ an **inverse scale parameter** or $\delta := 1/\lambda$ a **scale parameter**, and α is called **shape parameter**. $\text{Wei}_{0,\lambda,1} = \text{Exp}_{\lambda}$ is the well-known exponential distribution with inverse scale parameter λ.

Another model of interest is the three-parameter Fréchet model. Here the random variables $X_1, \ldots, X_n$ are independent and $\text{Fre}_{\gamma,\lambda,\alpha}$-distributed with unknown location parameter $\gamma \in \mathbf{R}$, inverse scale parameter $\lambda \in (0, \infty)$ and shape parameter $\alpha \in (0, \infty)$ where $\text{Fre}_{\gamma,\lambda,\alpha}$ denotes the **Fréchet distribution** (for minima) on $(\mathbf{R}, \mathcal{B}^1)$, which has cumulative distribution function

$$F_{\gamma,\lambda,\alpha}(z) = \begin{cases} 1 & , z \geq \gamma \\ 1 - \exp\left(-(-\lambda(z-\gamma))^{-\alpha}\right) & , z < \gamma \end{cases}.$$

Let further denote $X_{1\uparrow n} \leq \ldots \leq X_{n\uparrow n}$ the ordered values of the random vector $X = (X_1, \ldots, X_n)$. Now suppose that we observe only some order statistics $x_{i\uparrow n}$ for $i \in I \subseteq \{1, \ldots, n\}$.

Note that it is possible to put the Weibull distribution, the Fréchet distribution and the Gumbel distribution (for minima) after reparameterization into one model — the so-called generalized extreme value model for minima — [see e.g. Embrechts, Klüppelberg and Mikosch (1997, p. 152)].

It is known that the likelihood approach is only possible for the restriction to $\alpha > 1$ (or better $\alpha > 2$) — see Johnson, Kotz and Balakrishnan (1994, p. 656) — and even the application of the generalized maximum likelihood method Scholz (1980, p. 194 f) has proposed does not help — see Offinger (1996, p. 41 ff).

Therefore, in Section 6.2 we deal with some least squares and minimum distance approaches for estimating the parameters of the three-parameter Weibull model and Fréchet model. A practical application to river drain data will be given in Section 6.3.

6.2 Least Squares and Minimum Distance Methods

6.2.1 General

Formulation of a Least Squares method

The following method goes back to Bain and Antle (1967, p. 622):

Let X_i, $i = 1, \ldots, n$, be independent real-valued random variables that are P_ϑ-distributed, where $\vartheta \in \Theta$ is an unknown parameter. Suppose we observe only some order statistics $X_{i\uparrow n}$ for $i \in I$, where for example in the case of type II censoring $I = \{r + 1, \ldots, s\}$.

The general idea of the method is to choose ϑ in that way that for some suitable function U_ϑ the values $U_\vartheta(X_{i\uparrow n})$ are close to their expected values, i.e. we want to minimize over $\vartheta \in \Theta$ or some reasonable — chosen regarding the observations — subset Θ_X of Θ

$$d\left((U_\vartheta(X_{i\uparrow n}))_{i \in I}, (E_\vartheta(U_\vartheta(X_{i\uparrow n})))_{i \in I}\right),$$

where d denotes a suitable metric.

The choice $U_\vartheta = F_\vartheta$, the cumulative distribution function of P_ϑ, has the advantage that $E_\vartheta(U_\vartheta(X_{i\uparrow n})) = i/(n+1)$ is independent from ϑ, hence we want to solve

$$\min_\vartheta d\left((F_\vartheta(X_{i\uparrow n}))_{i \in I}, (i/(n+1))_{i \in I}\right). \tag{6.1}$$

Remark 6.2.1 *An alternative Least Squares method.* Let F^- denote the quantile function of a cumulative distribution function F, i.e.,

$$F^- : (0,1) \to \mathbf{R}, \ F^-(z) = \inf\{x \in \mathbf{R} : F(x) \geq z\}.$$

Instead of fitting $(F_\vartheta(X_{i\uparrow n}))_{i \in I}$ to $(i/(n+1))_{i \in I}$ we may also try to fit the observations $(X_{i\uparrow n})_{i \in I}$ to $\left(F_\vartheta^-(i/(n+1))\right)_{i \in I}$, i.e., we may want to solve the minimizing problem

$$\min_\vartheta d\left((X_{i\uparrow n})_{i \in I}, \left(F_\vartheta^-(i/(n+1))\right)_{i \in I}\right). \tag{6.2}$$

For the special choice of $d = d_2$, the Euclidean metric, we want to minimize the Q-statistics $\sum_{i \in I}\left[X_{i\uparrow n} - F_\vartheta^-(i/(n+1))\right]^2$, which is related with the Mallows-metric, see Witting and Müller-Funk (1995, p. 710 f).

Next we show that the approaches (6.1) and (6.2) are very similar to special minimum distance methods [see Witting and Müller-Funk (1995, p. 183 f)].

Remark 6.2.2 *Minimum distance methods.* The aim of the **minimum distance method** is to minimize the discrepancy, measured by a discrepancy function ϱ, of the empirical distribution function $\hat{F}_n$ and F_ϑ over $\vartheta \in \Theta$, i.e., to consider the minimization problem

$$\min_{\vartheta \in \Theta} \varrho(\hat{F}_n, F_\vartheta), \tag{6.3}$$

or to minimize over some suitable Θ_X again. Note that, if we do not observe all order statistics, that is $I \neq \{1, \ldots, n\}$, the empirical distribution function is not fully determined and therefore the discrepancy function has to take this into account.

Example 6.2.1 Consider for a metric $d : \mathbf{R}^{|I|} \times \mathbf{R}^{|I|} \rightarrow \mathbf{R}_0^+$ or $d : (0,1)^{|I|} \times (0,1)^{|I|} \rightarrow \mathbf{R}_0^+$ the following discrepancies, where "+" denotes limits from right and "−" limits from left:

$$\varrho_1(G, F) \;:=\; d\left(\left(G^-\left(i/n\right)\right)_{i \in I}, \left(F^-\left(i/n\right)\right)_{i \in I}\right)$$

$$\tilde{\varrho}_1(G, F) \;:=\; d\left(\left(G^-\left((i-1)/n+\right)\right)_{i \in I}, \left(F^-\left((i-1)/n+\right)\right)_{i \in I}\right)$$

$$\varrho_2(G, F) \;:=\; d\left(\left(G\left(X_{i \uparrow n}\right)\right)_{i \in I}, \left(F\left(X_{i \uparrow n}\right)\right)_{i \in I}\right)$$

$$\tilde{\varrho}_2(G, F) \;:=\; d\left(\left(G\left(X_{i \uparrow n}-\right)\right)_{i \in I}, \left(F\left(X_{i \uparrow n}-\right)\right)_{i \in I}\right) .$$

Note that the discrepancies ϱ_2 and $\tilde{\varrho}_2$ are dependent on the observations. Choosing these discrepancies we consider the following problems in (6.3):

$$\min_{\vartheta} \varrho_1(\hat{F}_n, F_\vartheta) \;=\; \min_{\vartheta} d\left(\left(X_{i \uparrow n}\right)_{i \in I}, \left(F_\vartheta^-\left(i/n\right)\right)_{i \in I}\right) \tag{6.4}$$

$$\min_{\vartheta} \tilde{\varrho}_1(\hat{F}_n, F_\vartheta) \;=\; \min_{\vartheta} d\left(\left(X_{i \uparrow n}\right)_{i \in I}, \left(F_\vartheta^-\left((i-1)/n+\right)\right)_{i \in I}\right) \tag{6.5}$$

$$\min_{\vartheta} \varrho_2(\hat{F}_n, F_\vartheta) \;=\; \min_{\vartheta} d\left(\left(F_\vartheta(X_{i \uparrow n})\right)_{i \in I}, \left(i/n\right)_{i \in I}\right) \tag{6.6}$$

$$\min_{\vartheta} \tilde{\varrho}_2(\hat{F}_n, F_\vartheta) \;=\; \min_{\vartheta} d\left(\left(F_\vartheta(X_{i \uparrow n}-)\right)_{i \in I}, \left((i-1)/n\right)_{i \in I}\right), \tag{6.7}$$

where the minimization is over $\vartheta \in \Theta$ or $\vartheta \in \Theta_X$. We may also try to hit the golden mean between (6.4) and (6.5) or between (6.6) and (6.7) and consider the minimization problems

$$\min_{\vartheta} d\left(\left(X_{i \uparrow n}\right)_{i \in I}, \left(F_\vartheta^-\left(\xi_i\right)\right)_{i \in I}\right) \quad \text{and} \tag{6.8}$$

$$\min_{\vartheta} d\left(\left(F_\vartheta(X_{i \uparrow n})\right)_{i \in I}, \left(\xi_i\right)_{i \in I}\right) \tag{6.9}$$

with $\xi_i := \frac{i-1/2}{n}$.

Note that for continuous distribution functions the alternative LS method (6.2) and the minimum distance methods (6.4) and (6.5) also have the form (6.8) and the LS method (6.1) and the minimum distance methods (6.6) and (6.7) also have the form (6.9) for $\xi_i \in \{i/(n+1), i/n, (i-1)/n\}$.

Discrepancies similar to ϱ_1 and $\tilde{\varrho}_1$ have been studied by Carmody, Eubank and LaRiccia (1984, p. 70 f).

6.2.2   Least squares and minimum distance estimators for the three-parameter Weibull model

First, we consider the corresponding minimization problems of the form (6.9) with $\xi_i \in \{i/(n+1), i/n, (i-1)/n, (i-1/2)/n\}$ in the three-parameter Weibull model. Here, the metric

$$
d_{\ln} : \begin{cases} (0,1)^{|I|} \times (0,1)^{|I|} \to \mathbf{R}_0^+ \\ ((t_i)_{i \in I}, (z_i)_{i \in I}) \mapsto \left(\sum_{i \in I} [\ln(-\ln(1-t_i)) - \ln(-\ln(1-z_i))]^2 \right)^{1/2} \end{cases}
$$
(6.10)

has proven to be more suitable than the Euclidean metric $d = d_2$, because for $d = d_2$ the minimization problems will be very difficult to solve numerically and the estimators have greater absolute bias and mean square error [see Offinger (1996, p. 131 ff and p. 154 ff)]. Another motivation is that $t \mapsto \ln(-\ln(1-t))$ is the quantile function of the Gumbel distribution for minima, that is, essentially we transfer to a Gumbel scale. Hence, we restrict ourselves to the metric $d = d_{\ln}$.

A reasonable restriction of the admissible region of the minimization problem is $\Theta_X := \{(\gamma, \lambda, \alpha) : \gamma < X_{\min}, \lambda > 0, \alpha > 0\}$, where $X_{\min} := X_{\min(I)\uparrow n}$. It will prove that this causes no numerical problems. Then we get the following minimization problem with $k_i := \ln(-\ln(1-\xi_i))$:

$$
\min_{(\gamma, \lambda, \alpha) \in (-\infty, X_{\min}) \times (0, \infty)^2} \sum_{i \in I} \left[\alpha \ln(\lambda) + \alpha \ln(X_{(i)} - \gamma) - k_i \right]^2 .
$$
(6.11)

For fixed γ and at least two different observation values the minimization problem has a unique solution. Put

$$
\begin{aligned}
t_i(\gamma) &:= \ln(X_{i\uparrow n} - \gamma), \\
\bar{t}(\gamma) &:= \frac{1}{|I|} \sum_{i \in I} t_i(\gamma), \\
\bar{k} &:= \frac{1}{|I|} \sum_{i \in I} k_i, \\
s_{tk}(\gamma) &:= \frac{1}{|I|} \sum_{i \in I} (t_i(\gamma) - \bar{t}(\gamma))(k_i - \bar{k}) = \frac{1}{|I|} \sum_{i \in I} t_i(\gamma) k_i - \bar{t}(\gamma) \bar{k} \quad \text{and} \\
s_{t^2}(\gamma) &:= \frac{1}{|I|} \sum_{i \in I} (t_i(\gamma) - \bar{t}(\gamma))^2,
\end{aligned}
$$

and we easily get the solution [compare Offinger (1996, p. 104 ff)]:

$$
\alpha(\gamma) = \frac{s_{tk}(\gamma)}{s_{t^2}(\gamma)} = \frac{\sum_{i \in I}(t_i(\gamma) - \bar{t}(\gamma))(k_i - \bar{k})}{\sum_{i \in I}(t_i(\gamma) - \bar{t}(\gamma))^2}
$$
(6.12)

$$
\lambda(\gamma) = \exp(\bar{k}/\alpha(\gamma)) / \exp(\bar{t}(\gamma)) = \frac{\prod_{i \in I}(-\ln(1-\xi_i))^{1/(n\alpha(\gamma))}}{\prod_{i \in I}(X_{(i)} - \gamma)^{1/n}}
$$
(6.13)

Substituting these solutions in (6.11) we get, because of

$$
\sum_{i \in I} \left[\alpha(\gamma) \ln(\lambda(\gamma)) + \alpha(\gamma) \ln(X_{(i)} - \gamma) - k_i \right]^2 = -|I| \frac{(s_{tk}(\gamma))^2}{s_{t^2}(\gamma)} + |I| s_{k^2}
$$

and the independence of s_{k^2} on γ, that we only have to solve the following one-dimensional problem:

$$\max_{\gamma \in (-\infty, X_{\min})} \frac{(s_{tk}(\gamma))^2}{s_{t^2}(\gamma)}. \qquad (6.14)$$

That is, we search for a $\gamma \in (-\infty, X_{\min})$ such that the correlation $\frac{s_{tk}(\gamma)}{\sqrt{[b]s_{t^2}(\gamma)s_{k^2}}}$ between t and k gets as high as possible.

For the numerical solution of the maximization problem a hybrid method by Brent [see Scales (1985, p. 46 ff)], that combines quadratic approximation and golden section, has proven to be very suitable [see Offinger (1996, p. 126 ff)], since numerical examinations showed that the function $\gamma \mapsto s_{tk}(\gamma)/s_{t^2}(\gamma)$ is strictly convex and unimodal or strictly increasing. Simulations showed that regarding bias and mean square error the choice $\xi_i = (i - 1/2)/n$ was better than $\xi_i = i/(n + 1)$, $\xi_i = i/n$ or $\xi_i = (i - 1)/n$. For obvious reasons we call this estimator with $\xi_i = (i - 1/2)/n$ **golden hit estimator I**, abbreviated GHE I. The results of a simulation study for this estimator can be found in Table 6.1. These show that the estimator, which is very simple to implement numerically, gives good estimation results: Absolute bias and mean square error are about the same as the modified moment estimator, the estimator by Wyckoff, Bain and Engelhardt (WBE) and the estimator by Zanakis examined in Cohen, Whitten and Ding (1984, p. 166), slightly better than maximum likelihood estimation[1] for $n = 50$ and $\alpha \in \{2.0, 3.0\}$ and about the same for $n = 1000$ and $\alpha \in \{2.0, 3.0\}$.

Next we take a look at the minimization problem (6.8) in our three-parameter Weibull model. After squaring and using the Euclidean metric $d = d_2$ we get:

$$\min_{(\gamma, \lambda, \alpha) \in \mathbf{R} \times (0, \infty)^2} \sum_{i \in I} \left[\gamma + \frac{1}{\lambda} \left(-\ln(1 - \xi_i) \right)^{1/\alpha} - X_{(i)} \right]^2 \qquad (6.15)$$

Only if we refrain from the restriction on Θ_X from above and if we look at the problem for fixed α, we can again reduce it to a one-dimensional one [compare Offinger (1996: p. 107 ff)] and implement it similarly. Putting

$$t_i(\alpha) := \left(-\ln(1 - \xi_i) \right)^{1/\alpha}$$

we get with analogous denotations as above

$$\lambda(\alpha) = \frac{s_{t^2}(\alpha)}{s_{tx}(\alpha)} \qquad (6.16)$$

$$\gamma(\alpha) = \bar{x} - \frac{s_{tx}(\alpha)}{s_{t^2}(\alpha)} \bar{t}(\alpha) \qquad (6.17)$$

---

[1]Maximum Likelihood estimation denotes the estimation obtained by searching for a **local** maximum of the likelihood function in the model with $\alpha > 1$, see Cheng and Taylor (1995, p. 6 ff).

Table 6.1: BIAS and MSE from a simulation study for the golden hit estimator I, implemented by (6.12), (6.13) and (6.14) with $\xi_i = (i - 1/2)/n$, for sample sizes of $n = 25, 50, 100, 1000$ and $I = \{1, \ldots, n\}$. The "true" parameters were $\gamma = 0$, $\delta = 1$ and $\alpha \in \{0.5, 1.0, 2.0, 3.0\}$. 1000 data sets of size n from $\text{Wei}_{\gamma, 1/\delta, \alpha}$ were generated

n=25	α							
	0.5		1.0		2.0		3.0	
	BIAS	MSE	BIAS	MSE	BIAS	MSE	BIAS	MSE
$\hat{\gamma}$	0.0021	0.0001	0.0132	0.0025	0.0033	0.0446	-0.2008	2.0222
$\hat{\alpha}$	0.0074	0.0103	0.0004	0.0483	0.0522	0.5564	0.8381	26.3019
$\hat{\delta}$	0.0696	0.2023	-0.0060	0.0498	-0.0072	0.0713	0.1966	2.0818

n=50	α							
	0.5		1.0		2.0		3.0	
	BIAS	MSE	BIAS	MSE	BIAS	MSE	BIAS	MSE
$\hat{\gamma}$	0.0005	0.0000	0.0054	0.0006	0.0011	0.0151	-0.0455	0.1240
$\hat{\alpha}$	0.0021	0.0045	-0.0014	0.0211	0.0254	0.2232	0.2019	2.2887
$\hat{\delta}$	0.0489	0.0924	0.0048	0.0226	0.0004	0.0253	0.0431	0.1353

n=100	α							
	0.5		1.0		2.0		3.0	
	BIAS	MSE	BIAS	MSE	BIAS	MSE	BIAS	MSE
$\hat{\gamma}$	0.0001	0.0000	0.0031	0.0001	0.0098	0.0034	0.0059	0.0153
$\hat{\alpha}$	0.0056	0.0022	0.0045	0.0098	0.0009	0.0731	0.0275	0.3473
$\hat{\delta}$	0.0257	0.0487	0.0022	0.0116	-0.0095	0.0070	-0.0061	0.0180

n=1000	α							
	0.5		1.0		2.0		3.0	
	BIAS	MSE	BIAS	MSE	BIAS	MSE	BIAS	MSE
$\hat{\gamma}$	0.0000	0.0000	0.0007	0.0000	0.0088	0.0002	0.0191	0.0008
$\hat{\alpha}$	0.0078	0.0002	0.0081	0.0014	-0.0085	0.0096	-0.0445	0.0326
$\hat{\delta}$	0.0031	0.0013	0.0015	0.0004	-0.0085	0.0003	-0.0192	0.0010

Table 6.2: BIAS and MSE from a simulation study for the maximum likelihood estimator in the model with $\alpha > 1$ for sample sizes of $n = 50, 1000$ and $I = \{1, \ldots, n\}$. The "true" parameters were $\gamma = 0$, $\delta = 1$ and $\alpha \in \{2.0, 3.0\}$. 1000 data sets of size $n = 50$ and 100 data sets of size $n = 1000$ were generated from $\text{Wei}_{\gamma, 1/\delta, \alpha}$

	$n = 50$				$n = 1000$			
	$\alpha = 2.0$		$\alpha = 3.0$		$\alpha = 2.0$		$\alpha = 3.0$	
	BIAS	MSE	BIAS	MSE	BIAS	MSE	BIAS	MSE
$\hat{\gamma}$	0.0359	0.0150	0.0037	0.1797	0.0134	0.0002	0.0060	0.0008
$\hat{\alpha}$	-0.0324	0.1931	0.1114	2.7989	-0.0163	0.0037	-0.0208	0.0160
$\hat{\delta}$	-0.0448	0.0259	-0.0107	0.1983	-0.0145	0.0004	-0.0068	0.0010

and as the resulting one-dimensional problem

$$\max_{\alpha\in(0,\infty)} \frac{(s_{tx}(\alpha))^2}{s_{t^2}(\alpha)} . \tag{6.18}$$

Here again the choice of $\xi_i = \frac{i-1/2}{n}$ proved to be better regarding bias and mean square error than $\xi_i \in \{i/(n+1), i/n, (i-1)/n\}$ We call this estimator the **golden hit estimator II**. The results of a simulation study for this estimator are in Table 6.3. These show that the golden hit estimator I has smaller absolute bias and mean square error for $\alpha \in \{0.5, 1.0, 2.0\}$ but greater absolute bias and mean square error for $\alpha = 3.0$. So, if we expect higher values for α it might be better to use GHE II.

Table 6.3: BIAS and MSE from a simulation study for the golden hit estimator II, implemented by (6.16), (6.17) and (6.18) with $\xi_i = (i-1/2)/n$, sample sizes $n = 25, 50, 100, 1000$ and $I = \{1, \ldots, n\}$; "true" parameters: $\gamma = 0$, $\delta = 1$ and $\alpha \in \{0.5, 1.0, 2.0, 3.0\}$; 1000 data sets of size n from $\mathrm{Wei}_{\gamma,1/\delta,\alpha}$ were generated

	α							
n=25	0.5		1.0		2.0		3.0	
	BIAS	MSE	BIAS	MSE	BIAS	MSE	BIAS	MSE
$\hat\gamma$	-0.1039	0.1422	-0.0301	0.0421	-0.0367	0.0799	-0.1189	0.3505
$\hat\alpha$	0.1016	0.0703	0.1199	0.1727	0.2293	0.8564	0.6385	6.1001
$\hat\delta$	0.3271	0.9486	0.0419	0.1589	0.0323	0.1150	0.1138	0.3844

	α							
n=50	0.5		1.0		2.0		3.0	
	BIAS	MSE	BIAS	MSE	BIAS	MSE	BIAS	MSE
$\hat\gamma$	-0.0854	0.1199	-0.0266	0.0232	-0.0253	0.0304	-0.0518	0.0897
$\hat\alpha$	0.0712	0.0381	0.0803	0.0765	0.1359	0.3074	0.2881	1.5368
$\hat\delta$	0.2628	0.5910	0.0431	0.0817	0.0266	0.0456	0.0512	0.1009

	α							
n=100	0.5		1.0		2.0		3.0	
	BIAS	MSE	BIAS	MSE	BIAS	MSE	BIAS	MSE
$\hat\gamma$	-0.0523	0.0662	-0.0075	0.0092	0.0002	0.0105	0.0037	0.0210
$\hat\alpha$	0.0382	0.0166	0.0331	0.0296	0.0438	0.1016	0.0520	0.3310
$\hat\delta$	0.1444	0.2668	0.0127	0.0377	-0.0008	0.0181	-0.0049	0.0258

	α							
n=1000	0.5		1.0		2.0		3.0	
	BIAS	MSE	BIAS	MSE	BIAS	MSE	BIAS	MSE
$\hat\gamma$	-0.0690	0.0141	-0.0063	0.0009	0.0047	0.0011	0.0122	0.0022
$\hat\alpha$	0.0274	0.0027	0.0171	0.0025	0.0089	0.0073	-0.0114	0.0247
$\hat\delta$	0.1185	0.0378	0.0111	0.0009	-0.0043	0.0008	-0.0122	0.0020

Remark 6.2.1 A program for personal computer that implements the mentioned and many more estimators that have been suggested in the three-parameter Weibull model is available by the author.

Remark 6.2.3 *Three-parameter Fréchet model.* The estimation methods we have treated here can very easily be transferred to a three-parameter Fréchet model. We consider the minimization problem (6.9) with metric $d = d_{\ln}$ (see formula (6.10)) and the restriction to $\tilde{\Theta}_x = \{(\gamma, \lambda, \alpha) : \gamma > X_{\max}, \lambda > 0, \alpha > 0\}$ with $X_{\max} := X_{\max(I)\uparrow n}$. This leads to

$$\alpha(\gamma) = s_{\tau\kappa}(\gamma)/s_{\tau^2}(\gamma) \text{ and}$$
$$\lambda(\gamma) = \exp(\bar{\kappa}/\alpha(\gamma))/\exp(\bar{\tau}(\gamma)),$$

with $\kappa_i := -\ln(-\ln(\xi_i))$ and $\tau_i(\gamma) := \ln(\gamma - X_{i\uparrow n})$, and the one-dimensional problem

$$\max_{\gamma \in (X_{\max}, -\infty)} \frac{(s_{\tau\kappa}(\gamma))^2}{s_{\tau^2}(\gamma)}.$$

We call this estimator the golden hit estimator I again. The derivation of an analogue to the golden hit estimator II is also very simple.

Remark 6.2.2 As with all minimum distance methods, the discrepancy provides a measure of goodness-of-fit in a natural way, but one also can use other test statistics, e.g. the EDF tests developed in Lockhart and Stephens (1994, p. 496 f) for the three-parameter Weibull model, where one of the test statistics is the Cramér-von Mises statistic

$$W^2(X) = d_2((F_{\gamma,\lambda,\alpha}(X_{i\uparrow n}))_{i=1}^n, (\xi_i)_{i=1}^n) + \frac{1}{12n}$$

with $\xi_i = \frac{i-1/2}{n}$, that is, essentially the objective function from (6.9) with Euclidian metric $d = d_2$.

6.3 Modelling of River Drain Data

6.3.1 The data

The river Danube has up to Ulm a typical low mountain range character, which gets Alpine influences by the river Iller at Ulm. Up to Donauwörth these influences are counterbalanced by other tributaries. On its way from roughly west to east the river Danube absorbs many waters coming from north and south, for example, the river Lech, which flows into the Danube from south a few kilometres after Donauwörth, or at Donauwörth the river Wörnitz from the north. For further details we refer to the hydrological yearbooks from München (second reference in the list), which most of the following data come from. First of all we should make clear that by a year the hydrologist means the time from November of the year before till the October of the referred year. Thus, for example, the year 1990 lasted from 1st November 1989 to 31th October 1990 and the minimum daily average value of 90.3 m^3/s for the year 1987 (see Table 6.5) was reached at 13th December 1986. Thereby a year contains a complete winter half year and a complete summer half year.

The values of the drain of the river Danube are measured every hour by level indicators. In Table 6.4 are the maximum values over a year of the maximum values over a day of the drain measurements of the river Danube at level indicator Schäfstall or (since 1971) at level indicator Donauwörth, 3.5 kilometres from Schäfstall. But both level indicators are before the entry of the river Lech and no power station is between them. On the other hand in Table 6.5 are the minimum values over a year of the daily average values for the river Danube at the same level indicator. For the measurements of the river Danube before 1956 it is not totally clear if really the yearly minimum of the daily average and not of the daily minimum values were listed. Apart from that the values of the river Danube in the years 1943-1945 are – caused by the turmoils of the war – uncertain.

Another data set can be found in Figure 6.1 Here the maximum drain values of the river Main at Würzburg from 1824 till 1995 are plotted and not printed number by number — for the sake of brevity; refer to third reference in the list. The river Main, which is the greatest tributary of the river Rhine, has low mountain character.

Table 6.4: Maximum drain of the river Danube at Donauwörth in m^3/s in the years 1941-1990

1941	1942	1943	1944	1945	1946	1947	1948	1949	1950
860	958	419	582	964	571	765	741	491	358
1951	1952	1953	1954	1955	1956	1957	1958	1959	1960
760	624	613	747	915	973	875	807	493	394
1961	1962	1963	1964	1965	1966	1967	1968	1969	1970
544	500	436	566	1002	784	627	849	673	1150
1971	1972	1973	1974	1975	1976	1977	1978	1979	1980
828	347	694	662	758	679	788	827	877	947
1981	1982	1983	1984	1985	1986	1987	1988	1989	1990
804	1110	774	669	718	603	973	1110	853	1150

Table 6.5: Minimum drain of the river Danube at Donauwörth in m^3/s in the years 1941-1990

1941	1942	1943	1944	1945	1946	1947	1948	1949	1950
104	75.0	54.2	45.8	70.7	92.2	42.6	43.8	42.6	39.8
1951	1952	1953	1954	1955	1956	1957	1958	1959	1960
55.5	53.7	69.2	53.0	100	85.4	84.4	70.9	54.6	63.2
1961	1962	1963	1964	1965	1966	1967	1968	1969	1970
67.9	50.9	50.6	64.2	65.0	80.5	84.2	76.8	83.0	78.0
1971	1972	1973	1974	1975	1976	1977	1978	1979	1980
72.1	49.7	64.6	85.6	103	65.4	78.7	99.0	94.6	68.8
1981	1982	1983	1984	1985	1986	1987	1988	1989	1990
88.7	97.4	77.6	73.3	64.7	71.6	90.3	98.6	75.6	72.6

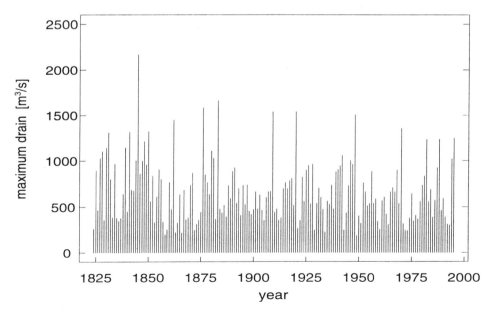

Figure 6.1: Maximum drain of the river Main at Würzburg in the years 1824-1995

6.3.2 General

As we have tested the data for serial dependence by various methods and could not reject the null hypothesis that there is no dependence and motivated by appropriate theorems from extreme value statistics [see Witting and Müller-Funk (1995, p. 594) or Castillo (1994, p. 27)] it is well-grounded to model the minimum annual drain by a generalized extreme value model for minima and therefore in essence by our three-parameter Weibull model or by our three-parameter Fréchet model (for minima). Analogously we tried to model the maximum annual drain by a three-parameter model with the Weibull distribution for maxima or by a three-parameter Fréchet model for maxima. Note that the Weibull distribution for maxima is obtained by reflecting the Weibull distribution (for minima) $\mathrm{Wei}_{\gamma,\lambda,\alpha}$. Instead of transferring the estimation methods to the model with the Weibull distribution for maxima, that is, the reflected Weibull distributions $\mathrm{rWei}_{\gamma,\lambda,\alpha}$, which have cumulative distribution function

$$F_{\gamma,\lambda,\alpha}(z) = \begin{cases} 1 & , z \geq \gamma \\ \exp\left(-(-\lambda(z-\gamma))^{\alpha}\right) & , z < \gamma \end{cases} ,$$

it is simpler to transfer the data by multiplying with -1, since it holds: If X is $\mathrm{rWei}_{\gamma,\lambda,\alpha}$-distributed, then $-X$ is $\mathrm{Wei}_{-\gamma,\lambda,\alpha}$-distributed. Thus we get the corresponding estimators for the model with the reflected distributions at once from the estimators for our original model. The same works for the Fréchet distributions.

6.3.3 Analysis of river Danube data

We take a three-parameter Weibull model for maxima — as indicated above — for the drain data of the river Danube from Table 6.4. The estimated parameter values of a reflected Weibull distribution are given in Table 6.6 for a number of different estimators. Figure 6.2 and the value $W^2(x) = 0.0173$ of the Cramér-von Mises statistic for the golden hit estimation I $\mathrm{rWei}_{1400,1/728,3.52}$ indicate a very good fit. Only the WBE-estimation with a value of $W^2(x) = 0.0553$, is out of the ordinary. For further details on the methods not treated here see Johnson, Kotz and Balakrishnan (1994, p. 641 ff), Cohen, Whitten and Ding (1984, p. 160 ff) and Cheng and Taylor (1995, p. 6 ff).

Problems arise with the minimum drain data of the river Danube from Table 6.5. Even if we leave out the uncertain values before 1956, the estimators still give very different results, see Table 1.7. The improvement of the W^2-value in case of the golden hit estimation I is remarkable, it decreases from 0.0406 for the complete data set to 0.0279 for the reduced data set. This good fit of the golden hit estimator I for the smaller data set is also illustrated in Figure 6.3. For the golden hit estimation II the Cramér-von Mises statistic is $W^2(x) = 0.0344$ for the complete data set and $W^2(x) = 0.0261$ for the reduced data set.

Table 6.6: Estimated parameter values of a reflected Weibull distribution $\text{rWei}_{\gamma,\lambda,\alpha}$ for maximum drain data of the river Danube from Table 6.4

estimator	γ	$\delta = \frac{1}{\lambda}$	α
golden hit estimator I	$1.40 \cdot 10^3$	728	3.52
golden hit estimator II	$1.39 \cdot 10^3$	723	3.53
maximum likelihood	$1.34 \cdot 10^3$	670	3.29
moment estimator	$1.43 \cdot 10^3$	755	3.69
modif. moment estimator	$1.36 \cdot 10^3$	683	3.28
WBE	$1.23 \cdot 10^3$	537	2.45
max. product of spacings	$1.42 \cdot 10^3$	749	3.49

Table 6.7: Estimated parameter values of a Weibull distribution $\text{Wei}_{\gamma,\lambda,\alpha}$ for the minimum drain data of the river Danube with and without the uncertain data before 1956

estimator	total Danube data			Danube data since '56		
	γ	$\delta = \frac{1}{\lambda}$	α	γ	$\delta = \frac{1}{\lambda}$	α
golden hit estimator I	30.6	47.0	2.44	35.8	45.0	3.26
golden hit estimator II	19.5	58.5	3.38	38.4	42.2	3.04
maximum likelihood	27.6	50.0	2.87	40.5	39.9	2.90
moment estimator	15.9	62.3	3.61	36.1	44.6	3.22
modif. moment estimator	27.6	50.0	2.78	29.4	51.6	3.82
WBE	34.6	41.7	2.27	42.6	37.9	2.56
max. product of spacings	21.0	57.2	3.15	34.6	46.4	3.13

6.3.4 Analysis of river Main data

At last we take a look at the river Main data from Figure 6.1, where some interesting problems occur.

A QQ-Plot of the negative observations versus k_i, the quantiles of the Gumbel distribution for minima, gave a weak indication that a Frechét model should be used. Another sign is the empirical skewness -1.20 of the negative observations (see also the histogram and the kernel estimation in Figure 6.4): Recall that the Weibull distributions $\text{Wei}_{\gamma,\lambda,\alpha}$ have skewness greater than the skewness of a Gumbel distribution, which is $-2 \cdot 6^{3/2} \zeta(3)/\pi^3 \approx -1.13955$, and the infimum is for shape parameter $\alpha \to \infty$ and that the Fréchet distribution have skewness smaller than that and the supremum is again for shape parameter $\alpha \to \infty$ (for $\alpha \leq 3$ the first three moments do no exist). So we expect a high estimated value of α in the Fréchet model. Indeed the golden hit estimator gives $\text{Fre}_{6107,1/6620,26.00}$ for the distribution of the negative maximum drain, that is, we estimate a Fréchet distribution for maxima with location parameter

-6107 m^3/s, scale parameter 6620 m^3/s and shape parameter 26.00, which is essentially a distribution on $(0, \infty)$ indeed.

Actually all estimators given in Table 6.8 fail in the case that the negative observations are modelled by a three-parameter Weibull distribution. For example, the function $\gamma \mapsto (s_{tk}(\gamma))^2/s_{t^2}(\gamma)$ from (6.14) is strictly increasing and the function $\gamma \mapsto \alpha(\gamma)$ from (6.12) tends to ∞ for $\gamma \to -\infty$, while the function $\alpha \mapsto (s_{tx}(\alpha))^2/s_{t^2}(\alpha)$ from (6.18) is strictly decreasing, which is another hint that a Fréchet model is more appropriate.

We also have fitted a three-parameter Weibull model for the original data and obtained the estimates from Table 6.8. Recall that this model is **not** motivated by extreme value considerations! Also note that the moment estimator and the golden hit estimator II give γ-estimates which are larger than the minimum observation value of 190 m^3/s. This is due to the fact that the optimization was not restricted to Θ_X, indicating that a restriction is necessary in general.

Finally, Figure 6.4 shows a histogram, a Gaussian kernel estimation, the density of the $\mathrm{Fre}_{-6107,1/6620,26.00}$-distribution for maxima and the density of the $\mathrm{Wei}_{179,1/543,1.51}$-distribution and we can see a good fit to the data since the histogram and the estimated nonparametric density are very similar to the estimated parametric density. This is backed up by the Cramér-von Mises statistic, which gives

$$W^2(X) = d_2((F_{179,1/543,1.51}(X_{i\uparrow n}))_{i=1}^n, ((i-1/2)/n)_{i=1}^n) + \frac{1}{12n} = 0.0216$$

for the Weibull estimation, indicating a very good fit, as the critical point for W^2 at a 50%-level is about 0.05, see Lockhart and Stephens (1994, p. 497).

Table 6.8: Estimated parameter values of a Weibull distribution $\mathrm{Wei}_{\gamma,\lambda,\alpha}$ for the Main data from Figure 6.1 with usage of different estimation methods

estimator	γ	$\delta = \frac{1}{\lambda}$	α
golden hit estimator I	179	543	1.51
golden hit estimator II	205	509	1.38
maximum likelihood	185	535	1.46
moment estimator	198	517	1.40
modified moment estimator	173	550	1.49
WBE	176	549	1.52
maximum product of spacings	175	550	1.46

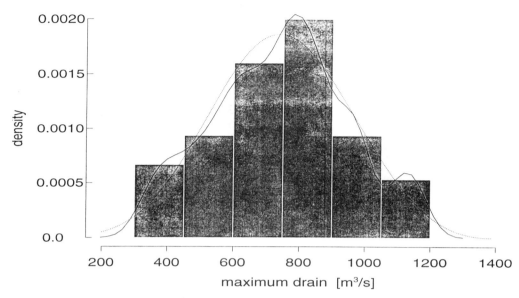

Figure 6.2: Histogram, Gaussian kernel estimation with bandwidth 200 m^3/s (solid line) and density of a estimated reflected Weibull distribution rWei$_{1400,1/728,3.52}$ (dotted line) for the maximum drain data of the river Danube, see Table 6.4

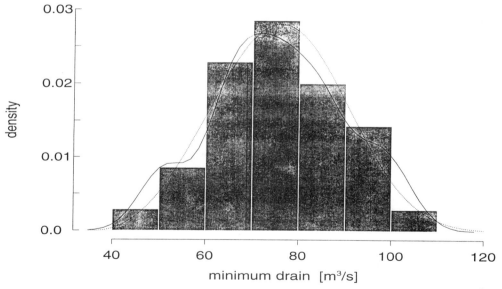

Figure 6.3: Histogram, Gaussian kernel estimation with bandwidth 20 m^3/s (solid line) and density of estimated Weibull distribution Wei$_{35.8,1/45.0,3.26}$ (dotted line) for the minimum drain data of the river Danube without the uncertain data before 1956, compare Table 6.5

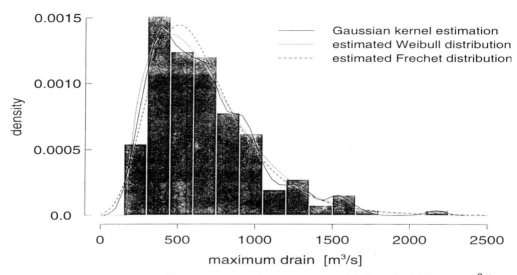

Figure 6.4: Histogram, Gaussian kernel estimation with bandwidth $250 \, m^3/s$, density of estimated Fréchet distribution for maxima $\text{rFre}_{6107,6620,26.00}$ and of estimated Weibull distribution $\text{Wei}_{179,543,1.51}$ for Main data from Table 6.1

For practical applications not only estimated parameter values of the distribution but confidence intervals are of special interest. If one wants to use bootstrap methods for calculations of such intervals, it is important to have good estimators of the parameter values that can be calculated rather quickly and numerically stable, since at least 1000 estimation processes have to be done, then. Here, our golden hit estimators do better than, for example, maximum likelihood estimation.

References

1. Bain, L. J. and Antle, C. E. (1967). Estimation of parameters in the Weibull distribution, *Technometrics*, **9**, 621–627.

2. Bayerisches Landesamt für Wasserwirtschaft München, *Deutsche gewässerkundliche Jahrbücher 1941-1990, Donaugebiet,* Bayerische Landesstelle für Gewässerkunde, München.

3. Bayerisches Staatsministerium für Landesentwicklung und Umweltfragen (1995), *Umwelt & Entwicklung Bayern 6/95: Hochwasser hausgemacht?* München.

4. Carmody, T. J., Eubank, R. L. and LaRiccia, V. N. (1984), A family of minimum quantile distance estimators for the three-parameter Weibull distribution *Statistische Hefte*, **25**, 69–82.

5. Castillo, E. (1994). Extremes in engineering applications, In *Extreme Value Theory and Applications, Proceedings of the Conference on Extreme Value Theory and Applications*, Vol. 1 (Eds., J. Galambos, J. Lechner and E. Simiu), Gaithersburg, Maryland 1993, pp. 15–42, Dordrecht: Kluwer.

6. Cheng, R. C. and Taylor, L. (1995). Non-regular maximum likelihood problems, *Journal of the Royal Statistical Society, B*, **57**, 3–44.

7. Cohen, A. C., Whitten, B. J. and Ding, Y. (1984). Modified moment estimation for the three-parameter Weibull distribution, *Journal of Quality Technology*, **16**, 159–167.

8. Embrechts, P., Klüppelberg, C. and Mikosch, T. (1997). *Modelling Extremal Events*, Berlin: Springer-Verlag.

9. Johnson, N. L., Kotz, S. and Balakrishnan, N. (1994). *Distributions in Statistics: Continuous Univariate Distributions, I*, New York: John Wiley & Sons.

10. Lockhart, R. A. and Stephens, M. A. (1994). Estimation and tests of fit for the three-parameter Weibull distribution, *Journal of the Royal Statistical Society B*, **56**, 491–500.

11. Offinger, R. (1996). *Schätzer in dreiparametrigen Weibull-Modellen und Untersuchung ihrer Eigenschaften mittels Simulation*, Diplomarbeit, Universität Augsburg.

12. Scales, L. E. (1985). *Introduction to Non-Linear Optimization* London: Macmillan.

13. Scholz, F. W. (1980). Towards a unified definition of maximum likelihood, *The Canadian Journal of Statistics*, **8**, 193–203.

14. Witting, H. and Müller-Funk, U. (1995). *Mathematische Statistik II: Parametrische Modelle und nichtparametrische Funktionale*, Stuttgart: Teubner.

PART II
Reliability Analysis

7

Maximum Likelihood Estimation With Different Sequential k-out-of-n Systems

Erhard Cramer and Udo Kamps

Aachen University of Technology, Aachen, Germany
University of Oldenburg, Oldenburg, Germany

Abstract: Sequential order statistics have been introduced as a more flexible model than ordinary order statistics to describe (sequential) k-out-of-n systems, where, after each failure, the remaining components possess a possibly different failure rate than before. We consider the situation of a sample of sequential k-out-of-n systems, which are allowed to have different structures. Explicit expressions of the maximum likelihood estimators of both the model parameters and the distribution parameters are derived in specific classes of distributions. Moreover, several useful properties of the estimators are presented.

Keywords and phrases: Sequential k-out-of-n system, sequential order statistics, maximum likelihood estimation

7.1 Introduction

Ordinary k-out-of-n systems consist of n components of the same kind which are supposed to have independent and identically distributed (iid) life-lengths. This approach postulates that the failure of some component does not affect the remaining ones. A concept weakening the iid assumption is based on the observation that the life-length of an ordinary k-out-of-n system is described by the $(n - k + 1)$-th order statistic in a sample of size n. Kamps (1995) proposes sequential order statistics as an extension of ordinary order statistics to model damage effects in the situation of a k-out-of-n structure. The failure of any component is allowed to influence the other components such that their underlying failure rate is parametrically adjusted with respect to the number of preceding failures. On the one hand this can be thought of as a damage caused by some component failure in the system. On the other hand it could mean that in case of a component failure increased weight is put on the remaining components. In this situation we refer to the system as *sequential k-out-of-n system*. In the following we consider the parametrization of a k-out-of-n system as $(n - r + 1)$-out-of-n system for some $1 \leq r \leq n$, where the r-th ordinary order statistic or the r-th sequential order statistic describes the life-length of the underlying system, respectively.

To be more precise, let F be an absolutely continuous distribution function with density function f and let $\alpha_1, \ldots, \alpha_n$ be positive real numbers. When starting the system, the n components are described by iid random variables with failure rate $\alpha_1 \frac{f}{1-F}$. The damage caused by the i-th failure is modelled in such a way that the remaining $n - i$ components are now supposed to have the failure rate

$$\alpha_{i+1} \frac{f}{1 - F}, \quad 1 \leq i \leq r - 1.$$

For further details and a precise description we refer to Kamps (1995, Chapter I.1) and to Cramer and Kamps (1996).

As mentioned above, the life-length of a sequential $(n - r + 1)$-out-of-n system is described by the r-th sequential order statistic in a sample of size n. For some $1 \leq r \leq n$, the random variables $X_*^{(1)}, \ldots, X_*^{(r)}$ are called *sequential order statistics* (based on F and $\alpha_1, \ldots, \alpha_r$) if their joint density function is given by

$$f^{X_*^{(1)}, \ldots, X_*^{(r)}}(x_1, \ldots, x_r)$$

$$= \frac{n!}{(n-r)!} \left(\prod_{j=1}^{r} \alpha_j \right) \left(\prod_{j=1}^{r-1} (1 - F(x_j))^{m_j} f(x_j) \right) (1 - F(x_r))^{\alpha_r(n-r+1)-1} f(x_r),$$

$$x_1 < \ldots < x_r, \tag{7.1}$$

with $m_j = (n - j + 1)\alpha_j - (n - j)\alpha_{j+1} - 1$, $1 \leq j \leq r - 1$, $r \leq n$.

Choosing $r = n$ and $\alpha_1 = \ldots \alpha_n = 1$, we obtain the joint density function of ordinary order statistics $X_{1,n} \leq \ldots \leq X_{n,n}$ based on iid random variables $X_1, \ldots, X_n$ with distribution function F. An important example of a distribution specified by a density function of the type (7.1) is Weinman's multivariate exponential distribution [see Block (1975)]. The connection to sequential order statistics is demonstrated in Kamps (1995) and Cramer and Kamps (1997).

In a previous paper [Cramer and Kamps (1996)], the authors derive maximum likelihood estimators of the model parameters. The maximum likelihood estimators $\alpha_1^*, \ldots, \alpha_r^*$ of $\alpha_1, \ldots, \alpha_r$ turn out to be independent and inverted gamma distributed random variables. Moreover, it has been shown that the estimators are sufficient, strongly consistent and asymptotically normal with respect to an increasing number of independent observations of some sequential $(n - r + 1)$-out-of-n system. We also considered simultaneous maximum likelihood estimation of both the model parameters and the distribution parameters for specific distributions, including Weibull distributions.

In the paper on hand we consider a more general situation with s sequential k-out-of-n systems. Each system is allowed to have a different structure, i.e. we have independent observations of a number of s

$$(n_i - r_i + 1)\text{-out-of-}n_i \text{ systems}, \quad 1 \leq r_i \leq n_i, 1 \leq i \leq s,$$

with r_i (dependent) observations, respectively, and with model parameters $(\alpha_{ij})_{1 \leq i \leq s, 1 \leq j \leq r_i}$. This leads to the data set

$$(x_{ij})_{1 \leq i \leq s, 1 \leq j \leq r_i} \quad \text{with} \quad x_{i1} \leq \ldots \leq x_{ir_i}, 1 \leq i \leq s.$$

The corresponding sequential order statistics are denoted by

$$(X_{*i}^{(j)})_{1 \leq i \leq s, 1 \leq j \leq r_i},$$

which, by assumption, are independent with respect to the index i.

Throughout this paper we assume without loss of generality that the r_i's are arranged in descending order:

$$r_1 \geq r_2 \geq \ldots \geq r_s \geq 1.$$

For $\nu = 1, \ldots, r_1$ let c_ν be the number of r_i's with $r_i \geq \nu$:

$$c_\nu = |\{i : r_i \geq \nu, 1 \leq i \leq s\}| = \max\{i : r_i \geq \nu, 1 \leq i \leq s\}.$$

Obviously, we have $c_1 = s$.

The paper is organized as follows. In Section 7.2 we suppose to have identical model parameters $\alpha_1, \alpha_2, \ldots$ in all systems, i.e.

$$\alpha_{ij} = \alpha_j, \quad 1 \leq i \leq s.$$

We obtain explicit expressions of the maximum likelihood estimators $\alpha_1^*, \ldots, \alpha_{r_1}^*$ of $\alpha_1, \ldots, \alpha_{r_1}$ which are shown to be independent and inverted gamma distributed random variables. The estimation of the model parameter α_ν is based on c_ν observations. Section 7.3 contains maximum likelihood estimators of $\lambda \alpha_1, \ldots, \lambda \alpha_{r_1}$ and simultaneous maximum likelihood estimators of η and $\lambda \alpha_1, \ldots,$ $\lambda \alpha_{r_1}$ when the specific classes of distributions are given by $F(t) = 1 - \exp(-\lambda g(t))$, $t \geq 0$, and $F(t) = 1 - \exp(-\lambda(g(t) - \eta))$, $t \geq g^{-1}(\eta)$, respectively, with an appropriate function g. In Section 7.4 we examine the maximum likelihood estimator λ^* of λ in the model based on $F(t) = 1 - \exp(-\lambda g(t))$, $t \geq 0$, where the α_{ij}'s are assumed to be known. The estimator λ^* is shown to be inverted gamma distributed and sufficient for λ. With respect to an increasing number of observed systems we obtain strong consistency and asymptotic normality for the maximum likelihood estimator λ^*. In the final section we present an example and apply the results of Section 7.4. We consider a sample of 2-out-of-4 systems to compare the estimators of λ when modelling the system either by ordinary order statistics or by sequential ones.

7.2   Sequential k-out-of-n Systems With Unknown Model Parameters

According to the model description and the joint density $f^{X_*^{(1)},\dots,X_*^{(r)}}$ stated in (7.1) we obtain the likelihood function

$$
L(\alpha_{ij}, x_{ij}; 1 \le i \le s, 1 \le j \le r_i)
$$

$$
= \left(\prod_{i=1}^{s} \frac{n_i!}{(n_i - r_i)!} \right) \left(\prod_{i=1}^{s} \prod_{j=1}^{r_i} \alpha_{ij} \right) \left(\prod_{i=1}^{s} \prod_{j=1}^{r_i - 1} (1 - F(x_{ij}))^{m_{ij}} f(x_{ij}) \right)
$$

$$
\times \prod_{i=1}^{s} (1 - F(x_{ir_i}))^{\alpha_{r_i}(n_i - r_i + 1) - 1} f(x_{ir_i}) = L((\alpha_{ij})_{i,j}), \quad \text{say}, \quad (7.2)
$$

where $m_{ij} = (n_i - j + 1)\alpha_{ij} - (n_i - j)\alpha_{i,j+1} - 1,\, 1 \le j \le r_i - 1,\, r_i \le n_i,\, 1 \le i \le s.$

Theorem 7.2.1 *Let $\alpha_{ij} = \alpha_j$, $1 \le i \le s$, $1 \le j \le r_i$. The maximum likelihood estimators of the model parameters $\alpha_1, \dots, \alpha_{r_1}$ are given by*

$$
\alpha_1^* = -s \left(\log \prod_{i=1}^{s} \bar{F}^{n_i}(x_{i1}) \right)^{-1} \quad \text{and}
$$

$$
\alpha_\nu^* = -c_\nu \left(\log \prod_{i=1}^{c_\nu} \left(\frac{\bar{F}(x_{i\nu})}{\bar{F}(x_{i,\nu-1})} \right)^{n_i - \nu + 1} \right)^{-1}, \quad 2 \le \nu \le r_1,
$$

where $\bar{F}(t) = 1 - F(t)$, $t \in \mathbf{R}$, denotes the survival function of F.

PROOF. Expressing the log-likelihood function $l((\alpha_\nu)_{1 \le \nu \le r_1}) = \log L((\alpha_\nu)_{1 \le \nu \le r_1})$ in terms of the hazard function

$$
h(t) = \frac{f(t)}{1 - F(t)}, \quad t \in \mathbf{R},
$$

and the cumulative hazard function

$$
H(t) = -\log(1 - F(t)), \quad t \in \mathbf{R},
$$

leads to the representation

$$
l((\alpha_\nu)_{1 \le \nu \le r_1}) = \sum_{i=1}^{s} \log \frac{n_i!}{(n_i - r_i)!} + \sum_{i=1}^{s} \sum_{j=1}^{r_i} \log \alpha_j - \alpha_1 \sum_{i=1}^{s} n_i H(x_{i1})
$$

$$
- \sum_{i=1}^{s} \sum_{j=2}^{r_i} (n_i - j + 1)\alpha_j \left[H(x_{ij}) - H(x_{i,j-1}) \right]
$$

$$
+ \sum_{i=1}^{s} \sum_{j=1}^{r_i} \log h(x_{ij}). \tag{7.3}
$$

$(\sum\limits_{i=p}^{q} \cdots = 0$ if $p > q$.) Introducing the abbreviations

$$A_1 = \sum_{i=1}^{s} n_i H(x_{i1}) \quad \text{and} \quad A_\nu = \sum_{i=1}^{c_\nu}(n_i - \nu + 1)\left[H(x_{i\nu}) - H(x_{i,\nu-1})\right],$$

$2 \leq \nu \leq r_1$, and interchanging the order of summation we obtain

$$
\begin{aligned}
l((\alpha_\nu)_{1\leq\nu\leq r_1}) &= \sum_{i=1}^{s} \log \frac{n_i!}{(n_i - r_i)!} + \sum_{\nu=1}^{r_1} c_\nu \log \alpha_\nu \\
&\quad - \sum_{\nu=1}^{r_1} \alpha_\nu A_\nu + \sum_{i=1}^{s}\sum_{j=1}^{r_i} \log h(x_{ij}).
\end{aligned}
$$

If we take the equation

$$\sum_{\nu=1}^{r_1} c_\nu \frac{\alpha_\nu}{\alpha_\nu^*} = \sum_{\nu=1}^{r_1} \alpha_\nu A_\nu$$

into account, we get the following upper bound of the log-likelihood function

$$l((\alpha_\nu)_{1\leq\nu\leq r_1}) \leq \sum_{i=1}^{s} \log \frac{n_i!}{(n_i - r_i)!} + \sum_{\nu=1}^{r_1} c_\nu(\log \alpha_\nu^* - 1) + \sum_{i=1}^{s}\sum_{j=1}^{r_i} \log h(x_{ij}).$$

$$(7.4)$$

Since (7.4) is based on the inequality

$$\log \alpha_\nu = \log \frac{\alpha_\nu}{\alpha_\nu^*} + \log \alpha_\nu^* \leq \frac{\alpha_\nu}{\alpha_\nu^*} - 1 + \log \alpha_\nu^*,$$

equality holds iff $\alpha_\nu = \alpha_\nu^*$, $\nu = 1, \ldots, r_1$, which yields the assertion. ∎

Remark 7.2.1 If we assume that the considered sequential k-out-of-n systems have an identical structure, i.e. $r_i = r$ and $n_i = n$, $1 \leq i \leq s$, the result is stated in Cramer and Kamps (1996).

Expressing the maximum likelihood estimators $\alpha_1^*, \ldots, \alpha_r^*$ in terms of the underlying random variables, i.e.

$$\alpha_1^* = -s \left(\sum_{i=1}^{s} n_i \log \bar{F}(X_{*i}^{(1)})\right)^{-1} \quad \text{and}$$

$$\alpha_\nu^* = -c_\nu \left(\sum_{i=1}^{c_\nu}(n_i - \nu + 1) \log \left(\frac{\bar{F}(X_{*i}^{(\nu)})}{\bar{F}(X_{*i}^{(\nu-1)})}\right)\right)^{-1}, \quad 2 \leq \nu \leq r_1,$$

we get the following results. The proofs are along the lines of those in Cramer and Kamps (1996, Theorem 3.2) and therefore they are omitted.

Theorem 7.2.2 *The maximum likelihood estimators* $\alpha_1^*, \ldots, \alpha_{r_1}^*$ *of* $\alpha_1, \ldots, \alpha_{r_1}$ *have the following properties:*

1. $\alpha_1^*, \ldots, \alpha_{r_1}^*$ *are independent.*

2. α_ν^* *has an inverted gamma distribution with parameters* c_ν *and* $c_\nu \alpha_\nu$, *i.e.*
   $\alpha_\nu^* \sim \alpha_\nu (\frac{1}{c_\nu} \sum\limits_{i=1}^{c_\nu} V_{i\nu})^{-1}$, $1 \leq \nu \leq r_1$ *where* $(V_{i\nu})_{1 \leq \nu \leq r_1, 1 \leq i \leq c_\nu}$ *are iid standard exponential random variables.*

3. $E(\alpha_\nu^*)^k = \frac{(c_\nu - k - 1)!}{(c_\nu - 1)!}(c_\nu \alpha_\nu)^k$, *if* $k \leq c_\nu - 1$; *hence* $E\alpha_\nu^* = \frac{c_\nu}{c_\nu - 1}\alpha_\nu$, $c_\nu > 1$;
   $Var(\alpha_\nu^*) = \frac{c_\nu^2}{(c_\nu - 1)^2(c_\nu - 2)}\alpha_\nu^2$, $c_\nu > 2$; $MSE(\alpha_\nu^*) = \frac{c_\nu + 2}{(c_\nu - 1)(c_\nu - 2)}\alpha_\nu^2$, $c_\nu > 2$,
   $1 \leq \nu \leq r_1$.

4. *The statistic* $(\alpha_1^*, \ldots, \alpha_{r_1}^*)$ *is sufficient for* $(\alpha_1, \ldots, \alpha_{r_1})$.

5. *The sequences of estimators* (α_ν^*) *are strongly consistent w.r.t.* $c_\nu \to \infty$
   $(1 \leq \nu \leq r_1)$.

6. α_ν^* *is asymptotically normal,* $1 \leq \nu \leq r_1$, *i.e.* $\sqrt{c_\nu}(\alpha_\nu^*/\alpha_\nu - 1) \xrightarrow{d} \mathcal{N}(0,1)$
   w.r.t. $c_\nu \to \infty$.

7.3   Estimation in Specific Distributions

In the situation of Section 7.2 we assume first that the underlying distribution function F is given by a one-parameter exponential family, i.e.

$$F(t) = 1 - \exp(-\lambda g(t)), \quad t \geq 0, \tag{7.5}$$

with an unknown parameter $\lambda > 0$ and a known increasing and differentiable function g on $[0, \infty)$ satisfying the regularity conditions $g(0) = 0$ and $\lim_{t \to \infty} g(t) = \infty$. Several distributions important in lifetime analysis are included in this setting, e.g. standard Weibull distributions $(g(t) = t^\beta, \beta > 0)$, exponential distributions $(g(t) = t)$, and Pareto distributions $(g(t) = \log(\frac{t+\vartheta}{\vartheta})$, $\vartheta > 0)$. By analogy with Theorem 7.2.1 we obtain the maximum likelihood estimators of the parameters.

Lemma 7.3.1 *In the situation (7.5), the maximum likelihood estimators of* $\tilde{\alpha}_\nu = \lambda \alpha_\nu$, $1 \leq \nu \leq r_1$, *are given by* $\tilde{\alpha}_\nu^* = \lambda \alpha_\nu^*$ *with* α_ν^* *as in Theorem 7.2.1, i.e.*

$$\tilde{\alpha}_1^* = -s \left(\sum_{i=1}^{s} n_i g(x_{i1}) \right)^{-1} \quad \text{and}$$

$$\tilde{\alpha}_\nu^* = -c_\nu \left(\sum_{i=1}^{c_\nu} (n_i - \nu + 1)[g(x_{i\nu}) - g(x_{i,\nu-1})] \right)^{-1}, \quad 2 \leq \nu \leq r_1.$$

An analogous result is valid for (7.5) with an additional shift parameter, i.e. for the family of distributions given by

$$F(t) = 1 - \exp(-\lambda(g(t) - \eta)), \quad t \geq g^{-1}(\eta), \tag{7.6}$$

with $\lambda > 0$, $\eta \in \mathbf{R}$ unknown and some increasing and differentiable function g on $[g^{-1}(\eta), \infty)$ satisfying $\lim_{t\to\infty} g(t) = \infty$. This parametrization includes some models often used in lifetime analysis, e.g. two-parameter exponential and Pareto distributions ($g(t) = t$ and $g(t) = \log(t)$, respectively). The special case $r_i = r$, $n_i = n$, $i = 1, \ldots, s$, leads to the results stated in Cramer and Kamps (1996). Varde (1970) considers the case of a two-component series system where the lifetime distribution is exponential. The proof is carried out as in Cramer and Kamps (1996) [see also Lawless (1982) and Epstein (1957)].

Lemma 7.3.2 *In the situation (7.6), the maximum likelihood estimators of η and $\tilde{\alpha}_\nu = \lambda \alpha_\nu$, $1 \leq \nu \leq r_1$, are given by*

$$\eta^* = \min_{1 \leq i \leq s} g(x_{i1})$$

$$\tilde{\alpha}_1^* = -s \left(\sum_{i=1}^{s} n_i [g(x_{i1}) - \eta^*] \right)^{-1} \quad and$$

$$\tilde{\alpha}_\nu^* = -c_\nu \left(\sum_{i=1}^{c_\nu} (n_i - \nu + 1)[g(x_{i\nu}) - g(x_{i,\nu-1})] \right)^{-1}, \quad 2 \leq \nu \leq r_1.$$

Moreover, the statistic $(\eta^*, \tilde{\alpha}_1^*, \ldots, \tilde{\alpha}_{r_1}^*)$ is sufficient for $(\eta, \lambda\alpha_1, \ldots, \lambda\alpha_{r_1})$.

7.4   Sequential *k*-out-of-*n* Systems With Known Model Parameters and Underlying One-Parameter Exponential Family

In the subsequent part we assume that the model parameters $\alpha_{i1}, \ldots, \alpha_{in_i}$, $1 \leq i \leq s$, are possibly different, but known and that the underlying distribution function F is given by (7.5) with an unknown parameter $\lambda > 0$.

Theorem 7.4.1 *The maximum likelihood estimator λ^* of λ is given by*

$$\lambda^* = \frac{\displaystyle\sum_{i=1}^{s} r_i}{\displaystyle\sum_{i=1}^{s} \sum_{j=1}^{r_i} (n_i - j + 1)\alpha_{ij}[g(x_{ij}) - g(x_{i,j-1})]} \tag{7.7}$$

where $x_{i0} = 0$, $1 \leq i \leq s$.

PROOF. Using the distributional assumptions we obtain the hazard function $h(t) = \lambda g'(t)$ and the cumulative hazard function $H(t) = \lambda g(t)$. The representation of the log-likelihood function given in (7.3) leads to

$$
\begin{aligned}
l(\lambda) \;=\; & \sum_{i=1}^{s} \log \frac{n_i!}{(n_i - r_i)!} + \sum_{i=1}^{s}\sum_{j=1}^{r_i} \log \alpha_{ij} \\
& - \lambda \sum_{i=1}^{s}\sum_{j=1}^{r_i}(n_i - j + 1)\alpha_{ij}[g(x_{ij}) - g(x_{i,j-1})] \\
& + \left(\sum_{i=1}^{s} r_i\right) \log \lambda + \sum_{i=1}^{s}\sum_{j=1}^{r_i} \log g'(x_{ij}).
\end{aligned}
$$

Maximizing the function $l(\cdot)$ with respect to λ yields the assertion.   ■

Remark 7.4.1

1. In the situation of Type II censoring, i.e. $\alpha_{ij} = 1$, $r_i = r$, $n_i = n$, $1 \le i \le s$, $1 \le j \le r_i$, some special cases of (7.7) appear in the literature. Apparently the case $s = 1$ is only considered. We refer to Lawless (1982, p. 102), Johnson, Kotz and Balakrishnan (1994, p. 514) for the exponential distribution $(g(t) = t)$, to Johnson, Kotz and Balakrishnan (1994, p. 656/7) for the Weibull distribution $(g(t) = t^\beta,\ \beta > 0)$ and to Johnson, Kotz and Balakrishnan (1994, p. 593) for the Pareto distribution $(g(t) = \log(\frac{t+\vartheta}{\vartheta}),\ \vartheta > 0)$.

2. In the particular case $\alpha_{ij} = \alpha_i$ for all j, the denominator in (7.7) can be written as

$$
(n_i - r_i + 1)\alpha_i g(x_{ir_i}) + \sum_{j=1}^{r_i-1} \alpha_i g(x_{ij}),
$$

which, for $s = 1$ and $\alpha_i = 1$ for all i, leads to the representation usually found in the literature (see e.g., Epstein 1957, eq. (3) for the exponential distribution).

We close this section by showing some properties of the estimator λ^*.

Theorem 7.4.2

1. $\lambda^* \sim V^{-1}$ *where V is a gamma distributed random variable with parameters $R = \sum\limits_{i=1}^{s} r_i$ and λR.*

2. $E(\lambda^*)^k = \frac{(R-k-1)!}{(R-1)!}(\lambda R)^k$, $k \le R - 1$; *in particular*, $E\lambda^* = \frac{R}{R-1}\lambda$, $R > 1$; $MSE(\lambda^*) = \frac{R+2}{(R-1)(R-2)}\lambda^2$, $Var(\lambda^*) = \frac{R^2}{(R-1)^2(R-2)}\lambda^2$, $R > 2$.

   Hence, $\lambda_u^* = \frac{R-1}{R}\lambda^*$ *is an unbiased estimator of λ with* $Var(\lambda_u^*) = \frac{\lambda^2}{R-2}$, $R > 2$.

3. λ^* is sufficient for λ.

4. λ^* is strongly consistent for λ, i.e. $\lambda^* \longrightarrow \lambda$ a.e. w.r.t. $R \to \infty$.

5. $\sqrt{R}(\lambda^*/\lambda - 1) \xrightarrow{d} \mathcal{N}(0,1)$ w.r.t. $R \to \infty$.

7.5 Example: Sequential 2-out-of-4 System

Suppose that the s underlying systems have a common sequential 2-out-of-4 structure with a distribution function specified by (7.5). Without loss of generality we assume that $\alpha_1 = 1$ and that α_2 is an arbitrary, but known positive real number. Applying the results of Section 7.4, the maximum likelihood estimator λ^* of λ is given by

$$\lambda^* = \frac{2s}{\sum_{i=1}^{s} \left(4g(X_{*i}^{(1)}) + 3\alpha_2[g(X_{*i}^{(2)}) - g(X_{*i}^{(1)})] \right)}$$

in terms of the random variables $X_{*i}^{(1)}, X_{*i}^{(2)}$, $i = 1, \ldots, s$. In case of an underlying ordinary k-out-of-n structure, i.e. $\alpha_2 = 1$, the maximum likelihood estimator λ_o^* of λ is given by

$$\lambda_o^* = \frac{2s}{\sum_{i=1}^{s} \left(4g(X_{*i}^{(1)}) + 3[g(X_{*i}^{(2)}) - g(X_{*i}^{(1)})] \right)}.$$

To analyze the different impact of the respective modelling of the system, we consider the ratio $Q = \lambda^*/\lambda_o^*$. The distribution of Q is obtained as follows. Since the random variables $4g(X_{*i}^{(1)})$ and $3\alpha_2[g(X_{*i}^{(2)}) - g(X_{*i}^{(1)})]$ are iid according to $F(t) = 1 - \exp(-t)$, $t \geq 0$, for all $i = 1, \ldots, s$ (cf. Cramer and Kamps 1996) we obtain

$$Q = \frac{Y_s + \frac{1}{\alpha_2} Z_s}{Y_s + Z_s} = 1 - (1 - \frac{1}{\alpha_2}) \frac{Z_s}{Y_s + Z_s}$$

by putting $Y_s = \sum_{i=1}^{s} 4g(X_{*i}^{(1)})$ and $Z_s = \sum_{i=1}^{s} 3\alpha_2[g(X_{*i}^{(2)}) - g(X_{*i}^{(1)})]$. Since Y_s and Z_s are independent and gamma distributed with parameters s and 1 we obtain that $Z_s/(Y_s + Z_s)$ has a standard beta distribution with both parameters equal to s. Therefore the expectation of Q is given by $E[Q] = 1 - (1 - 1/\alpha_2)/2 = 1/2 + 1/(2\alpha_2)$. A plot of $E[Q]$ as a function of α_2 is shown in Figure 7.1.

In case of an underlying exponential lifetime distribution, the expectation $E[Q]$ can be interpreted as follows. After the failure of the first component the hazard rate of the remaining components has changed to α_2. If this value is large, the survival time of these components tends to be short. Therefore the lifetime of the whole system is essentially determined by the first time to failure. In the classical model it is assumed that such effects can be neglected. Consequently, the lifetime of the system is estimated too optimistically.

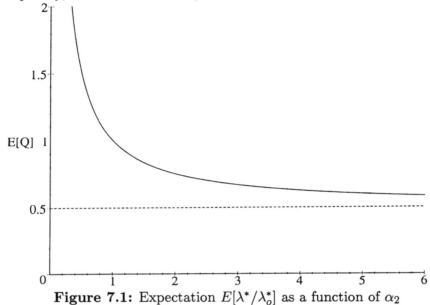

Figure 7.1: Expectation $E[\lambda^*/\lambda_o^*]$ as a function of α_2

Acknowledgement. The authors are thankful to a referee for helpful remarks.

References

1. Block, H. W. (1975). Continuous multivariate exponential extensions, In *Reliability and Fault Tree Analysis* (Eds., R. E. Barlow, J. B. Fussel and N. D. Singpurwalla), pp. 285–306, Philadelphia: SIAM.

2. Cramer, E. and Kamps, U. (1996). Sequential order statistics and k-out-of-n systems with sequentially adjusted failure rates, *Annals of the Institute of Statistical Mathematics*, **48**, 535–549.

3. Cramer, E. and Kamps, U. (1997). The UMVUE of $P(X < Y)$ based on Type-II censored samples from Weinman multivariate exponential distributions, *Metrika*, **46**, 93–121.

4. Epstein, B. (1957). Simple estimators of the parameters of exponential distributions when samples are censored, *Annals of the Institute of Statistical Mathematics*, **8**, 15–26.

5. Johnson, N. L., Kotz, S. and Balakrishnan, N. (1994). *Continuous Univariate Distributions*, vol. 1, Second edition, New York: John Wiley & Sons.

6. Kamps, U. (1995). *A Concept of Generalized Order Statistics*, Stuttgart: Teubner.

7. Lawless, J. F. (1982). *Statistical Models and Methods for Lifetime Data*, New York: John Wiley & Sons.

8. Varde, S. D. (1970). Estimation of reliability of a two exponential component series system, *Technometrics*, **12**, 867–875.

8

Stochastic Models for the Return of Used Devices

Berthold Heiligers and Jürgen Ruf

Institut für Mathematische Stochastik, Magdeburg, Germany

Abstract: We present stochastic models for describing the number of returns of previously sold products. These models may be used e.g. for estimating the total number of returns. Thereby they provide a basis for deciding how to handle returns.

Keywords and phrases: Additive models, coupling, (total) in-service time, increasing failure rate

8.1 Introduction

In many different countries political issues about an obligation to take back used products are presently discussed, [see e.g. the German 'Elektronikschrottver-ordnung', [ESV] (1992)]. For a company's decision on how to handle returns it is therefore of increasing importance to have reliable data on the total number of returns, expected in any period of time in the future.

Taking back used devices may be an advantage for a company, regarding not only its image on environmental protection. For example, it might be an important goal of producers to supply their customers with spare parts to the products, even when the regular production has already ended. On the other hand, the main components of returned devices are often of excellent quality. Thus a systematic "cannibalization" of returns may reduce the total number of spare parts to be produced and stored only for possible later use, resulting in decreased expenses, e.g. for stock keeping. Of course, this has to be compared carefully with the increased costs caused by all of the involved departments.

This paper aims at providing a mathematical basis for a company's decision on handling returned devices. The prerequisite is a reliable estimation of the total number of returns in any period of time in the future. This calls for a stochastic model, as these numbers will depend on the random return times.

Actually, we started our investigations from a practical situation, in which an additional problem arose: We were concerned with High-Tech products of a fairly limited production. This is why no relevant data was available, which is necessary for applying standard regression or extrapolation techniques. To overcome these difficulties, we model the in-service times of the devices at the individual customers.

8.2   Additive Models for Returns

With any of the n produced devices, we associate its (nonrandom) time t_k of being sold, and its (random) total in-service time V_k staying at the customers before being returned, $1 \leq k \leq n$. The quantity of interest is the number $N_{a,b}$ of returns within a time period $(a, b]$, ($a < b$), i.e.

$$N_{a,b} = \#\{1 \leq k \leq n : a < t_k + V_k \leq b\}.$$

Often, total in-service times are supposed to be exponentially distributed; in the context we were concerned with, however, such an assumption turned out to be inadequate. For, we had to handle durations of more than two years, and thus history cannot be neglected. Actually, practice suggested that *the longer the device is already used the higher the probability of a return in the near future.* Mathematically, this is reflected by an increasing failure rate (IFR) of the V_k's, that is, the conditional probability

$$P(V_k \leq t + h \mid V_k > t) \qquad (8.1)$$

is an increasing function of t, for any fixed $h > 0$. For example, Weibull distributions with shape parameter exceeding 1 have this property, (here, additionally, the conditional probabilities from (8.1) have the desirable property of tending to 1 with increasing t). Convolutions of certain independent exponential distributions may serve as another example for IFR distributions, see Corollary 8.2.1.

The model we worked with came from the following considerations. Often devices can be resold several times (as second-hand). We therefore take into account the in-service times of the devices at the different customers. After device k has been sold for the first time, at (known) time t_k, it will be in use for the (random) time $\tau_{k,1}$. Then the owner will try to resell the device as second-hand; it is reasonable to suppose that the chance p for a successful resale depends on the age of the particular device, measured relatively to the beginning of the serial production, that is, $p = p(t_k + \tau_{k,1})$. At time $t_k + \tau_{k,1}$ the device is either resold or returned. If the device is resold, the second owner will utilize it for the random time $\tau_{k,2}$, reselling it thereafter with probability $p(t_k + \tau_{k,1} + \tau_{k,2})$, and so on. The owner, say the l-th one, who cannot resell the device, will return it to the company—this happens with probability $1 - p(t_k + T_{k,l})$, where

$$T_{k,l} = \sum_{i=1}^{l} \tau_{k,i} \tag{8.2}$$

is the total in-service time of device k staying at the first l users. Figure 8.1 illustrates the resale process.

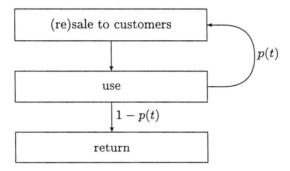

Figure 8.1: (Re)sale or return of the k-th device at age t

There is a constructive way for modeling the resale process (which can easily be implemented for simulation purposes): Consider for fixed k independent and uniformly distributed random variables $U_{k,l}$ over $[0, 1]$, being also independent of the $\tau_{k,l}$'s; then the zero-one variable

$$B_{k,l} = \begin{cases} 1, & \text{if } U_{k,l} \leq p(t_k + T_{k,l}); \\ 0, & \text{otherwise}; \end{cases} \tag{8.3}$$

describes success of the l-th attempt of reselling ($B_{k,l} = 1$) or non-success ($B_{k,l} = 0$), and $P(B_{k,l} = 1) = p(t_k + T_{k,l})$, as desired. The total number of users of the k-th device is consequently

$$C_k = \min\{l \in \mathbb{N} : B_{k,l} = 0\}, \tag{8.4}$$

where $\min \emptyset := \infty$, and

$$V_k = T_{k,C_k} = \sum_{i=1}^{C_k} \tau_{k,i} \tag{8.5}$$

is the total in-service time of that device when returned.

With the above settings (8.2)–(8.5), we say that $V_1, \ldots, V_n$ form an *additive model*, provided that

- the involved (positive) individual in-service times $\tau_{k,l}, l \in \mathbb{N}, 1 \leq k \leq n$, are independent and for any fixed k identically distributed and, as a further mathematical assumption,

- p is a piecewise continuous function with $\limsup_{t\to\infty} p(t) < 1$.

In an additive model, the total number of users of a device is finite almost surely, i.e. for every k

$$P(C_k < \infty) = 1. \tag{8.6}$$

For, by Fubini's theorem and dominated convergence it follows that

$$P(C_k = \infty) = \int \lim_{l\to\infty} \prod_{i=1}^{l} p(t_k + T_{k,i})\, dP;$$

here the strong law of large numbers yields $\lim_{l\to\infty} T_{k,l} = \infty$ almost surely, finally implying $\lim_{l\to\infty} \prod_{i=1}^{l} p(t_k + T_{k,i}) = 0$ almost surely.

It can be shown that in an additive model the total in-service times $V_1, \ldots, V_n$ are independent. Thus, for the total number of returns within any period $(a, b]$ we find

$$P(N_{a,b} = \ell) = \sum \prod_{k=1}^{n} q_k^{j_k}(1 - q_k)^{1-j_k}, \tag{8.7}$$

where the sum is taken over all $j_1, \ldots, j_n \in \{0, 1\}$ summing up to ℓ, $0 \le \ell \le n$, and $q_k = P(a < t_k + V_k \le b)$, for abbreviation. In particular, expectation and variance of $N_{a,b}$ are given by

$$\mathrm{E}[N_{a,b}] = \sum_{k=1}^{n} q_k, \tag{8.8}$$

$$\mathrm{Var}[N_{a,b}] = \sum_{k=1}^{n} q_k(1 - q_k). \tag{8.9}$$

It might be helpful to remark that often explicit computation of the probabilities (8.7) can be avoided, as excellent approximations are available. For example, the Poisson distribution with parameter $\sum_{k=1}^{n} q_k$ works well for small values of the q_k's [see Peköz and Ross (1994, pp. 449–462)]. The central limit theorem provides another surrogate to (8.7) [see e.g. Feller (1971)].

In any case, analyzing $N_{a,b}$ requires knowledge of the distributions of the V_k's from (8.5). The following lemma gives a basis for these investigations.

Lemma 8.2.1 *In an additive model we have for all $0 < v < \infty$ (and all $1 \le k \le n$),*

$$P(V_k \le v) = \sum_{l\in\mathbb{N}} \int_{\{T_{k,l}\le v\}} (1 - p(t_k + T_{k,l})) \prod_{i=1}^{l-1} p(t_k + T_{k,i})\, dP.$$

PROOF. Let $0 < v < \infty$. Observing equation (8.6) we get

$$
\begin{aligned}
P(V_k \leq v) &= \sum_{l \in \mathbb{N}} P(T_{k,l} \leq v, C_k = l) \\
&= \sum_{l \in \mathbb{N}} P(T_{k,l} \leq v, U_{k,l} - p(t_k + T_{k,l}) > 0, \\
&\qquad \text{and } U_{k,i} - p(t_k + T_{k,i}) \leq 0 \text{ for all } 1 \leq i \leq l-1),
\end{aligned}
$$

and the assertion follows by applying Fubini's theorem to each term involved in the latter sum. ∎

Explicit formulae for $P(V_k \leq v)$ can be obtained when specifying the distributions of the in-service times $\tau_{k,l}$. Next we consider two special cases of exponential and geometrical distributions, respectively.

Theorem 8.2.1 *Assume that, for all k, the individual in-service times $\tau_{k,l}$ are exponentially distributed, with the same parameter $\alpha_k > 0$. Then, for all $v > 0$,*

$$
P(V_k \leq v) = 1 - \exp\Big\{ - \alpha_k \int_0^v (1 - p(t_k + x))\, dx \Big\}.
$$

PROOF. Abbreviating

$$
\widetilde{p}(y) = \int_0^y p(t_k + z)\, dz, \quad y > 0,
$$

we get from Lemma 8.2.1 with some algebra

$$
\begin{aligned}
P(V_k \leq v) &= \sum_{l=0}^{\infty} \int_0^v \frac{(1 - p(t_k + y))\alpha_k^{l+1}\widetilde{p}^l(y)}{\exp\{\alpha_k y\}\, l!}\, dy = \int_0^v \frac{\alpha_k(1 - p(t_k + y))}{\exp\{\alpha_k(y - \widetilde{p}(y))\}}\, dy \\
&= \int_0^v \Big[\frac{d}{dy}(y - \widetilde{p}(y)) \Big] \frac{\alpha_k}{\exp\{\alpha_k(y - \widetilde{p}(y))\}}\, dy \\
&= 1 - \exp\{-\alpha_k(v - \widetilde{p}(v))\} \\
&= 1 - \exp\{-\alpha_k \int_0^v (1 - p(t_k + x))\, dx\}.
\end{aligned}
$$

∎

Corollary 8.2.1 *Consider the exponential additive model from Theorem 8.2.1, and suppose additionally that p is nonincreasing with $\lim_{t \to \infty} p(t) = 0$. Then the V_k have the IFR property.*

PROOF. By straightforward computation, the failure rate of V_k is found to be

$$\alpha_k(1 - p(t_k + t)), \ t > 0,$$

and the assertion follows since p is nonincreasing. ∎

Next we state a result similar to Theorem 8.2.1 for the discrete analogue to the exponential distribution.

Theorem 8.2.2 *Consider an additive model with $t_k \in \mathbb{N}$ for all k, and geometrically distributed in-service times $\tau_{k,l}$, with parameter $0 < \theta_k < 1$. Then, for all $v \in \mathbb{N}$,*

$$P(V_k = v) = (1 - p(t_k + v))(1 - \theta_k)^v \sum_{l=1}^{v} \left(\frac{\theta_k}{1 - \theta_k}\right)^l g_k(l, v),$$

where the $g_k(l, v)$ are recursively defined as follows independent of θ_k). Let

$$g_k(l, v) = g_k(l, v - 1) + p(t_k + v - 1)g_k(l - 1, v - 1), \quad \text{for } 2 \leq l \leq v,$$

where

$$g_k(l, v) = \begin{cases} 0, & \text{if} \quad v < l; \\ 1, & \text{if} \quad l = 1. \end{cases}$$

PROOF. Firstly, reasoning as in the proof of Lemma 8.2.1 yields

$$P(V_k = v) = \sum_{l \in \mathbb{N}} \int_{\{T_{k,l}=v\}} (1 - p(t_k + T_{k,l})) \prod_{i=1}^{l-1} p(t_k + T_{k,i}) \, dP.$$

Thus, from

$$\{T_{k,l} = v\} = \{1 \leq T_{k,1} < \ldots < T_{k,l-1} < v = T_{k,l}\},$$

and (for $x_0 = 0 < x_1 < \ldots x_{l-1} < v = x_l$),

$$P(T_{k,i} = x_i \text{ for all } 1 \leq i \leq l) = \prod_{i=1}^{l} P(\tau_{k,i} = x_i - x_{i-1}) = \left(\frac{\theta_k}{1 - \theta_k}\right)^l (1 - \theta_k)^v,$$

we find that all terms in the above sum equal zero if $l > v$, and otherwise

$$\int_{\{T_{k,l}=v\}} (1 - p(t_k + T_{k,l})) \prod_{i=1}^{l-1} p(t_k + T_{k,i}) \, dP$$

$$= \ (1 - p(t_k + v))\left(\frac{\theta_k}{1 - \theta_k}\right)^l (1 - \theta_k)^v g_k(l, v),$$

with

$$g_k(l, v) = \sum_{1 \le x_1 < \ldots < x_l = v} \prod_{i=1}^{l-1} p(t_k + x_i).$$

The stated recurrence relation for g_k is easily verified, observing that

$$g_k(l, v) = \sum \prod_{i=1}^{l-1} p(t_k + x_i) + \widetilde{\sum} \prod_{i=1}^{l-1} p(t_k + x_i),$$

where the first and second sum $\sum$ and $\widetilde{\sum}$ are taken over all $1 \le x_1 < \ldots < x_{l-1} < v - 1$ and $1 \le x_1 < \ldots < x_{l-1} = v - 1$, respectively. ∎

8.3 Model Fit

One of the fundamentals for achieving a satisfactory model fit is the justification of identical distributions for the in-service times $\tau_{k,l}$ for any fixed k. This may call for grouping the customers into homogeneous clusters according to their behavior, all of which can be treated thereafter by the methods described above. We will not discuss in this paper adequate clustering techniques, but instead we refer the reader to literature on multivariate analysis, [see e.g. Mardia, Kent and Bibby (1979) or Kaufman and Rousseeuw (1990)]. For our purposes we assume that homogeneity of the customers population is guaranteed.

The next crucial step is the choice of an adequate function p, reflecting the resale chances varying with the time t. Of course, at this stage of the modeling phase, additional knowledge on the customer's behavior is required. In this context, marketing departments should be involved in the set-up process.

There is no one–for–all rule for finding p, but some general hints might be helpful. For example, technical progress may cause products to become outdated, and thus may suggest to choose a nonincreasing function p, (ignoring the possibility of describing rarities). It appears to be reasonable to assume $p(t)$ tends to zero, when the time t increases. Besides that, a frequent observation in practice is that there are perfect resale chances ($p = 1$) as long as the serial production continues, but after termination, at time t_0, say, they tend (more or less rapidly) to zero. The following three families (8.10)–(8.12) of p-functions are expected to serve good in real-life situations. Here we choose parameters $a > 0$, $c > 1$ and $\gamma \in [0, 1]$; in all three cases, $\gamma < 1$ describes a lack in the resale chances at the termination time t_0 of the serial production.

The function

$$p_1(t) = \begin{cases} 1, & \text{if } t < t_0; \\ \gamma \left(\frac{1}{t - t_0 + 1} \right)^a, & \text{otherwise;} \end{cases} \tag{8.10}$$

reflects, depending on the parameter a, a rather rapid decrease of the chances immediately after termination at time t_0. More flexibility is achieved by

$$p_2(t) = \begin{cases} 1, & \text{if } t < t_0; \\ \frac{\gamma s (t+t_m-t_0)^{c-1}}{\exp\{a(t+t_m-t_0)^c\}}, & \text{otherwise}; \end{cases} \qquad (8.11)$$

where $s = \frac{act_m \exp\{(c-1)/c\}}{c-1}$ and $t_m = (\frac{c-1}{ac})^{\frac{1}{c}}$. As limiting case $c = 1$ we obtain the exponential-type p-function

$$p_3(t) = \begin{cases} 1, & \text{if } t < t_0; \\ \frac{\gamma}{\exp\{a(t-t_0)\}}, & \text{otherwise}; \end{cases} \qquad (8.12)$$

Figure 8.2 displays the shapes of these functions for selected parameters.

Finally, for the above p-functions we take up the additive model from Theorem 8.2.1, and give explicit formulae for the distribution of V_k. Since in all three cases $p(t) = 1$ if $t \leq t_0$, but $p(t) < 1$ if $t > t_0$, the probabilities $P(V_k \leq v)$ will essentially depend not only on the difference $\Delta_k = t_0 - t_k$, but particularly on the *sign* of Δ_k. In order to state closed form expressions we utilize the following notations

$$\Delta_k^+ = \begin{cases} \Delta_k, & \text{if } \Delta_k \geq 0; \\ 0, & \text{otherwise}; \end{cases}$$

$$\Delta_k^- = \begin{cases} 0, & \text{if } \Delta_k \geq 0; \\ -\Delta_k, & \text{otherwise}. \end{cases}$$

The following corollary is easily verified, and therefore we omit the proof.

Corollary 8.3.1 *Consider the exponential additive model from Theorem 8.2.1, with $p = p_1$, $p = p_2$, or $p = p_3$. Then*

$$P(V_k \leq v) = \begin{cases} 0, & \text{if } v \leq \Delta_k^+; \\ 1 - \exp\{-\alpha_k(v - \Delta_k^+ - \gamma u(v))\}, & \text{if } v > \Delta_k^+; \end{cases}$$

where $u(v)$ is defined as follows:

for $p = p_1$ with $a \neq 1$,

$$u(v) = ((v - \Delta_k + 1)^{1-a} - (\Delta_k^- + 1)^{1-a})/(1 - a),$$

for $p = p_1$ with $a = 1$,

$$u(v) = \ln\left(\frac{v - \Delta_k + 1}{\Delta_k^- + 1}\right),$$

for $p = p_2$,

$$u(v) = -s(\exp\{-a(v - \Delta_k + t_m)^c\} - \exp\{-a(\Delta_k^- + t_m)^c\}),$$

for $p = p_3$,

$$u(v) = -(\exp\{-a(v - \Delta_k)\} - \exp\{-a\Delta_k^-\})/a.$$

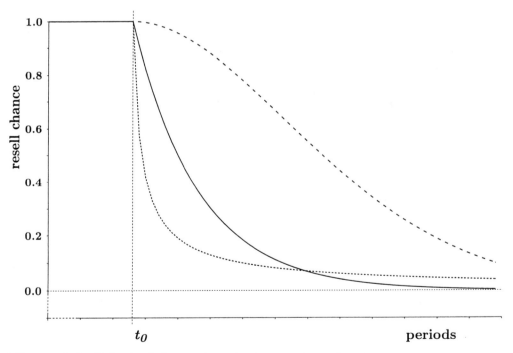

Figure 8.2: Principal shape of resale chances from $(8.10) - (8.12)$ (with $\gamma = 1$); the dotted, dashed and solid lines correspond to p_1, p_2, and p_3, respectively

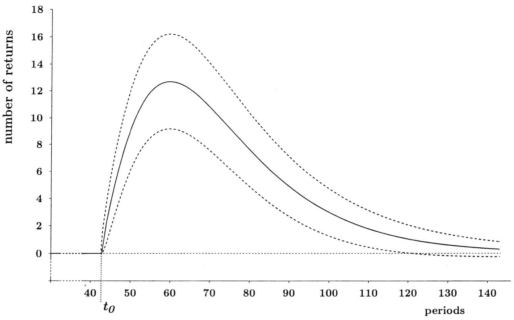

Figure 8.3a: Expected number of returns in one-month periods in an additive
model. The in-service times have identical exponential distributions with
parameter α, and the resell chance is described by p_3 from above with
the same parameter $a = \alpha$ (and $\gamma = 1$). The solid line gives the expected
number of returns; the dotted lines represent the expectations plus/minus
standard deviation

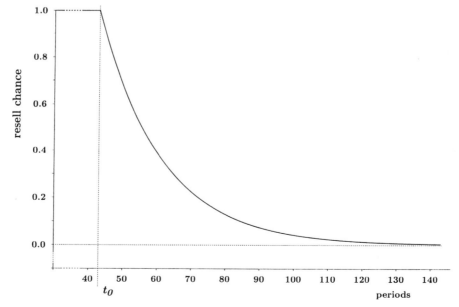

Figure 8.3b: Resell chance p, corresponding to Figure 8.3a

Our final example reports a practical situation, which arose at the Siemens-Nixdorf Company. In this case we dealt with a special kind of CPU, produced in a period of $t_0 = 43$ months, (starting at month 0, say). We found out that an exponential additive model as in Theorem 8.2.1 with the same parameter $\alpha_k = \alpha$ for all k was adequate for describing the returns. As p-function we took $p = p_3$, with parameter a equal to the model parameter α, and $\gamma = 1$. The idea was to describe some coupling between in-service times at the users and resale chances, observed for the CPU: *Longer in-service times go along with a smaller decrease of the resale chances.*

Figure 8.3a displays a typical example for the resulting expected number of returns from (8.8) in one-month periods in the future. The model parameter α could be estimated from data on previous productions (e.g. by ordinary least squares methods) or obtained by inquiries about the in-service times. For reference, Figure 8.3b shows the corresponding function $p = p_3$.

References

1. ESV-Elektronikschrottverordnung (1992). Working paper, Bundesministerin für Umwelt, Naturschutz und Reaktorsicherheit, Government of the Federal Republic of Germany.

2. Feller, W. (1971). *An Introduction to Probability Theory and its Applications, II*, New York: John Wiley & Sons.

3. Kaufman, L. and Rousseeuw, P. L. (1990). *Finding Groups in Data,* New York: John Wiley & Sons.

4. Mardia, K. V., Kent, J. T. and Bibby, J. M. (1979). *Multivariate Analysis,* London: Academic Press.

5. Peköz, E. and Ross, S. (1994). *Improving Poisson Approximations*, Probability in the Engineering and Informational Sciences, Vol. 8, No. 4.

9

Some Remarks on Dependent Censoring in Complex Systems

Tina Herberts and Uwe Jensen

University of Ulm, Ulm, Germany

Abstract: We consider a sample of independent binary complex systems, all having the same structure. Each system is continuously observed until it fails. The failure time of a component is observed and recorded if the component fails before the system does. If a component is still functioning at the time of system failure, its lifelength cannot be observed and the time of system failure is recorded as censoring time instead. Based on these failure and censoring times, Doss, Freitag and Proschan (1989) introduced an estimator of the distribution of the system lifelength in the case that the components of the system are independent. In this paper we discuss possibilities to weaken the assumption of independent component lifelengths. Under these weaker assumptions we introduce methods of estimating the distribution functions of the system lifelength and the component lifelengths.

Keywords and phrases: Reliability, binary complex systems, system life distribution, dependent censoring

9.1 Introduction and Summary

Suppose that we have a sample of n independent coherent systems, all of them consisting of m components and having the same structure ϕ. We refer to Barlow and Proschan (1981) for definitions and further details relating to coherent systems. Each system and each component is in either a functioning state or a failed state, and the state of the system only depends on the states of the components. Each system is continuously observed until it fails. For all components in each system, either a failure time or a censoring time is recorded. The failure time of the component is recorded if the component fails before or at the time of system failure. A censoring time is recorded if the component is still functioning at the time of system failure. In this case, one only knows that the component lifelength is greater than the system lifelength. The information contained in the sample comprises the failure times and censoring times of the components, and we also know whether a component failed or whether it was censored. From this information we wish to estimate the joint distribution function of the component lifelengths and the distribution F of the system lifelength.

Doss, Freitag and Proschan (1989) introduced an estimator of F under the assumption that the component lifelengths are independent. We use the following notation:

- For $i = 1, \ldots, n$ and $j = 1, \ldots, m$, X_{ij} is the lifelength of component j in system i,

- F_j is the distribution function of the lifelength of component j (i.e. $X_{ij} \sim F_j$ for all i),

- T_i is the lifelength of system i,

- $Z_{ij} = \min(X_{ij}, T_i)$, and $Z_{(1)j} \leq Z_{(2)j} \leq \ldots \leq Z_{(n)j}$ are the ordered values of $Z_{1j}, \ldots, Z_{nj}$,

- $\delta_{ij} = I(X_{ij} \leq T_i)$, where $I(A)$ denotes the indicator of the set A, and $\delta_{(i)j} = \delta_{kj}$ if $Z_{(i)j} = Z_{kj}$.

For each coherent structure ϕ of independent components, there corresponds a function h_ϕ (the *reliability function*) such that

$$\bar{F}(t) = h_\phi(\bar{F}_1(t), \ldots, \bar{F}_m(t)), \tag{9.1}$$

where $\bar{G}(t) = 1 - G(t)$ for a distribution function G. To construct the estimator $\hat{F}$, F_j is estimated by its Kaplan-Meier estimator $\hat{F}_j$, where

$$\hat{F}_j(t) = 1 - \prod_{i:Z_{(i)j} \leq t} \left(\frac{n-i}{n-i+1} \right)^{\delta_{(i)j}}. \tag{9.2}$$

Then $\hat{F}$ is defined by

$$\hat{F}(t) = \begin{cases} 1 - h_\phi(\hat{\bar{F}}_1(t), \ldots, \hat{\bar{F}}_m(t)) & \text{if } t < T_{(n)}, \\ 1 & \text{if } t \geq T_{(n)}, \end{cases} \tag{9.3}$$

where $T_{(n)} = \max(T_1, \ldots, T_n)$. In Doss, Freitag and Proschan (1989) the strong uniform consistency and the weak convergence of $\hat{F}$ to a Gaussian process with known covariance structure are proved (see p. 131, Theorem 9.2.1 in this article) and the asymptotic efficiency of $\hat{F}(t)$ versus the empirical estimator

$$\hat{F}^{emp}(t) = \frac{1}{n}\sum_{i=1}^{n} I(T_i \leq t) \tag{9.4}$$

is discussed. For all systems except for series systems, there exist $t \geq 0$ such that the asymptotic variance of $\hat{F}(t)$ is less than the asymptotic variance of $\hat{F}^{emp}(t)$. This is what one would expect since $\hat{F}$ uses all the information contained in the sample while $\hat{F}^{emp}$ only uses the information about the system lifelengths.

As mentioned above, the estimator $\hat{F}$ is defined under the assumption of independent component lifelengths. Frequently, however, the components are dependent since they are subject to the same operating environment. In this case, the distribution of the system lifelength usually cannot be written as a function of the distributions of the component lifelengths, i.e. formula (9.1) no longer holds true. In addition, the Kaplan-Meier estimator only estimates the distributions of the component lifelengths consistently under the *constant sum condition* [see Williams and Lagakos (1977)]. In Kalbfleisch and MacKay (1979) it is shown that under the assumption that X_{ij} and Z_{ij} are absolutely continuous, this condition is equivalent to the condition

$$h_j(u) = \lambda_j(u) \tag{9.5}$$

for all $u > 0$, where h_j and λ_j are defined by

$$h_j(u) = \lim_{h \downarrow 0} \frac{1}{h} P(u \leq X_{ij} < u + h | X_{ij} \geq u) \text{ and}$$

$$\lambda_j(u) = \lim_{h \downarrow 0} \frac{1}{h} P(u \leq Z_{ij} < u + h, \delta_{ij} = 1 | Z_{ij} \geq u)$$

$$= \lim_{h \downarrow 0} \frac{1}{h} P(u \leq X_{ij} < u + h, X_{ij} \leq T_i | X_{ij} \geq u, T_i \geq u).$$

h_j is the failure rate of X_{ij} and λ_j is the failure rate of X_{ij} in the presence of censoring. Independent censoring is a special case in which equation (9.5) holds.

In Doss, Freitag and Proschan (1989) it is shown that if the component lifelengths are independent, one can always define random variables $Y_{ij}, j = 1, \ldots, m$, such that

$$(Z_{ij}, \delta_{ij}) = (\min(X_{ij}, Y_{ij}), I(X_{ij} \leq Y_{ij})) \tag{9.6}$$

and

$$X_{ij} \text{ and } Y_{ij} \text{ are independent.} \tag{9.7}$$

Therefore the Kaplan-Meier estimator can be used to estimate the distributions of the component lifelengths in this case. If the component lifelengths are dependent the Kaplan-Meier estimator usually no longer estimates the distribution functions of the component lifelengths consistently. So in this case, the estimator (9.3) cannot be used to estimate the distribution function of the system lifelength.

In this article we consider the estimation of the distributions of the system lifelength and the component lifelengths in the case that the component lifelengths are dependent. In Section 9.2 we introduce methods for estimating the distribution of the system lifelength and the joint distribution function of the component lifelengths if components within parallel systems are dependent. In this case the Kaplan-Meier estimator and multivariate extensions of the Kaplan-Meier estimator, respectively, can be used to estimate the (joint) distribution function of the component lifelengths. In Section 9.3 we consider the case that all component lifelengths are dependent. We review a result of Langberg, Proschan and Quinzy (1978) which states a condition for converting a system having dependent component lifelengths into a system consisting of independent components. By means of this result the unobservable marginal distributions can be estimated consistently.

9.2 Dependence of the Components Within a Parallel System

In this paragraph component lifelengths may be dependent as long as the lifelengths and their censoring variables remain independent. This is best explained in terms of a simple example.

Consider the system shown in Figure 9.1.

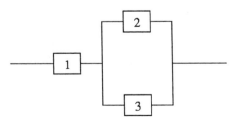

Figure 9.1: A parallel system with dependence of components

Let X_j denote the lifelength of component j. X_1 is censored by $Y_1 = X_2 \vee X_3$, where $x \vee y = \max(x, y)$, X_2 is censored by $Y_2 = X_1$, and X_3 is censored by $Y_3 = X_1$. If X_1 is independent of (X_2, X_3), but X_2 and X_3 are dependent random variables, then X_i and Y_i are independent pairs of random variables for

$i = 1, 2, 3$. So in this example the independence of the component lifelengths and their censoring variables is maintained even if the components of the parallel system are dependent.

This example can be generalized to series structures which are composed of parallel systems as shown in Figure 9.2.

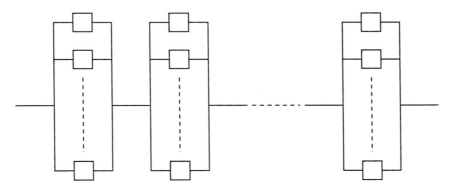

Figure 9.2: A series structure composing of parallel systems

Every system can be represented as a series system being composed of parallel systems, where the components of the jth parallel system are the elements of the jth minimal cut set of the system [see Barlow and Proschan (1981)]. Again, we consider a sample of n independent systems where each system now consists of l subsystems in series, the jth subsystem containing m_j components in parallel.

We use the following notation:

- Again, T_i is the lifelength of the ith system ($i = 1, \ldots, m$), and F is the distribution of this lifelength;

- for $j = 1, \ldots, l$, X_{ij} represents the lifelength of the jth parallel system of the ith system, and F_j is the corresponding distribution function;

- X_{ijk} represents the lifelength of component k of the jth parallel structure in system i, and F_{jk} is its distribution function.

Again, every component is observed until system failure, and a failure time or a censoring time is recorded. Since the system fails as soon as one of the parallel systems fails, X_{ijk} is censored by

$$Y_{ij} = \min_{\nu \neq j} X_{i\nu} = \min_{\nu \neq j} (\max_{1 \leq \mu \leq m_\nu} X_{i\nu\mu}). \tag{9.8}$$

Y_{ij} is a function of $(X_{i\nu\mu} : \mu = 1, \ldots, m_\nu, \nu \neq j)$, but it does not depend on $X_{ij\mu}$ for $\mu = 1, \ldots, m_j$. Thus, if the random vectors $(X_{i11}, \ldots, X_{i1m_1}), \ldots, (X_{il1}, \ldots, X_{ilm_l})$ are independent, the component lifelengths X_{ijk} and their censoring variables Y_{ij} will be independent for $k = 1, \ldots, m_j$ and $j = 1, \ldots, l$ even if the components within a parallel system are dependent. In the following we therefore consider systems which can be represented as l parallel systems in series such that the components of different parallel systems are independent. Components within the same parallel system may be dependent.

9.2.1   Estimation of F by means of $\hat{F}_{jk}$

For a system as described above the distributions of the component lifelengths can be estimated consistently by their Kaplan-Meier estimators $\hat{F}_{jk}$, $k = 1, \ldots,$ m_j, $j = 1, \ldots, l$, since the component lifelengths and their censoring variables are independent. However, the distribution function of the system lifelength cannot be estimated by means of the $\hat{F}_{jk}$ unless the dependence structure of the X_{ijk}, $k = 1, \ldots, m_j$, is known for each j. If the X_{ijk}, $k = 1, \ldots, m_j$, are positively dependent, we have

$$P(\max_{1 \leq k \leq m_j} X_{ijk} \leq t) \geq \prod_{k=1}^{m_j} P(X_{ijk} \leq t),$$

and so

$$1 - \prod_{j=1}^{l} \left(1 - \prod_{k=1}^{m_j} \hat{F}_{jk}(t) \right)$$

is an estimator of a lower bound for $F(t)$. If more information about the dependence structure of the X_{ijk}, $k = 1, \ldots, m_j$, is available, one might express $F(t)$ in terms of $F_{jk}(t)$, $k = 1, \ldots, m_j$, $j = 1, \ldots, l$, and obtain an estimator of $F(t)$ by substituting $F_{jk}(t)$ by its Kaplan-Meier estimator.

For example, consider the system shown in Figure 9.1. Assume that (X_2, X_3) follows a Marshall-Olkin-type distribution, i.e.

$$
\begin{aligned}
X_2 &= \min(U_2, U_{23}) \text{ and} \\
X_3 &= \min(U_3, U_{23}),
\end{aligned}
$$

where U_2, U_3 and U_{23} are nonnegative, independent random variables which are independent of X_1. U_j $(j = 2, 3)$ represents the time until X_j alone fails and U_{23} represents the time until X_2 and X_3 fail simultaneously. The distribution function of the system lifelength can be calculated as follows:

$$
\begin{aligned}
F(t) &= 1 - \bar{F}_1(t) P(X_2 \vee X_3 > t) \\
&= 1 - \bar{F}_1(t)(P(X_2 > t) + P(X_3 > t) - P(X_2 > t, X_3 > t)) \\
&= 1 - \bar{F}_1(t) P(U_{23} > t)(P(U_2 > t) + P(U_3 > t) - P(U_2 > t)P(U_3 > t)),
\end{aligned}
$$

where $\bar{F}_1(t) = P(X_1 > t)$. To estimate $\bar{F}_1(t)$ and $P(U_I > t)$ for $I \in \{2, 3, 23\}$, note that the system can be represented as shown in Figure 9.3, where now all components are independent. Therefore $\bar{F}_1(t)$ and $P(U_I > t)$, $I \in \{2, 3, 23\}$, can be estimated consistently by their Kaplan-Meier estimators, and thus we obtain a strongly consistent estimator $\hat{F}(t)$ of $F(t)$. Since we are in the independent case again, we obtain the weak convergence of $\hat{F}$ by the following result which is proved in Doss, Freitag and Proschan (1989):

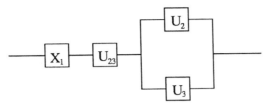

Figure 9.3: Representation of the system in Figure 9.1 with all components being independent

Theorem 9.2.1 *Consider a system which is composed of m independent components, and suppose that the distribution functions $F_1, \ldots, F_m$ of the component lifelengths are continuous and that $\tau > 0$ is such that $F_j(\tau) < 1$ for $j = 1, \ldots, m$. Let h_ϕ be the corresponding reliability function and*

$$K_j(t) = \frac{\partial h_\phi}{\partial u_j}(u_1, \ldots, u_m)|_{u_k = \bar{F}_k(t), k = 1, \ldots, m},$$

and let $\bar{H}_j(u) = P(Y_{ij} > u)$. Then as $n \to \infty$,

$$n^{1/2}(\hat{F} - F) \Rightarrow W$$

weakly in $D[0, \tau]$, where W is a mean 0 Gaussian process with covariance structure given by

$$Cov\left(W(t_1), W(t_2)\right) = \sum_{j=1}^{m} K_j(t_1) K_j(t_2) \bar{F}_j(t_1) \bar{F}_j(t_2) \int_0^{t_1} \frac{dF_j(u)}{\bar{H}_j(u) \bar{F}_j(u)} \quad (9.9)$$

for $0 \leq t_1 \leq t_2 \leq \tau$.

Here, $D[0, \tau]$ is the space of all real-valued functions defined on $[0, \tau]$ that are right-continuous and have left limits, with the Skorokhod metric topology.

By means of Theorem 9.2.1, the asymptotic relative efficiency of $\hat{F}(t)$,

$$ARE(t) = \frac{\text{asymptotic variance of } \hat{F}^{emp}(t)}{\text{asymptotic variance of } \hat{F}(t)},$$

can be calculated. Doss, Freitag and Proschan (1989) show that it is monotonically decreasing and that it satisfies

$$ARE(t) > 1 \text{ for some } t \geq 0.$$

This example can be generalized to arbitrary systems. For example, one can consider the case that a system comprises several modules, components of different modules are independent and the component lifelengths of the jth module can be represented as

$$X_{jk} = \min(U_{jk}, U_j),$$

where U_{jk} and U_j are independent random variables representing the time until the component alone or all components of the module simultaneously fail, respectively. Similarly to the system considered above, this system can be converted into a system comprising independent components, and therefore the distribution function of the system lifelength can be estimated by an estimator similar to the one introduced by Doss, Freitag and Proschan (1989).

If nothing is known about the dependence structure of the $X_{ijk}, k = 1, \ldots, m_j$, the Kaplan-Meier estimators $\hat{F}_{jk}$ cannot be used in order to estimate the distribution function of the system lifelength. Instead, one has to estimate the distributions F_j of the lifelengths X_{ij} of the parallel systems in order to obtain an estimator of F. This can be done in different ways.

9.2.2   Estimation of F by means of the Kaplan-Meier estimators of F_j

Since the lifelength X_{ij} of the jth parallel system and its censoring variable Y_{ij} are independent, $F_j(t)$ can be estimated by the Kaplan-Meier estimator

$$\hat{F}_j(t) = 1 - \prod_{i: Z_{(i)j} \leq t} \left(\frac{n-i}{n-i+1} \right)^{\delta_{(i)j}},$$

where $Z_{ij} = \min(X_{ij}, Y_{ij})$, $\delta_{ij} = I(X_{ij} \leq Y_{ij})$ and $Z_{(i)j}$ and $\delta_{(i)j}$ are defined as before. That is, the parallel systems are treated as modules and the system is considered a series system consisting of l independent modules. Therefore we are in the independent case again, and by another result of Doss, Freitag and Proschan (1989) we have

$$n^{1/2}(\hat{F}_1 - F_1, \ldots, \hat{F}_l - F_l) \Rightarrow (W_1, \ldots, W_l)$$

weakly in $D^l[0, \tau]$ as $n \to \infty$, where $W_1, \ldots, W_l$ are independent mean 0 Gaussian processes with covariance structure given by

$$\text{Cov}\,(W_j(t_1), W_j(t_2)) = \bar{F}_j(t_1)\bar{F}_j(t_2) \int_0^{t_1} \frac{dF_j(u)}{\bar{H}_j(u)\bar{F}_j(u)} \tag{9.10}$$

for $0 \leq t_1 \leq t_2 \leq \tau$. Here, $D^l[0, \tau]$ denotes the product space of $D[0, \tau]$.

Since the asymptotic variance of $\hat{F}_j(t)$ can be estimated consistently by Greenwood's formula [see Hall and Wellner (1980)], this result enables the construction of simultaneous confidence intervals for $(F_1(t), \ldots, F_l(t))$ in large samples.

As in (9.3), the distribution function F of the system can be estimated by

$$\hat{F}^{(1)}(t) = \begin{cases} 1 - \prod_{j=1}^{l} \hat{\bar{F}}_j(t) & \text{if } t < T_{(n)}, \\ 1 & \text{if } t \geq T_{(n)}. \end{cases} \qquad (9.11)$$

However, since the system fails as soon as one of the parallel systems fails, we have

$$\delta_{(i)1} + \ldots + \delta_{(i)l} = 1 \quad \text{for all } i \in \{1, \ldots, n\}$$

and

$$Z_{(i)1} = \ldots = Z_{(i)l} = T_{(i)} \quad \text{for all } i.$$

Therefore

$$\begin{aligned} \prod_{j=1}^{l} \hat{\bar{F}}_j(t) &= \prod_{j=1}^{l} \left(\prod_{i:Z_{(i)j} \leq t} \left(\frac{n-i}{n-i+1} \right)^{\delta_{(i)j}} \right) \\ &= \prod_{i:T_{(i)} \leq t} \left(\frac{n-i}{n-i+1} \right)^{\delta_{(i)1} + \ldots + \delta_{(i)l}} \\ &= \prod_{i:T_{(i)} \leq t} \left(\frac{n-i}{n-i+1} \right), \end{aligned}$$

and $\hat{F}^{(1)}$ turns out to be the empirical distribution function $\hat{F}^{emp}$.

The disadvantage of considering the parallel systems as modules is that the information about the component lifelengths is not utilized in this case. The next method uses all the information contained in the sample.

9.2.3 Estimation of F by means of multivariate Kaplan-Meier estimators

Instead of estimating the distribution function of the lifelength of the jth parallel system, the joint distribution

$$G_j(t_1, \ldots, t_{m_j}) = P(X_{ij1} \leq t_1, \ldots, X_{ijm_j} \leq t_{m_j})$$

of the component lifelengths of the jth parallel system is estimated by a multivariate estimator of the distribution function of censored random variables. Several estimators of such distributions have been suggested, including those of Campbell and Földes (1982), Burke (1988), Tsai, Leurgans and Crowley (1986), Dabrowska (1988), Prentice and Cai (1992) and Lin and Ying (1993). All of these estimators are strongly uniformly consistent, and some of them converge weakly to a Gaussian process.

Let $\hat{G}_j$ be one of the multivariate estimators mentioned above. Then

$$\hat{F}^{(2)}(t) = \begin{cases} 1 - \prod_{j=1}^{l}(1 - \hat{G}_j(t, \ldots, t)) & \text{if } t < T_{(n)}, \\ 1 & \text{if } t \geq T_{(n)} \end{cases} \qquad (9.12)$$

is an estimator of F which is strongly uniformly consistent. For example, consider the estimator of the survival function of the bivariate random vector (X_{ij1}, X_{ij2}) which has been suggested by Lin and Ying (1993). This estimator can be applied in the case of univariate censoring, which means that X_{ij1} and X_{ij2} are censored by the same censoring variable Y_{ij}. This condition is satisfied in our case. Lin and Ying (1993) suggested the following estimator $\hat{\bar{G}}_j(t_1, t_2)$ of the survival function $\bar{G}_j(t_1, t_2) = P(X_{ij1} > t_1, X_{ij2} > t_2)$:

$$\hat{\bar{G}}_j(t_1, t_2) = \frac{1}{n} \sum_{i=1}^{n} I(Z_{ij1} > t_1, Z_{ij2} > t_2)/\hat{\bar{H}}_j(t_1 \vee t_2), \qquad (9.13)$$

where

$$Z_{ijk} = \min(X_{ijk}, Y_{ij}) \text{ and } \delta_{ijk} = I(X_{ijk} \leq Y_{ij})$$

are the observable data and $\hat{\bar{H}}_j(t)$ is the Kaplan-Meier estimator of the survival function $\bar{H}_j(t) = P(Y_{ij} > t)$. By setting

$$1 - \hat{G}_j(t, t) = \hat{\bar{G}}_j(t, 0) + \hat{\bar{G}}_j(0, t) - \hat{\bar{G}}_j(t, t) \qquad (9.14)$$

we obtain

$$1 - \hat{G}_j(t, t) = \frac{\frac{1}{n} \sum_{i=1}^{n} I(Z_{ij1} \vee Z_{ij2} > t)}{\hat{\bar{H}}_j(t)}.$$

But this is just the Kaplan-Meier estimator of $P(X_{ij1} \vee X_{ij2} > t)$ [see Shorack and Wellner (1986, p. 295)], i.e.

$$1 - \hat{G}_j(t, t) = \hat{\bar{F}}_j(t),$$

and therefore $\hat{F}^{(2)} = \hat{F}^{(1)} = \hat{F}^{emp}$. Again, we do not obtain a new estimator, at least not by using the multivariate estimator of Lin and Ying (1993).

The advantage of this method is that all information contained in the sample is used and that we obtain estimators of the joint distribution function of the component lifelengths. Since the random vectors

$$(X_{i11}, \ldots, X_{i1m_1}), \ldots, (X_{il1}, \ldots, X_{ilm_l})$$

are independent, $P(X_{i11} > t_{11}, \ldots, X_{ilm_l} > t_{lm_l})$ can be estimated consistently by

$$\hat{\bar{G}}_1(t_{11}, \ldots, t_{1m_1}) \cdot \ldots \cdot \hat{\bar{G}}_l(t_{l1}, \ldots, t_{lm_l}). \qquad (9.15)$$

Therefore, this gives a better result than the first method where only the joint distribution of $X_{ij}, j = 1, \ldots, l$, could be estimated.

9.3 Dependence of the Lifelengths and their Censoring Variables

In the last section we considered the case that the component lifelengths and their censoring variables are independent. Now we consider a system where all of the components are dependent. In this case the lifelengths and their censoring variables are dependent, and the Kaplan-Meier estimator usually does not estimate the distribution function of the component lifelengths consistently. Again, the system we consider is represented as m parallel systems in series, and the parallel systems are treated as modules. Therefore we confine ourselves to considering a series system being composed of m components where the component lifelengths X_j are dependent now. In Langberg, Proschan and Quinzy (1978) a necessary and sufficient condition for the existence of a set of *independent* random variables $\{H_I, I \subseteq \{1, \ldots, m\}\}$ is given such that the lifelength of the original series system and the occurrence of its failure pattern have the same joint distribution as the lifelength and the occurrence of the failure pattern of a series system of components having the independent lifelengths $\{H_I, I \subseteq \{1, \ldots, m\}\}$. Here, the failure pattern of the system is the set of components whose simultaneous failure coincides with that of the system, and we denote the failure pattern by $\xi(\mathbf{X})$:

$$\xi(\mathbf{X}) = \begin{cases} I & \text{if } T = X_j \ \forall \ j \in I, T \neq X_j \ \forall \ j \notin I, \\ \emptyset & \text{otherwise,} \end{cases}$$

where $\mathbf{X} = (X_1, \ldots, X_m)$.

The condition for the existence of the derived series system of independent components is that the sets of discontinuities of the F_I are pairwise disjoint, where $F_I(t) = P(T \leq t, \xi(\mathbf{X}) = I)$. The survival functions of the $H_I, I \subseteq \{1, \ldots, m\}$, are given by

$$P(H_I > t) = \bar{G}_I(t) = \exp\left(-\int_0^t \frac{dF_I^c}{\bar{F}}\right) \prod_{a(I,j) \leq t} \left(\frac{\bar{F}(a(I,j))}{\bar{F}(a(I,j)-)}\right), \tag{9.16}$$

where F_I^c is the continuous part of F_I and $\{a(I, j)\}_j$ is the set of discontinuities of F_I.

Now we consider a sample of n independent systems with system lifelengths T_i and component lifelengths $X_{ij}, j = 1, \ldots, m, i = 1, \ldots, n$. Substituting F_I and F in formula (9.16) by their empirical counterparts $\hat{F}_{I,n}^{emp}$ and $\hat{F}_n^{emp}$, respectively, where

$$\hat{F}_{I,n}^{emp}(t) = \frac{1}{n} \sum_{i=1}^n I(T_i \leq t, \xi(\mathbf{X}_i) = I)$$

and

$$\hat{F}_n^{emp}(t) = \frac{1}{n} \sum_{i=1}^n I(T_i \leq t),$$

we obtain an estimator $\hat{\bar{G}}_{I,n}$ of G_I. Since

$$F(t) = P(T \leq t) = P(\min_{I \subseteq \{1,\ldots,m\}} H_I \leq t) = 1 - \prod_{I \subseteq \{1,\ldots,m\}} \bar{G}_I(t),$$

a natural estimator of $F(t)$ is $1 - \prod_{I \subseteq \{1,\ldots,m\}} \bar{G}_I(t)$. More generally, Langberg, Proschan and Quinzy (1981) derive an estimator of

$$\bar{M}_I(t) = P(T_I > t), \tag{9.17}$$

where $I \subseteq \{1,\ldots,m\}$ and $T_I = \min_{j \in I} X_j$. Let

$$\mathcal{J}_I = \{J \subseteq \{1,\ldots,m\} : J \cap I \neq \emptyset\}.$$

Then it is shown in Langberg, Proschan and Quinzy (1981) that

$$\hat{\bar{M}}_{I,n}(t) = \prod_{J \in \mathcal{J}_I} \hat{\bar{G}}_{J,n}(t) \tag{9.18}$$

is a strongly consistent estimator of $\bar{M}_I$ if the following conditions hold:

(i) the functions $F_I, I \subseteq \{1,\ldots,m\}$, have no common discontinuities;

(ii)
$$\frac{\bar{M}_I(a)}{\bar{M}_I(a-)} = \begin{cases} \bar{F}(a)/\bar{F}(a-) & \text{if } a \in D(F_{\mathcal{J}_I}), \\ 1 & \text{otherwise,} \end{cases}$$

where $F_{\mathcal{J}_I}(t) = P(T \leq t, \xi(\mathbf{X}) \in \mathcal{J}_I)$ and $D(F_{\mathcal{J}_I})$ denotes the set of discontinuities of $F_{\mathcal{J}_I}$; and

(iii)
$$P(T_{I^c} \geq t | T_I = t) = P(T_{I^c} > t | T_I > t),$$

where I^c denotes the complement of I.

Conditions (i)–(iii) are satisfied if the $F_I, I \subseteq \{1,\ldots,m\}$, are continuous and the component lifelengths are independent. But there are also examples of dependent random variables satisfying those conditions:

Example 9.3.1 Suppose that the vector (X_1, X_2) has the Marshall-Olkin bivariate exponential distribution, where now (in contrast to the previous example) X_1 and X_2 are the component lifelengths of a series system. The corresponding survival probability is

$$P(X_1 > t_1, X_2 > t_2) = \exp(-\lambda_1 t_1 - \lambda_2 t_2 - \lambda_{12}(t_1 \vee t_2))$$

for $t_1, t_2 \geq 0$. Then conditions (i) and (ii) are satisfied since the marginal distributions M_1, M_2 and M_{12} are continuous. Besides,

$$P(X_2 \geq t | X_1 = t) = e^{-\lambda_2 t} = P(X_2 > t | X_1 > t).$$

Therefore condition (iii) is satisfied for $I = \{1\}$, and $\bar{M}_I$ can be estimated consistently by $\hat{\bar{M}}_{I,n}$ although X_1 and X_2 are dependent for $\lambda_{12} > 0$.

References

1. Barlow, R. E. and Proschan, F. (1981). *Statistical Theory of Reliability and Life Testing*, Silver Spring, MD: To Begin With.

2. Burke, M. D. (1988). Estimation of a bivariate distribution function under random censorship, *Biometrika*, **75**, 379–382.

3. Campbell, G. and Földes, A. (1982). Large sample properties of nonparametric bivariate estimators with censored data, In *Nonparametric Statistical Inference, Colloquia Mathematica-Societatis János Bolyai* (Eds. B. V. Gnedenko, M. L. Puri and I. Vincze), pp. 103–122, Amsterdam: North Holland.

4. Dabrowska, D. M. (1988). Kaplan-Meier estimate on the plane, *Annals of Statistics*, **16**, 1475–1489.

5. Doss, H., Freitag, S. and Proschan, F. (1989). Estimating jointly system and component reliabilities using a mutual censorship approach, *Annals of Statistics*, **17**, 764–782.

6. Hall, W. J. and Wellner, J. A. (1980). Confidence bands for a survival curve from censored data, *Biometrika*, **67**, 133–143.

7. Kalbfleisch, J. D. and MacKay, R. J. (1979). On constant-sum models for censored survival data, *Biometrika*, **66**, 87–90.

8. Langberg, N., Proschan, F. and Quinzy, A. J. (1978). Converting dependent models into independent ones, preserving essential features, *Annals of Probability*, **6**, 174–181.

9. Langberg, N., Proschan, F. and Quinzy, A. J. (1981). Estimating dependent life lengths, with applications to the theory of competing risks, *Annals of Statistics*, **9**, 157–167.

10. Lin, D. Y. and Ying, Z. (1993). A simple nonparametric estimator of the bivariate survival function under univariate censoring, *Biometrika*, **80**, 573–581.

11. Prentice, R. L. and Cai, J. (1992). Covariance and survival function estimation using censored multivariate failure time data, *Biometrika*, **79**, 495–512.

12. Shorack, G. R. and Wellner, J. A. (1986). *Empirical Processes with Applications to Statistics*, New York: John Wiley & Sons.

13. Tsai, W.-Y., Leurgans, S. and Crowley, J. (1986). Nonparametric estimation of a bivariate survival function in the presence of censoring, *Annals of Statistics*, **14**, 1351–1365.

14. Williams, J. S. and Lagakos, S. W. (1977). Models for censored survival analysis: constant-sum and variable-sum models, *Biometrika*, **64**, 215–224.

10

Parameter Estimation in Damage Processes: Dependent Observation of Damage Increments and First Passage Time

Waltraud Kahle and Axel Lehmann

Otto-von-Guericke-Universität Magdeburg, Magdeburg, Germany

Abstract: In this paper we describe statistical methods for estimating the parameters of damage processes if in one realization both process increments and a failure time are observable. The likelihood function for such observations is developed and point estimates are compared with those for other models.

Keywords and phrases: Process observation at discrete time points, likelihood function, parameter estimation, damage processes

10.1 Introduction

Investigating the reliability of products that are affected by damage processes such as wear, corrosion, and crack-growth it is often necessary to observe the development of damage processes which are characterized by a gradual drift of the mean value. The choise of the mathematical model for these damage processes is based on the assumption of an additive accumulation of damage with constant wear intensity. Regarding every damage increment as an additive superposition of a large number of small effects, we can assume the damage process to be normally distributed. Therefore the damage measure $Z(t)$ can be described by the following model [see Kahle (1994)]

$$Z(t) = z_0 + \sigma W(t - t_0) + \mu \cdot (t - t_0), \quad t \geq t_0 \tag{10.1}$$

with z_0 -   constant initial damage ($z_0 \in \mathbf{R}$),

           t_0 -   beginning of the damage ($t_0 \in \mathbf{R}$),

           μ -   drift parameter ($\mu \in \mathbf{R}$),

σ -   variance parameter ($\sigma > 0$),

$W(t)$ -   standard Wiener process on $[0, \infty)$.

We suppose that a failure of a product will occur if the damage process arrives at a certain boundary damage level, which is unknown. For a given boundary level h, the lifetime T_h of the product is then determined as the instant at which the damage process $Z(t)$ exceeds the level h for the first time:

$$T_h = \inf\{t \geq t_0 : Z(t) \geq h\} . \tag{10.2}$$

It is well known that for $z_0 < h$ the lifetime T_h follows an inverse Gaussian distribution with the Lebesgue density function

$$f_{T_h}(t) = \frac{h - z_0}{\sqrt{2\pi\sigma^2(t - t_0)^3}} \exp\left(-\frac{(h - z_0 - \mu(t - t_0))^2}{2\sigma^2(t - t_0)}\right) I_{\{t>t_0\}}, \tag{10.3}$$

where $I_{\{.\}}$ denotes an indicator variable, assuming the value 1 if the relation in brackets is true, and the value 0 otherwise. Obviously, P^{T_h} is the Dirac measure ε_{t_0} concentrated at t_0 if $z_0 \geq h$.

The lifetime distribution depends on the parameters of the damage process z_0, t_0, μ, σ^2, and on the boundary level h. Observing only the failure times of a sample of products we can estimate the parameters μ, σ^2, and t_0, for known $h - z_0$, according to the probability density (10.3).

The observation of the underlying damage process at fixed times t_j gives an alternative approach to parameter estimation which is particular useful for products with high reliability since failure times have not to be observed. If we ignore the boundary level h, the parameters μ, σ^2, t_0, and z_0 may be estimated by use of date of the damage process before a failure occurs. Supposed that a sample is given by

- n independent realizations $Z_i(t)$, $i = 1, \ldots, n$, of the damage process,

- m_i observations $z_{ij} = Z_i(t_{ij})$ of the damage measure at discrete time points t_{ij}, $j = 1, \ldots, m_i$, in each realization $Z_i(t)$, $i = 1, \ldots, n$; ,

the resulting likelihood function has the form

$$L = \prod_{i=1}^{n} \prod_{j=1}^{m_i} \frac{1}{\sqrt{\sigma^2(t_{ij} - t_{ij-1})}} \phi\left(\frac{(z_{ij} - z_{ij-1}) - \mu(t_{ij} - t_{ij-1})}{\sqrt{\sigma^2(t_{ij} - t_{ij-1})}}\right) \tag{10.4}$$

where $t_{i0} = t_0$ and $z_{i0} = z_0$, $i = 1, \ldots, n$. In this case , which is formally obtained on setting $h = \infty$, we make neither use of information on failure times nor of the fact that the damage process does not exceed the level h between two observation points. Such assumptions were chosen by Kahle (1994) and Pieper and Tiedge (1983). Note that the number of observations and the observation points may be different in each realization.

Parameter estimation procedures based either on process increments or on failure times or on increments and failure times, *independent* of each other, are investigated in Kahle (1994).

In the present paper we give the likelihood function for the case that in each realization of the damage process both process increments and a failure time are observable and we estimate simultaneously the parameters μ, σ^2, t_0, z_0, and h. By contrast with the models considered in Kahle (1994), here we observe for each sample path of the damage process either a failure time or the process increments at fixed observation times under the condition, that the process has not yet exceeded the damage level h. Hence, these conditional process increments and the failure time T_h are dependent random variables. To compute the likelihood function, we have to find in Section 10.2 the conditional distribution of the process under the condition that the level h is not exceeded and the joint distribution of conditional process increments and the lifetime T_h.

10.2   The Likelihood Function if Both Damage Increments and Failure Times are Observed

Let $Z(t) = z_0 + \sigma W(t - t_0) + \mu \cdot (t - t_0)$, $t \geq t_0$, be a damage process on a probability space $(\Omega, \mathcal{F}, P)$ with values in $(\mathbf{R}, \mathcal{B})$ as defined in (10.1). Further, for $m \in \mathbf{N}$, let $t_1, \ldots, t_m$ be fixed observation points with $t_0 < t_1 < \ldots < t_m < \infty$. To simplify the notation we consider only one realization of $Z(t)$ and drop the subscript i. We assume that a failure is observable at any time $t > t_0$ and that we stop observing the damage process after a failure has occured. Hence, in each time interval $(t_{j-1}, t_j]$, $j = 1, \ldots, m$, we observe either a failure at time $\tau \in (t_{j-1}, t_j]$, i.e. a realization τ of the lifetime T_h in $(t_{j-1}, t_j]$, or we observe the damage measure $z_j = Z(t_j)$ at t_j under the condition that the process has not yet exceeded the level h until the time t_j. The observable stopped process is given by

$$\widetilde{Z}(t) = \begin{cases} Z(t) & , \quad t_0 \leq t \leq T_h \\ \infty & , \quad t > T_h \end{cases} \quad , \quad t \geq t_0.$$

Thus, $\widetilde{Z}(t)$ is a functional of $\{Z(s); t_0 \leq s \leq t\}$,

$$\widetilde{Z}(t) = F_t(T_h, Z(t)), \quad t \geq t_0 \tag{10.5}$$

where

$$F_t(\tau, z) = \begin{cases} z & , \quad \tau \geq t \\ \infty & , \quad \tau < t \end{cases} \quad , \quad t, \tau, z \in \mathbf{R},$$

and measurable with respect to the σ-Algebra $\mathcal{F}_t = \sigma\{Z(s); t_0 \le s \le t\}$.

A sample of censored observations has the structure

$$X = F(T_h, Z(t_1), \ldots, Z(t_m)) = (\min(T_h, t_m), \widetilde{Z}(t_1), \ldots, \widetilde{Z}(t_m)) \qquad (10.6)$$

with

$$F(\tau, z_1, \ldots, z_m) = (F_0(\tau), F_{t_1}(\tau, z_1), \ldots, F_{t_m}(\tau, z_m))$$

and

$$F_0(\tau) = \min(\tau, t_m),$$

and the statistical model is given by

$$(\mathbf{R} \times \overline{\mathbf{R}}^m, \mathcal{B} \otimes \overline{\mathcal{B}}^m, (\mathrm{P}_\theta^F)_{\theta = (\mu, \sigma^2, z_0, t_0, h) \in \Theta \subset \mathbf{R}^5}),$$

where $\overline{\mathbf{R}} = \mathbf{R} \cup \{-\infty, +\infty\}$ and $\overline{\mathcal{B}}$ is the σ-algebra of Borel sets in $\overline{\mathbf{R}}$.

To compute the likelihood function of X we have to find the Radon-Nikodym derivative of P_θ^F with respect to a dominating measure ν

$$L(\tau, \mathbf{z}; \theta) = \frac{\mathrm{d}\mathrm{P}_\theta^F}{\mathrm{d}\nu}(\tau, \mathbf{z}), \quad \tau \in \mathbf{R}, \ \mathbf{z} = (z_1, \ldots, z_m) \in \overline{\mathbf{R}}^m. \qquad (10.7)$$

A crucial step in the derivation of $L(\tau, \mathbf{z}; \theta)$ is the calculation of the conditional density $f(s, x, t, z, h)$ of the process $Z(t)$ at time t starting at point $x = Z(s)$ under the condition that the level h is not exceeded in $[s, t]$. For that purpose we consider the Markov family $(Z(t), P_{s,x})$ with the transition function

$$P(s, x, t, B) = (2\pi\sigma^2(t - s))^{-1/2} \int_B \exp\left(-\frac{(y - x - \mu(t - s))^2}{2\sigma^2(t - s)}\right) \mathrm{d}y$$

for $t > s \ge t_0$ and $B \in \mathcal{B}$, i.e., for all $s \ge t_0$ and $x \in \mathbf{R}$, $P_{s,x}$ is a probability measure on $\mathcal{F}_{\ge s} = \sigma\{Z(u); u \ge s\}$ with $P_{s,x}(Z(t) \in B) = P(s, x, t, B)$. Let T_h^s denote the first passage time of $Z(t)$ to the upper boundary h on $[s, \infty)$

$$T_h^s = \inf\{t \ge s : Z(t) \ge h\}.$$

In case $s = t_0$ we have $T_h^{t_0} = T_h$ according to definition (10.2). For every fixed $s \ge t_0$ and $x \in \mathbf{R}$, T_h^s is a random variable on the probability space $(\Omega, \mathcal{F}_{\ge s}, P_{s,x})$. Hence, for $x < h$ the induced probability measure $P_{s,x}^{T_h^s}$, i.e. the conditional distribution of T_h^s under the condition $\{Z(s) = x\}$, has the Lebesgue density

$$g(s, x, t, h) = \frac{h - x}{\sqrt{2\pi\sigma^2(t - s)^3}} \exp\left(-\frac{(h - x - \mu(t - s))^2}{2\sigma^2(t - s)}\right) I_{\{t > s\}}, \qquad (10.8)$$

whereas for $x \geq h$, $P_{s,x}^{T_h^s}$ is the Dirac measure ε_s concentrated at s. Using total probability and the Markov property for Markov families [cf. Wentzell (1979, pp. 109)] we get for $x < h$ and $s < t$

$$P_{s,x}(Z(t) \leq z, T_h^s \geq t)$$

$$= P_{s,x}(Z(t) \leq z) - \int_s^t P_{s,x}(Z(t) \leq z \,|\, T_h^s = u)\, P_{s,x}^{T_h^s}(rmdu)$$

$$= P(s,x,t,(-\infty,z]) - \int_s^t P(u,h,t,(-\infty,z])\, g(s,x,u,h)\, rmdu$$

and hence, applying Fubinis theorem and passing to densities

$$f(s,x,t,z,h) = p(s,x,t,z) - \int_s^t p(u,h,t,z)\, g(s,x,u,h)\, rmdu \qquad (10.9)$$

where $p(s,x,t,z)$ is the transition density of $(Z(t), P_{s,x})$. Using (10.9), the following lemma provides an explicit representation for the density $f(s,x,t,z,h)$.

Lemma 10.2.1 *Let $t_0 \leq s < t$, $x < h$, and $z \in \mathbf{R}$. The density (10.9) is given by*

$$f(s,x,t,z,h) = \frac{1}{\sigma\sqrt{t-s}}\, \phi\left(\frac{z - x - \mu(t-s)}{\sigma\sqrt{t-s}}\right)$$

$$\times \left[1 - \exp\left(-\frac{2(h-x)(h-z)}{\sigma^2(t-s)}\right)\right] I_{\{z \leq h\}}, \qquad (10.10)$$

where ϕ is the density of the standard normal distribution.

The proof may be found in the appendix. The next lemma gives the density of the finite dimensional distributions of $Z(t)$ starting at $x = Z(s)$ under the condition that h is not exceeded in terms of f. Let λ^k be the k-dimensional Lebesgue measure on $\overline{\mathcal{B}}^k$ ($k \in \mathbf{N}$). We use the convention $\prod_{j=j_1}^{j_2} \alpha_j = 1$ if $j_2 < j_1$.

Lemma 10.2.2 *Let $t_0 \leq s < t_1$, $\mathbf{t}_{1,k} = (t_1,\ldots,t_k)$, $\mathbf{z}_{1,k} = (z_1,\ldots,z_k) \in \mathbf{R}^k$, and $\mathbf{Z}_{\mathbf{t}_{1,k}} = (Z(t_1),\ldots,Z(t_k))$ for $1 \leq k \leq m$. Then, the probability measure $P_{s,x}(\mathbf{Z}_{\mathbf{t}_{1,k}} \in B, T_h^s \geq t_k)$, $B \in \mathcal{B}^k$, has the λ^k-density*

$$f_k(s,x,\mathbf{t}_{1,k},\mathbf{z}_{1,k},h) = f(s,x,t_1,z_1,h) \prod_{j=2}^k f(t_{j-1},z_{j-1},t_j,z_j,h). \qquad (10.11)$$

PROOF. For $k = 1$, clearly $f_1(s,x,\mathbf{t}_{1,1},\mathbf{z}_{1,1},h) = f(s,x,t_1,z_1,h)$ by the definition of f. For simplicity we will prove (10.11) only for $k = 2$. The general case can be shown analogously by induction. Let $B_1, B_2 \in \mathcal{B}$. Then, applying total probability and making use of the Markov property, we have

$$P_{s,x}(\mathbf{Z_{t_{1,2}}} \in B_1 \times B_2, T_h^s \geq t_2)$$

$$= P_{s,x}(Z(t_2) \in B_2, T_h^{t_1} \geq t_2, Z(t_1) \in B_1, T_h^s \geq t_1)$$

$$= \int_{B_1} P_{s,x}(Z(t_2) \in B_2, T_h^{t_1} \geq t_2 \,|\, Z(t_1) = z_1) \, f(s, x, t_1, z_1, h) \, \mathrm{d}z_1$$

$$= \int_{B_1} P_{t_1,z_1}(Z(t_2) \in B_2, T_h^{t_1} \geq t_2) \, f(s, x, t_1, z_1, h) \, \mathrm{d}z_1$$

$$= \int_{B_2 \times B_1} f(t_1, z_1, t_2, z_2, h) \, f(s, x, t_1, z_1, h) \, \mathrm{d}z_2 \, \mathrm{d}z_1 \;.$$

$$\blacksquare$$

Based on these preliminary lemmata we can derive the likelihood function $L(\tau, \mathbf{z}; \cdot)$. We shall show that P_θ^F is for all $\theta \in \Theta$ absolutely continuous with respect to the measure

$$\nu = (\lambda + \varepsilon_{t_m}) \otimes (\lambda + \varepsilon_\infty)^m \tag{10.12}$$

where λ is the Lebesgue measure on $\overline{B}$ and, for every $x \in \mathbf{R}$, ε_x is the Dirac measure on $\overline{B}$ concentrated at x.

Theorem 10.2.1 *Let ν be defined as in (10.12) on $\mathcal{B} \otimes \overline{\mathcal{B}}^m$. Then, for all $\theta \in \Theta$, P_θ^F is absolutely continuous with respect to ν and has the Radon-Nikodym derivative*

$$L(\tau, \mathbf{z}; \theta) = \frac{\mathrm{d}\mathrm{P}_\theta^F}{\mathrm{d}\nu}(\tau, \mathbf{z})$$

$$= \prod_{k=1}^m \left(g(t_{k-1}, z_{k-1}, \tau, h) \prod_{j=1}^{k-1} f(t_{j-1}, z_{j-1}, t_j, z_j, h) \prod_{j=k}^m I_{\{z_j = \infty\}} \right)^{I_{\{t_{k-1} \leq \tau < t_k\}}}$$

$$\times \left(\prod_{j=1}^m f(t_{j-1}, z_{j-1}, t_j, z_j, h) \right)^{I_{\{\tau = t_m\}}} I_{\{t_0 \leq \tau \leq t_m\}}, \tag{10.13}$$

where $\tau \in \mathbf{R}$, and $\mathbf{z} = (z_1, \ldots, z_m) \in \overline{\mathbf{R}}^m$.

PROOF. Let $1 \leq i \leq j \leq n$, $(y_i, \ldots, y_j) \in \overline{\mathbf{R}}^{j-i+1}$, and $t \in \mathbf{R}$. For the sake of brevity we define $\mathbf{t}_{i,j} = (t_i, \ldots, t_j)$, $\mathbf{y}_{i,j} = (y_i, \ldots, y_j)$, $(-\infty, \mathbf{y}_{i,j}] = (-\infty, y_i] \times \cdots \times (-\infty, y_j]$, and $E = (-\infty, t] \times (-\infty, \mathbf{y}_{1,m}]$. Further, we use the abbreviation $\{\mathbf{Z_{t_{i,j}}} \leq \mathbf{y}_{i,j}\} = \{Z(t_i) \leq y_i, \ldots, Z(t_j) \leq y_j\}$ and an analogous one for $\widetilde{\mathbf{Z}}(t)$. Recalling (10.6) we have for all $\theta \in \Theta$

$$\mathrm{P}_\theta^F(E) = \mathrm{P}_{t_0,z_0}(\min(T_h, t_m) \leq t, \widetilde{\mathbf{Z}}_{\mathbf{t}_{1,m}} \leq \mathbf{y}_{1,m})$$

$$= \sum_{k=1}^m I_{\{t \geq t_{k-1}\}} \mathrm{P}_{t_0,z_0}(t_{k-1} \leq T_h \leq \min(t, t_k), \widetilde{\mathbf{Z}}_{\mathbf{t}_{1,m}} \leq \mathbf{y}_{1,m})$$

$$+ I_{\{t \geq t_m\}} \mathrm{P}_{t_0,z_0}(T_h \geq t_m, \widetilde{\mathbf{Z}}_{\mathbf{t}_{1,m}} \leq \mathbf{y}_{1,m}). \tag{10.14}$$

To calculate the first m terms of (10.14) take $t_{k-1} \leq t < t_k$ with $1 \leq k \leq m$ and define $\bigcap_{k=k_1}^{k_2} A_k = \Omega$ if $k_2 < k_1$. Then, for $s \geq t_{k-1}$ by the definition of $\tilde{\mathbf{Z}}(s)$

$$
P_{t_0,z_0}(t_{k-1} \leq T_h \leq s, \ \tilde{\mathbf{Z}}_{\mathbf{t}_{1,m}} \leq \mathbf{y}_{1,m})
$$
$$
= \ P_{t_0,z_0}(\{t_{k-1} \leq T_h \leq s, \ \mathbf{Z}_{\mathbf{t}_{1,k-1}} \leq \mathbf{y}_{1,k-1})\} \cap \{\tilde{\mathbf{Z}}_{\mathbf{t}_{k,m}} \leq \mathbf{y}_{k,m})\})
$$
$$
= \ P_{t_0,z_0}(T_h^{t_{k-1}} \leq s, \ T_h^{t_0} \geq t_{k-1}, \ \mathbf{Z}_{\mathbf{t}_{1,k-1}} \leq \mathbf{y}_{1,k-1}) \prod_{j=k}^{m} I_{\{y_j=\infty\}}. \quad (10.15)
$$

Applying total probability, the Markov property, and Lemma 10.2.2 yields

$$
P_{t_0,z_0}(T_h^{t_{k-1}} \leq s, \ T_h^{t_0} \geq t_{k-1}, \ \mathbf{Z}_{\mathbf{t}_{1,k-1}} \leq \mathbf{y}_{1,k-1})
$$
$$
= \int_{(-\infty,\mathbf{y}_{1,k-1}]} P_{t_0,z_0}(T_h^{t_{k-1}} \leq s \,|\, Z(t_{k-1}) = z_{k-1})
$$
$$
\times \ f_{k-1}(t_0, z_0, \mathbf{t}_{1,k-1}, \mathbf{z}_{1,k-1}, h) \, \lambda^{k-1}(\mathrm{d}\mathbf{z}_{1,k-1})
$$

and, moreover, recalling the distribution of $P_{s,x}^{T_h^s}$

$$
P_{t_0,z_0}(T_h^{t_{k-1}} \leq s \,|\, Z(s_{k-1}) = z_{k-1}) \ = \ P_{t_{k-1},z_{k-1}}(T_h^{t_{k-1}} \leq s)
$$
$$
= \int_{t_{k-1}}^{s} g(t_{k-1}, z_{k-1}, \tau, h) \, \lambda(\mathrm{d}\tau)
$$

if $z_{k-1} < h$. Finally, the product of indicator functions in (10.15) can be written as

$$
\prod_{j=k}^{m} I_{\{y_j=\infty\}} = \int_{(-\infty,\mathbf{y}_{k,m}]} \prod_{j=k}^{m} I_{\{z_j=\infty\}} \, \varepsilon_\infty^{m-k+1}(\mathrm{d}\mathbf{z}_{k,m}).
$$

Combining the above equations we obtain

$$
P_{t_0,z_0}(t_{k-1} \leq T_h \leq s, \ \tilde{\mathbf{Z}}_{\mathbf{t}_{1,m}} \leq \mathbf{y}_{1,m})
$$
$$
= \int_{(-\infty,\mathbf{y}_{1,m}]} \int_{t_{k-1}}^{s} g(t_{k-1}, z_{k-1}, \tau, h) \, \lambda(\mathrm{d}\tau) \, f_{k-1}(t_0, z_0, \mathbf{t}_{1,k-1}, \mathbf{z}_{1,k-1}, h)
$$
$$
\times \prod_{j=k}^{m} I_{\{z_j=\infty\}} \, (\lambda^{k-1} \otimes \varepsilon_\infty^{m-k+1})(\mathrm{d}\mathbf{z}_{1,m}) \quad . \quad (10.16)
$$

Now, from (10.10) and (10.11) it is obvious, that the measure $\lambda^{k-1} \otimes \varepsilon_\infty^{m-k+1}$ in (10.16) can be replaced by

$$
(\lambda + \varepsilon_\infty)^m = \sum_{(\mu_1,\ldots,\mu_m) \in \{\lambda, \varepsilon_\infty\}^m} \bigotimes_{i=1}^{m} \mu_i
$$

without altering the value of the integral, since integrating with respect to the summands of $(\lambda + \varepsilon_\infty)^m$ gives zero except for the case $\mu_i = \lambda^{k-1} \otimes \varepsilon_\infty^{m-k+1}$. Hence, going back to (10.14) and using (10.15) and Lemma 10.2.2 we obtain

$$
\begin{aligned}
\mathrm{P}_\theta^F(E) &= \int\limits_{(-\infty, \mathbf{y}_{1,m}]} \left(\int_{t_0}^t \sum_{k=1}^m \left[I_{\{t_{k-1} \le \tau < t_k\}} g(t_{k-1}, z_{k-1}, \tau, h) \, \lambda(\mathrm{d}\tau) \right. \right. \\
&\qquad \times f_{k-1}(t_0, z_0, \mathbf{t}_{1,k-1}, \mathbf{z}_{1,k-1}, h) \prod_{j=k}^m I_{\{z_j = \infty\}} \bigg] \\
&\qquad \left. + \int_{t_0}^t I_{\{\tau = t_m\}} \, \varepsilon_{t_m}(\mathrm{d}\tau) \, f_m(t_0, z_0, \mathbf{t}_{1,m}, \mathbf{z}_{1,m}, h) \right) (\lambda + \varepsilon_\infty)^m(\mathrm{d}\mathbf{z}_{1,m}) \\
&= \int\limits_{(-\infty, t] \times (-\infty, \mathbf{y}_{1,m}]} \left(\sum_{k=1}^m \left[I_{\{t_{k-1} \le \tau < t_k\}} g(t_{k-1}, z_{k-1}, \tau, h) \right. \right. \\
&\qquad \times \prod_{j=1}^{k-1} f(t_{j-1}, z_{j-1}, t_j, z_j, h) \prod_{j=k}^m I_{\{z_j = \infty\}} \bigg] \\
&\qquad \left. + I_{\{\tau = t_m\}} \prod_{j=1}^m f(t_{j-1}, z_{j-1}, t_j, z_j, h) \right) \nu(\mathrm{d}(\tau, \mathbf{z}_{1,m})),
\end{aligned}
$$

where $\nu = (\lambda + \varepsilon_{t_m}) \otimes (\lambda + \varepsilon_\infty)^m$.

Thus, the integrand of the last equation is a version of the Radon-Nikodym derivative of P_θ^F with respect to ν, since $\mathcal{B} \otimes \overline{\mathcal{B}}^m$ is generated by the sets E. To complete the proof note that this integrand and the right hand side of (10.13) are equivalent. ∎

In general we observe not only one realization of the damage process but the realizations of n independent damage processes $Z_i(t)$, $1, \ldots, n$, corresponding to n independent items. Let $t_{i1}, \ldots, t_{im_i}$ with $t_0 < t_{i1} < \ldots < t_{im_i} < \infty$ be the m_i fixed observation points of the realization of $Z_i(t)$, $M = \sum_{i=1}^n m_i$, and $T_h^{(i)}$ the lifetime of $Z_i(t)$. The censored observations in this realization have the form

$$
\begin{aligned}
&(\tau_i, z_{i1}, \ldots, z_{im_i}) \\
&= (\min(T_h^{(i)}, t_{im_i}), F_{t_{i1}}(T_h^{(i)}, Z_i(t_{i1})), \ldots, F_{t_{im_i}}(T_h^{(i)}, Z_i(t_{im_i}))).
\end{aligned}
$$

Then the statistical model is given by

$$
(\mathbf{R}^n \times \overline{\mathbf{R}}^M, \mathcal{B}^n \otimes \overline{\mathcal{B}}^M, (\otimes_{i=1}^n \mathrm{P}_\theta^F)_{\theta = (\mu, \sigma^2, z_0, t_0, h) \in \Theta \subset \mathbf{R}^5})
$$

with the likelihood function

$$
L(\boldsymbol{\tau}, \mathbf{z}^{(1)}, \ldots, \mathbf{z}^{(n)}; \theta) = \prod_{i=1}^n L(\tau_i, \mathbf{z}^{(i)}; \theta)
$$

where $\boldsymbol{\tau} = (\tau_1, \ldots, \tau_n) \in \mathbf{R}^n$ and $\mathbf{z}^{(i)} = (z_{i1}, \ldots, z_{im_i}) \in \overline{\mathbf{R}}^{m_i}$, $i = 1, \ldots, n$.

10.3   An Example

Substituting formulae (10.8) and (10.10) for the functions g and f in equation (10.13) and setting $t_{i0} = t_0$ and $z_{i0} = z_0$ we get the likelihood function of the ith realization $(i = 1, \ldots, n)$

$$
\begin{aligned}
L(\tau_i, \mathbf{z}^{(i)}; \theta) &= \prod_{j=1}^{k-1} \left(\frac{1}{\sqrt{\sigma^2(t_{ij} - t_{ij-1})}} \phi \left(\frac{(z_{ij} - z_{ij-1}) - \mu(t_{ij} - t_{ij-1})}{\sqrt{\sigma^2(t_{ij} - t_{ij-1})}} \right) \right. \\
&\quad \left. \cdot \left[1 - \exp \left(-\frac{2(h - z_{ij-1})(h - z_{ij})}{\sigma^2(t_{ij} - t_{ij-1})} \right) \right] \right) \prod_{j=k}^{m_i} I_{\{z_j = \infty\}} \\
&\quad \times \frac{h - z_{ik-1}}{\sqrt{\sigma^2(\tau_i - t_{ik-1})^3}} \phi \left(\frac{h - z_{ik-1} - \mu(\tau_i - t_{ik-1})}{\sqrt{\sigma^2(\tau_i - t_{ik-1})}} \right) \quad (10.17)
\end{aligned}
$$

if $t_{k-1} \leq \tau_i < t_k$ and $1 \leq k \leq m_i$ and

$$
\begin{aligned}
L(\tau_i, \mathbf{z}^{(i)}; \theta) &= \prod_{j=1}^{m_i} \left(\frac{1}{\sqrt{\sigma^2(t_{ij} - t_{ij-1})}} \phi \left(\frac{(z_{ij} - z_{ij-1}) - \mu(t_{ij} - t_{ij-1})}{\sqrt{\sigma^2(t_{ij} - t_{ij-1})}} \right) \right. \\
&\quad \left. \cdot \left[1 - \exp \left(-\frac{2(h - z_{ij-1})(h - z_{ij})}{\sigma^2(t_{ij} - t_{ij-1})} \right) \right] \right) \quad (10.18)
\end{aligned}
$$

if $\tau_i = t_{m_i}$. To compare our estimates based on the likelihood function in (10.17) and (10.18) with previous results we consider the case that only process increments are observed. As shown in Kahle (1994), in this case the likelihood function of the ith realization is given by

$$
L(\mathbf{z}^{(i)}; \theta) = \prod_{j=1}^{m_i} \frac{1}{\sqrt{\sigma^2(t_{ij} - t_{ij-1})}} \phi \left(\frac{(z_{ij} - z_{ij-1}) - \mu(t_{ij} - t_{ij-1})}{\sqrt{\sigma^2(t_{ij} - t_{ij-1})}} \right) . \quad (10.19)
$$

We notice, that the likelihood function (10.19) can be obtained from (10.18) if h tends to infinity, that means if a boundary level h does not exist and, consequently, exceeding the boundary is impossible.

As example, we have simulated a sample of $n = 5$ process realizations with $m_i = 10$ observation points t_{ij} in each realization, which have equal distances $t_{ij} - t_{ij-1} = 1$ and with true parameter values $\mu = 1.0$, $\sigma^2 = 0.25$, $t_0 = 0$, $z_0 = 0$, and $h = 5.0$. The simulated data are shown in the following table:

i	1	2	3	4	5
$z_{ij} - z_{ij-1}$	1.290	0.397	0.744	0.984	1.223
	1.096	1.619	1.125	0.904	0.769
	0.715	0.427	1.047	1.384	1.775
	0.339	1.574	0.530	1.000	0.567
	0.028		0.123		0.426
	1.231		1.346		
τ_i	6.559	4.967	6.018	4.461	5.138

Supposed that the parameter values $t_0 = 0$ and $z_0 = 0$ are known, we first estimate the parameters μ and σ^2 maximizing the likelihood function (10.19), that means, we ignore that the damage process does not exceed the boundary level between observation points. Using maximum likelihood estimators given by Kahle (1994) we get

$$\hat{\mu} = 0.9198, \qquad \hat{\sigma}^2 = 0.1983.$$

In Figure 10.1, 0.95- and 0.99-confidence regions for μ and σ^2 based on the asymptotic χ^2-distribution of the likelihood ratio are shown for this model.

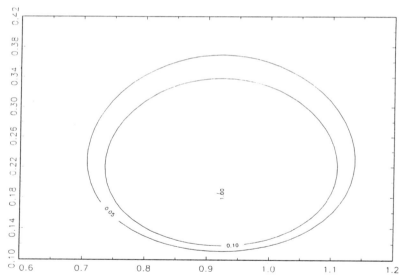

Figure 10.1: Confidence estimates for (μ, σ^2)

Now, again supposing that $t_0 = z_0 = 0$ are known, we compute maximum likelihood estimates for the parameters μ, σ^2, and h based on the likelihood function (10.17) of the exact model. The maximization of the likelihood function and also the calculation of confidence regions was done by the computer program Gauss (1984-1996) and yielded the point estimates

$$\hat{\mu} = 0.9222, \qquad \hat{\sigma}^2 = 0.1915, \qquad \hat{h} = 5.0047.$$

The estimates turned out to be difficult to find so that a good initial value of the parameters have to be known. Such initial values may be the estimates of μ and σ^2 from the likelihood function (10.19) and an initial value of h can be found from a likelihood function based on the inverse Gaussian density (10.3). The Figures 10.2, 10.3 and 10.4 show 0.95- and 0.99-confidence regions based on the asymptotic χ^2-distribution of the likelihood ratio for the parameters (μ, σ^2), (μ, h), and (σ^2, h), respectively. In all examples considered both the estimates of the parameter σ^2 and the confidence regions for the exact model are smaller then the estimates based on (10.19). Note that in this case it is also possible to estimate the parameter h.

A simulation study to investigate the bias and the variance of the point estimates and also the exact confidence level of asymptotic confidence regions will be done in forthcoming works.

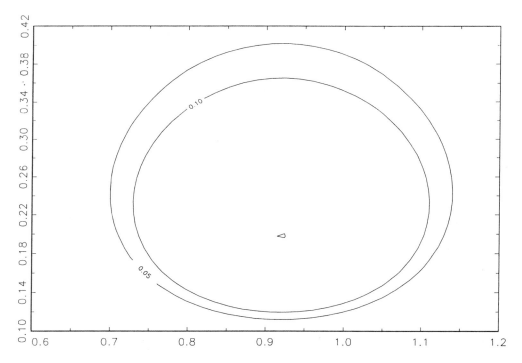

Figure 10.2: Confidence estimates for (μ, σ^2)

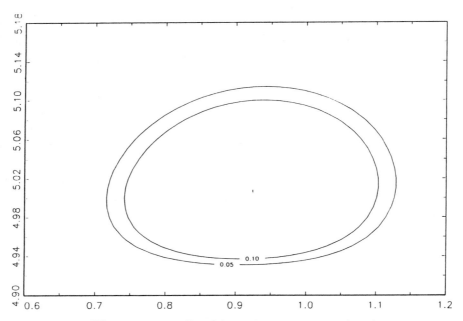

Figure 10.3: Confidence estimates for (μ, h)

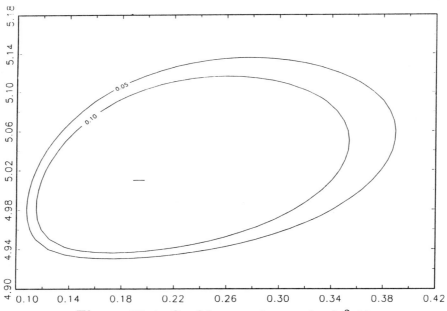

Figure 10.4: Confidence estimates for (σ^2, h)

10.4 Appendix: Proof of Lemma 10.2.1

PROOF. Recalling equation (10.9) we have to compute the integral

$$
\begin{aligned}
I &= \int_s^t p(u,h,t,z)\, g(s,x,u,h)\, \mathrm{d}u \\
&= \int_s^t \frac{h-x}{2\pi\sigma^2\sqrt{(t-u)(u-s)^3}} \\
&\qquad \times \exp\left[-\frac{1}{2\sigma^2}\left(\frac{(z-h-\mu(t-u))^2}{t-u} + \frac{(h-x-\mu(u-s))^2}{u-s}\right)\right]\mathrm{d}u
\end{aligned}
$$

for $t > s \geq t_0$ and $x < h$. Applying the substitution $y = ((t-u)/(u-s))^{1/2}$ for $s < u < t$, i.e., $u = (t+sy^2)/(1+y^2)$ with $+\infty > y > 0$, we obtain after some algebra

$$
\begin{aligned}
I &= \frac{h-x}{\pi\sigma^2(t-s)}\exp\left[-\frac{(z-h)^2+(h-x)^2-2\mu(t-s)(z-x)+\mu^2(t-s)^2}{2\sigma^2(t-s)}\right] \\
&\qquad \times \int_0^\infty \exp\left[-\left(\frac{\beta^2}{y^2}+\alpha^2 y^2\right)\right]\mathrm{d}y
\end{aligned}
$$

with $\beta^2 = (z-h)^2/(2\sigma^2(t-s))$ and $\alpha^2 = (h-x)^2/(2\sigma^2(t-s))$. Denoting the last integral with I_1 it can be easily checked that for $\alpha \neq 0$

$$
\begin{aligned}
I_1 &= \frac{\sqrt{\pi}}{4\alpha}\left[e^{2\alpha\beta}\left(\mathrm{erf}\left(\alpha z+\frac{\beta}{z}\right)-1\right)+e^{-2\alpha\beta}\left(\mathrm{erf}\left(\alpha z-\frac{\beta}{z}\right)+1\right)\right]\Bigg|_{z=0}^{z=\infty} \\
&= \frac{\sqrt{\pi}}{2|\alpha|}e^{-2|\alpha\beta|},
\end{aligned}
$$

where $\mathrm{erf}(x) = 2\pi^{-1/2}\int_0^x e^{-t^2}\,\mathrm{d}t$.

Hence, as $|\alpha| = |h-x|(2\sigma^2(t-s))^{-1/2}$ and $x < h$, we get

$$
I = \frac{1}{\sqrt{2\pi\sigma^2(t-s)}}\exp\left[-\frac{Q-2\mu(t-s)(z-x)+\mu^2(t-s)^2}{2\sigma^2(t-s)}\right],
$$

with

$$
\begin{aligned}
Q &= (z-h)^2+2|z-h|(h-x)+(h-x)^2 \\
&= \begin{cases}(z-2h+x)^2 = (z-x)^2+4(h-x)(h-z), & z \leq h \\ (z-x)^2, & z > h\end{cases},
\end{aligned}
$$

and, finally

$$
I = \frac{1}{\sigma\sqrt{t-s}}\phi\left(\frac{z-x-\mu(t-s)}{\sigma\sqrt{t-s}}\right)\times\begin{cases}\exp\left(-\dfrac{2(h-x)(h-z)}{\sigma^2(t-s)}\right), & z \leq h \\ 1, & z > h\end{cases},
$$

where ϕ is the density of the standard normal distribution. This completes the proof.   ■

References

1. Gauss, C. F. (1984-1996). *Mathematical and Statistical System*, Maple Valley, WA: Aptech Systems.

2. Kahle, W. (1994). Simultaneous confidence regions for the parameters of damage processes, *Statistical Papers*, **35**, 27–41.

3. Pieper, V. and Tiedge, J. (1983). Zuverlässigkeitsmodelle auf der Grundlage stochastischer Modelle von Verschleißprozessen, *Mathemathsche Operationsforschung und Statistik, Series Statistics*, **14**, 485–502.

4. Wentzell, A. D. (1979). *Theorie zufälliger Prozesse*, Berlin: Akademie-Verlag.

11

Boundary Crossing Probabilities of Poisson Counting Processes with General Boundaries

Axel Lehmann

Otto-von-Guericke-Universität Magdeburg, Magdeburg, Germany

Abstract: This paper gives exact boundary crossing probabilities for finite time intervals associated with Poisson processes and general moving boundaries in both one and two boundary cases. These probabilities are relevant in various applications of sequential stopping in statistics, reliability analysis, risk theory, dam engineering, etc. We deal with both homogeneous and nonhomogeneous Poisson processes as well as mixed Poisson processes where the Pólya-Lundberg process is considered in detail.

Keywords and phrases: Poisson processes, Pólya-Lundberg process, first passage time, general boundaries, stopping time, sequential analysis

11.1 Introduction

Let $\{N(t); t \geq 0\}$ be a homogeneous or nonhomogeneous Poisson counting process with increasing right continuous step functions as sample paths. A basic problem in many fields such as reliability theory, storage theory, queueing theory, sequential analysis, and optimal stopping is that of calculating the probability that $N(\cdot)$ does not cross a general upper or lower boundary before a time t

$$P(N(s) < f(s) \text{ for } 0 \leq s \leq t)$$

$$P(N(s) > g(s) \text{ for } 0 \leq s \leq t) \tag{11.1}$$

$$P(g(s) < N(s) < f(s) \text{ for } 0 \leq s \leq t) .$$

Provided a separable version of $\{N(t); t \geq 0\}$ is taken, the probabilities in (11.1) are well defined. In the present paper we give these probabilities for finite t with

respect to nondecreasing upper and lower boundaries f and g by deriving the
exact distribution function of first passage times τ of $\{N(t)\}$

$$
\begin{aligned}
\tau^f &= \inf\{t > 0 : N(t) \geq f(t)\,\} \\
\tau_g &= \inf\{t > 0 : N(t) \leq g(t)\} \\
\tau_{g,f} &= \inf\{t > 0 : N(t) \notin (g(t), f(t))\}
\end{aligned}
\qquad (11.2)
$$

where $\inf \emptyset = \infty$.

The problem of computing boundary crossing probabilities of Poisson count-
ing processes for infinite time intervals, i.e. calculating the probabilities $\mathrm{P}(\tau = \infty)$ that $N(\cdot)$ never crosses the boundaries, is not covered here and will be
considered in forthcoming works.

While a substantial literature on boundary crossing problems for continuous-
time continuous-state processes exists, the literature on boundary crossing probabilities of continuous-time discrete-state processes is more sparse and mainly
concentrated on processes with independent increments and constant boundaries [see for instance Baxter and Donsker (1957), Borovkov (1965) and Durbin
(1971, 1985)]. For the homogeneous Poisson process Pyke (1959) and recently
Zacks (1991) gave expressions of the first passage time distribution with respect
to a linear upper or lower boundary, and Franz (1977) for certain linear upper
boundaries. Some results for specific upper boundaries were given by Daniels
(1963), Whittle (1993), and Gallot (1966). Durbin (1971) obtained boundary
crossing probabilities for strictly increasing boundaries as solution of a system
of linear equations. Recently, Stadje (1993) gave the exact first passage time
distribution for continuous and strictly increasing upper boundaries and Picard
and Lefèvre (1996) for continuous lower boundaries. Solutions for the homogeneous and especially the nonhomogeneous Poisson process with respect to
general boundaries, in particular in the two boundary case, are not available in
the literature.

Boundary crossing probabilities are useful for sequential stopping methods in reliability testing [Genedenko, Belyayev and Solovyen (1969), Lisek and
Hochschild (1983)], queueing theory, storage theory and dam models [Prabhu
(1980), Pyke (1959)], risk theory [Schmidt (1996)], and sequential analysis [Sieg-
mund (1986)]. In reliability analysis consider e.g. the failure times $(T_n)_{n \geq 1}$ of an
item having a continuous random lifetime with distribution function F and being immediately repaired such that the failure rate is the same as just before the
loss (minimal repair). Then the renewal counting process is a nonhomogeneous
Poisson process with mean value function $\Lambda(t) = -\ln(1 - F(t))$. If n independent items with lifetime distribution functions F_i are considered the renewal
counting process is nonhomogeneous Poisson with $\Lambda(t) = -\sum_{i=1}^{n} \ln(1 - F_i(t))$.
Given boundaries f and g the above first passage times specify different sequential test procedures for reliability testing.

The assumption of nondecreasing boundary functions means no loss of generality as the following argument shows. Suppose that f and g are arbitrary real

valued upper and lower boundary functions on $[0, \infty)$ with existing one-sided limits. To ensure that the random variables τ in (11.2) are stopping times, i.e $\{\tau \leq t\} \in \mathcal{F}_t$, where $\mathcal{F}_t$ is the σ-algebra generated by $\{N(s); 0 \leq s \leq t\}$, we may assume that the first entrance sets $\{(x, y) \in \mathbf{R}_+ \times \mathbf{R} : y \geq f(x)\}$ and $\{(x, y) \in \mathbf{R}_+ \times \mathbf{R} : y \leq g(x)\}$ are closed. This is equivalent to the conditions

$$f(t) \leq \min\{f(t-0), f(t+0)\}, \quad t \geq 0 \tag{11.3}$$

and

$$g(t) \geq \max\{g(t-0), g(t+0)\}, \quad t \geq 0. \tag{11.4}$$

By the nondecreasing nature of the sample paths a lower boundary function g satisfying relation (11.4) can obviously be replaced by the nondecreasing boundary

$$g^*(s) = \sup\{g(u) : 0 \leq u \leq s\}, \quad s \geq 0$$

without altering the crossing probabilities in (11.1) for all t. Actually the valleys of g are cut off to yield g^*. For example, the arcs $\overparen{GH}$ and $\overparen{IK}$ in Figure 11.1 are not eligible as first crossing places and can be replaced by the segments $\overline{GH'}$ and $\overline{IK}$.

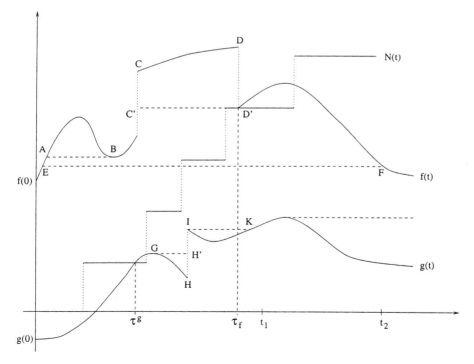

Figure 11.1: Functions $N(t)$, $f(t)$ and $g(t)$, and boundaries

In case of an upper boundary f that satisfies condition (11.3) the humps of f are cut off to obtain a nondecreasing boundary function with the same boundary crossing probabilities (11.1). In contrast to the lower boundary case these substitutions depend on the time interval $[0,t]$ so that different values of t imply different modified nondecreasing boundaries. For example, in Figure 11.1 the segments $\overline{AB}$ and $\overline{C'D'}$ can be substituted for the arcs $\overset{\frown}{AB}$ and $\overset{\frown}{CD}$ if $t = t_1$, whereas for $t = t_2$ the whole arc $\overset{\frown}{EF}$ has to be replaced by $\overline{EF}$. For fixed $t > 0$, the equivalent nondecreasing upper boundary on $[0,t]$ is given by

$$f_t^*(s) = \inf\{f(u) : s \le u \le t\}, \quad 0 \le s \le t.$$

Note that f_t^* (g^*) is left (right) continuous if f and g fulfill the conditions (11.3) and (11.4).

Deriving exact boundary crossing probabilities for nondecreasing boundaries we start in Section 11.2 with the homogeneous Poisson process using its independent increments and the nondecreasing nature of its sample paths. Based on these results we solve in Section 11.3 boundary crossing problems for nonhomogeneous Poisson processes by nonrandom time transformations. Finally, in Section 11.4 boundary crossing probabilities for a special mixed Poisson process, the Pólya-Lundberg process which has many applications in risk theory, are derived by conditioning on the random intensity of the process.

11.2   The Homogeneous Poisson Process

Let $\{N(t); t \ge 0\}$ be a time homogeneous Poisson process with events occurring at intensity $\lambda > 0$ and with nondecreasing right continuous sample paths starting at the point $(0,0)$ with probability one. Further, let f and g denote real valued nondecreasing upper and lower boundary functions on $[0,\infty)$ with $g(0) < 0 < f(0)$. To fulfill the conditions (11.3) and (11.4) we assume an upper boundary f to be left continuous and a lower boundary g to be right continuous.

Set $n(t) = \lfloor f(t) \rfloor$, $m(t) = \max(\lceil g(t) \rceil, -1)$, $t \ge 0$ and $n_0 = n(0) + 1$ where $\lfloor x \rfloor = \max\{k \in \mathbf{Z} : k < x\}$ and $\lceil x \rceil = \max\{k \in \mathbf{Z} : k \le x\}$ denote left and right continuous versions of the integer part. We define inverse functions of f and g on $\mathbf{N}_0$ in different ways by

$$f_n = f^{-1}(n) = \begin{cases} 0, & 0 \le n < n_0 \\ \sup\{t \ge 0 : f(t) \le n\}, & n \ge n_0 \end{cases}$$

and

$$g_n = g^{-1}(n) = \inf\{t \ge 0 : g(t) \ge n\}, \quad n \ge 0,$$

which differ only if the boundary has flat sections (see Figure 11.2). For convenience let $g_{-1} = 0$. The sequences $(f_n)_{n \geq 0}$ and $(g_n)_{n \geq -1}$ are nonnegative and nondecreasing and are sufficient to define f and g in the sequel. Note that some of the points f_n (g_n) will coincide if f (g) has jumps of size greater than or equal to one.

By the nondecreasing nature of the Poisson sample paths a first crossing of a lower boundary g occurs necessarily at continuity points of the sample path, so that g may be crossed only at the points g_n, $n \geq 0$. Thus, the first passage time τ_g has a discrete distribution. In the upper case with a nondecreasing boundary f, the distribution function of τ^f is continuous since the event $\{\tau^f = t\}$ with $t \in [0, \infty)$ can only occur if $N(\cdot)$ jumps at t, implying $P(\tau^f = t) = 0$ for all $t \in [0, \infty)$. Note that this fact is no langer true if f is decreasing. Finally, in the two boundary case the distribution of $\tau_{f,g}$ is neither continuous nor discrete.

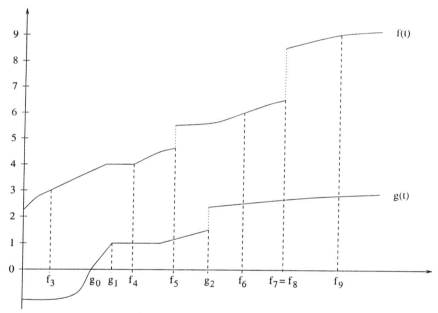

Figure 11.2: Functions $f(t)$ and $g(t)$, and their inverse

11.2.1 Upper boundary case

Let
$$P_n(t) = P(N(t) = n, \tau^f > t), \qquad t \geq 0, n \geq 0$$
denote the probability that a Poisson sample path reaches the level n at time t without crossing the upper boundary f on $[0, t]$. Since clearly $P_n(t) = 0$ if $n > n(t)$, the probability that $N(\cdot)$ does not cross the upper boundary f on $[0, t]$ yields to
$$P(\tau^f > t) = \sum_{n=0}^{n(t)} P_n(t), \qquad t \geq 0.$$

The functions $P_n(t)$ are given in Theorem 11.2.1. Recalling $P(N(0) = 0) = 1$ and using the notation $I_{\{\cdot\}}$ for an indicator variable that assumes the value 1 if the relation in brackets is true and the value 0 otherwise, we have obviously $P_n(0) = I_{\{n=0\}}$.

Theorem 11.2.1 *Let $t \in [0, \infty)$. Then*

$$P_0(t) = e^{-\lambda t} \tag{11.5}$$

and

$$P_n(t) = \begin{cases} e^{-\lambda t} \lambda^n \sum_{j=0}^{n} p_{j,n} \dfrac{(t - f_n)^{n-j}}{(n - j)!} \,, & t > f_n \\ 0 & , \quad t \leq f_n \end{cases} \quad , \quad n \geq 1 \tag{11.6}$$

where the constants $p_{j,n}$ are recursively defined by

$$p_{0,n} = 1, \qquad n \geq 0, \tag{11.7}$$

$$p_{j,n} = \sum_{i=0}^{j} p_{i,j} \dfrac{(f_n - f_j)^{j-i}}{(j - i)!} \,, \qquad 1 \leq j \leq n - 1; \; n \geq 0, \tag{11.8}$$

$$p_{j,n} = 0, \qquad j \geq n, j \neq 0; \; n \geq 0. \tag{11.9}$$

PROOF. Because of $P(\tau^f > 0) = 1$ formula (11.5) is obvious from $P_0(t) = P(N(t) = 0)$. Let $n \geq 1$. For $t \leq f_n$, clearly $P_n(t) = 0$ since $t = 0$ or $f(t) \leq n$. Decomposing the event $\{N(t) = n, \tau^f > t\}$ according to the value of $N(f_n)$ and using the Markov property we obtain for $t > f_n$

$$\begin{aligned} P_n(t) &= \sum_{j=0}^{n} P(N(f_n) = j, \tau^f > f_n, N(t) = n) \\ &= \sum_{j=0}^{n} P(N(t - f_n) = n - j) \, P_j(f_n) \end{aligned} \tag{11.10}$$

where $P_n(f_n) = 0$ if $f_n > 0$. Moreover, introducing the auxiliary functions $p_n(t) = e^{\lambda t} \lambda^{-n} P_n(t)$, $t \geq 0$ and setting $p_{j,n} = p_j(f_n)$ we have

$$P(N(t - f_n) = n - j) \, P_j(f_n) = e^{-\lambda t} \lambda^n p_{j,n} \dfrac{(t - f_n)^{n-j}}{(n - j)!}, \quad t > f_n, n \geq 0$$

and, hence, formula (11.6) follows immediately. If we replace in the above equations the index n by j and the variable t by f_n we get by the same arguments for all $n \geq 0$

$$P_0(f_n) = e^{-\lambda f_n}, \tag{11.11}$$

$$P_j(f_n) = \sum_{i=0}^{j} P(N(f_j) = i,\ \tau^f > f_j,\ N(f_n) = j)$$

$$= \sum_{i=0}^{j} P(N(f_n - f_j) = j - i)\, P_i(f_j)$$

$$= e^{-\lambda f_n} \lambda^j \sum_{i=0}^{j} p_{i,j} \frac{(f_n - f_j)^{j-i}}{(j-i)!}, \quad 1 \le j \le n-1, \quad (11.12)$$

and

$$P_j(f_n) = 0, \quad j \ge n,\, j \ne 0. \tag{11.13}$$

Now by the definition of the auxiliary functions $p_n(t)$, the recursion formulae (11.7) - (11.9) are simple consequences of the equations (11.11) - (11.13). ∎

Formula (11.6) shows that $F_{\tau f}$ is almost everywhere continuously differentiable in t (except for the points f_n) and thus absolutely continuous. Rewriting (11.6) and changing the order of summation we obtain

$$P_n(t) = e^{-\lambda t} \lambda^n \sum_{j=0}^{n} p_{j,n} \frac{(t - f_n)^{n-j}}{(n-j)!}$$

$$= e^{-\lambda t} \lambda^n \sum_{l=0}^{n} \left(\sum_{j=0}^{n-l} p_{j,n} \frac{(-f_n)^{n-j-l}}{(n-j-l)!} \right) \frac{t^l}{l!}$$

$$= e^{-\lambda t} \lambda^n \sum_{l=0}^{n} \rho_{n-l} \frac{t^l}{l!}, \quad t \ge f_n$$

with

$$\rho_{n-l} = \sum_{j=0}^{n-l} p_{j,n} \frac{(-f_n)^{n-j-l}}{(n-j-l)!}.$$

This is the expression of the first passage time distribution for the upper boundary case given in Stadje (1993). Since $P_n(f_n) = 0$, the coefficients ρ_n can be recursively calculated by

$$\rho_n = - \sum_{l=1}^{n} \rho_{n-l} \frac{(f_n)^l}{l!}, \quad n \ge 1 \tag{11.14}$$

and $\rho_0 = 1$. Although our recursion formula (11.8) seems not as simple as the recursion (11.14), it can be expected to be less vulnerable to rounding errors for large n, because it has only terms of the same sign whereas (11.14) has terms of alternating sign.

11.2.2   Lower boundary case

We have already mentioned that, in contrast to τ^f, the first passage time τ_g has a discrete distribution concentrated at the points g_n, $n \geq 0$ and possibly at infinity if the distribution of τ_g is defective. As some points g_n can coincide if g has jumps of size greater than or equal to one, the events $\{\tau_g = g_n\}$ may not be disjoint for different n. For this reason, we define for $n \in \mathbf{N}_0$ the index functions

$$\overline{n} = \max\{k \geq n : g_k = g_n\}$$

and

$$\underline{n} = \min\{0 \leq k \leq n : g_k = g_n\}.$$

Hence, $\underline{n} = n = \overline{n} \Leftrightarrow g_{n-1} < g_n < g_{n+1}$ for $n \geq 0$. The probability of first crossing the boundary at g_n is given by

$$P(\tau_g = g_n) = \sum_{k=\underline{n}}^{\overline{n}} P(\tau_g = g_k, N(g_k) = k), \quad n \geq 0$$

and the probability that the boundary is crossed on $[0, t]$ by

$$P(\tau_g \leq t) = \sum_{n=0}^{m(t)} P(\tau_g = g_n, N(g_n) = n), \quad t \geq 0$$

with disjoint events $\{\tau_g = g_n, N(g_n) = n\}$, $n \geq 0$.

Theorem 11.2.2 *Let $n \in \mathbf{N}$. Then*

$$P(\tau_g = g_n, N(g_n) = n) = e^{-\lambda g_n} \lambda^n \sum_{j=\underline{n}}^{n} q_{j,\underline{n}-1} \frac{(g_n - g_{\underline{n}-1})^{n-j}}{(n-j)!} \tag{11.15}$$

with constants $q_{j,n}$ recursively defined by

$$q_{j,-1} = I_{\{j=0\}}, \quad j \geq 0 \tag{11.16}$$

and

$$q_{j,n} = \begin{cases} 0 & , \quad 0 \leq j \leq \overline{n} \\ \sum_{i=n}^{j} q_{i,n-1} \dfrac{(g_n - g_{n-1})^{j-i}}{(j-i)!} & , \quad j > \overline{n} \end{cases} , \quad n \geq 0. \tag{11.17}$$

PROOF. Denote the probability that a Poisson sample path reaches the level n at time t without crossing the lower boundary g on $[0, t]$ by

$$Q_n(t) = P(N(t) = n, \tau_g > t), \quad t \geq 0, n \geq 0.$$

Again we define auxiliary functions $q_n(t) = e^{\lambda t} \lambda^{-n} Q_n(t)$, $t \geq 0$ and set $q_{j,n} = q_j(g_n)$, $j \geq 0$, $n \geq -1$. Remind that $g_{-1} = 0$ and $\underline{0} = 0$. A first crossing of g

at level n implies that the value of $N(g_{\underline{n}-1})$ is between $\underline{n}$ and n. Thus, by the Markov property we obtain

$$
\begin{aligned}
\mathrm{P}(\tau_g = g_n,\, N(g_n) = n) &= \sum_{j=\underline{n}}^{n} \mathrm{P}(N(g_{\underline{n}-1}) = j,\, \tau_g > g_{\underline{n}-1},\, N(g_n) = n) \\
&= \sum_{j=\underline{n}}^{n} \mathrm{P}(N(g_n - g_{\underline{n}-1}) = n - j)\, Q_j(g_{\underline{n}-1}) \\
&= \mathrm{e}^{-\lambda g_n} \lambda^n \sum_{j=\underline{n}}^{n} q_{j,\underline{n}-1} \frac{(g_n - g_{\underline{n}-1})^{n-j}}{(n-j)!}\,.
\end{aligned}
$$

The defintion (11.16) is clear since $Q_j(g_{-1}) = Q_j(0) = \mathrm{P}(N(0) = j)$. The recursion formula (11.17) follows for $n \geq 0$ from

$$
\begin{aligned}
Q_j(g_n) &= \sum_{i=n}^{j} \mathrm{P}(N(g_{n-1}) = i,\, \tau_g > g_{n-1},\, N(g_n) = n) \\
&= \sum_{i=n}^{j} \mathrm{P}(N(g_n - g_{n-1}) = j - i)\, Q_i(g_{n-1}) \\
&= \mathrm{e}^{-\lambda g_n} \lambda^j \sum_{i=n}^{j} q_{i,n-1} \frac{(g_n - g_{n-1})^{j-i}}{(j-i)!}
\end{aligned}
$$

for $j > \bar{n}$ and from $Q_j(g_n) = 0$ for $0 \leq j \leq \bar{n}$ since $g(g_n) \leq \bar{n}$. Note that in case $g_n = g_{n-1}$ formula (11.17) is still valid and yields $q_{j,n} = q_{j,n-1}$. ∎

11.2.3 Two boundary case

We denote by

$$
R_n(t) = \mathrm{P}(N(t) = n,\, \tau_{g,f} > t), \quad t \geq 0,\, n \geq 0
$$

the probability that a Poisson sample path has the value n at time t and crosses neither the upper boundary f nor the lower boundary g on $[0, t]$. The corresponding probability in the upper boundary case has been calculated on decomposing the event $\{N(t) = n,\, \tau^f > t\}$ according to the value of $N(f_n)$ since no crossing of f on $(f_n, t]$ is possible provided that f has not been crossed on $[0, f_n]$ and that $\{N(t) = n\}$ occurs. As some points g_j may lay in $(f_n, t]$, this is no longer true in the two boundary case. Therefore, we denote the ordered union of the points f_n and g_n by $(h_n)_{n \geq 0}$ and define for all $n \geq 0$ and $t \geq 0$ the point $h_{k(n,t)}$ in this sequence by

$$
h_{k(n,t)} = \max(f_n, g_{m(t)}), \qquad n \geq 0,\, t \geq 0.
$$

Although this definition does not uniquely determine the index $k(n, t)$, the point $h_{k(n,t)}$ is yet well defined. Hence, there are no points g_j between $h_{k(n,t)}$ and t

so that $R_n(t)$ can be calculated on decomposing the event $\{N(t) = n, \tau_{g,f} > t\}$ according to the value of $N(h_{k(n,t)})$ as done in Theorem 11.2.3. Then, the probability that neither f nor g are crossed on $[0, t]$ is given by

$$P(\tau_{g,f} > t) = \sum_{n=m(t)+1}^{n(t)} R_n(t). \qquad (11.18)$$

Theorem 11.2.3 *Let* $t \in [0, \infty)$. *Then*

$$R_0(t) = I_{\{0 \le t < g_0\}} e^{-\lambda t} \qquad (11.19)$$

and

$$R_n(t) = I_{\{f_n < t < g_n\}} e^{-\lambda t} \lambda^n \sum_{j=0}^{n} r_{j,k(n,t)} \frac{(t - h_{k(n,t)})^{n-j}}{(n-j)!}, \quad n \ge 1 \qquad (11.20)$$

where the constants $r_{j,n}$ *are recursively defined for all* $n \ge 0$ *by*

$$r_{0,n} = I_{\{g(h_n) < 0 < f(h_n)\}} \qquad (11.21)$$

and

$$r_{j,n} = I_{\{g(h_n) < j < f(h_n)\}} \sum_{i=0}^{j} r_{i,k(j,h_n)} \frac{(h_n - h_{k(j,h_n)})^{j-i}}{(j-i)!}, \quad j \ge 1. \qquad (11.22)$$

PROOF. Formula (11.19) is clear by the definition of g_0 and the positivity of f. With the settings $r_n(t) = e^{\lambda t} \lambda^{-n} R_n(t)$, $t \ge 0$ and $r_{j,n} = r_j(h_n)$, $j, n \ge 0$ we have for $n \ge 1$

$$
\begin{aligned}
R_n(t) &= \sum_{j=0}^{n} P(N(h_{k(n,t)}) = j, \ \tau_{g,f} > h_{k(n,t)}, \ N(t) = n) \\
&= \sum_{j=0}^{n} P(N(t - h_{k(n,t)}) = n - j) \, R_j(h_{k(n,t)}) \\
&= e^{-\lambda t} \lambda^n \sum_{j=0}^{n} r_{j,k(n,t)} \frac{(t - h_{k(n,t)})^{n-j}}{(n-j)!}
\end{aligned}
$$

for $f_n < t < g_n$ and $R_n(t) = 0$ otherwise. By analogy,

$$
\begin{aligned}
R_j(h_n) &= I_{\{g(h_n) < j < f(h_n)\}} \sum_{i=0}^{j} P(N(h_n - h_{k(j,h_n)}) = j - i) \, R_i(h_{k(j,h_n)}) \\
&= I_{\{g(h_n) < j < f(h_n)\}} \, e^{-\lambda h_n} \lambda^j \sum_{i=0}^{j} r_{i,k(j,h_n)} \frac{(h_n - h_{k(j,h_n)})^{j-i}}{(j-i)!}
\end{aligned}
$$

for $j \geq 1$ which implies the recursion formula (11.22). Finally, formula (11.21) is obvious. ∎

Analyzing the equations (11.18) and (11.20) shows that the distribution of $\tau_{g,f}$ is a mixture of an absolutely continuous and a discrete distribution with the discrete part probabilities

$$P(\tau_{g,f} = g_l) = e^{-\lambda g_l} \sum_{n=\underline{l}}^{\bar{l}} \lambda^n \sum_{j=0}^{n} r_{j,k(n,g_l)} \frac{(g_l - h_{k(n,g_l)})^{n-j}}{(n-j)!}, \quad l \geq 0.$$

11.3   The Nonhomogeneous Poisson Process

Let $\{Y(t); t \geq 0\}$ be a nonhomogeneous Poisson process with mean value function $\Lambda(t)$, i.e., $E(Y(t)) = \Lambda(t)$, $t \geq 0$, which is assumed to be nondecreasing and right continuous with $\Lambda(0) = 0$ and $\lim_{t \to \infty} \Lambda(t) = \infty$. Hence,

$$Y(t) = N_1(\Lambda(t)), \quad t \geq 0$$

where $\{N_1(t); t \geq 0\}$ is a homogeneous Poisson process with unit intensity. We will deal with first passage time problems of $\{Y(t)\}$ by transforming them to equivalent ones of $\{N_1(t)\}$. Given boundaries f and g which are defined as in Section 11.2, we denote with $\tau(Y)$ the first passage times of the nonhomogeneous Poisson process according to the definition (11.2). We need two versions of the pseudo inverse function of Λ which differ only when Λ has flat sections. Define

$$\Lambda^-(s) = \inf\{t \geq 0 : \Lambda(t) \geq s\}, \quad s \geq 0$$

and

$$\Lambda^+(s) = \sup\{t \geq 0 : \Lambda(t) \leq s\}, \quad s \geq 0.$$

Clearly, Λ^- is left continuous and Λ^+ is right continuous. By the monotonicity of the sample paths and the boundaries we have in the upper boundary case

$$\begin{aligned} P(\tau^f(Y) > t) &= P(Y(u) < f(u) \text{ for } 0 \leq u \leq t) \\ &= P(N_1(s) < f(\Lambda^-(s)) \text{ for } 0 \leq s \leq \Lambda(t)) \\ &= P(\tau^{\tilde{f}}(N_1) > \Lambda(t)), \quad t \geq 0 \end{aligned}$$

where $\tau^{\tilde{f}}(N_1)$ is the first passage time of $\{N_1(t)\}$ with respect to the upper boundary $\tilde{f}(s) = f(\Lambda^-(s))$, $s \geq 0$. Notice that

$$\{N_1(\Lambda(u)) < f(u) \; \forall u \in [t_1, t_2)\} = \{N_1(\Lambda(t_1)) < f(t_1)\}$$

if Λ is constant on $[t_1, t_2)$ and that

$$\{N_1(\Lambda(t)) < f(t)\} = \{N_1(s) < f(\Lambda^-(s)) \quad \forall s \in [\Lambda(t-0), \Lambda(t)]\}$$

if Λ has a jump at t. In the same way one can easily deduce

$$P(\tau_g(Y) > t) = P(\tau_{\tilde{g}}(N_1) > \Lambda(t)), \quad t \geq 0$$

and

$$P(\tau_{g,f}(Y) > t) = P(\tau_{\tilde{g},\tilde{f}}(N_1) > \Lambda(t)), \quad t \geq 0$$

where $\tilde{g}(s) = g(\Lambda^+(s))$, $s \geq 0$ is the transformed lower boundary. If Λ has flat sections then the transformed boundaries will possess jumps, whereas jumps of the mean value function imply flat sections of $\tilde{f}$ and $\tilde{g}$. Since the transformed boundaries satisfy the conditions set up in Section 11.2, the boundary crossing probabilities of $\{Y(t)\}$ can be calculated by the formulae given in Theorems 11.2.1–11.2.3.

We remark that, from a technical point of view, there is no need for the assumption $\Lambda(0) = 0$, since the above results remain valid in the case $\Lambda(0) > 0$ if we extend the defintion of Λ^+ by $\Lambda^+(s) = 0$ for all $0 \leq s < \Lambda(0)$.

11.4   A Special Mixed Poisson Process

A mixed Poisson process $\{X(t); t \geq 0\}$ is a conditionally Poisson process with mean value function Λt, $t \geq 0$, where Λ is a nonnegative random variable. It can be written as

$$X(t) = N_1(\Lambda t), \qquad t \geq 0$$

where $\{N_1(t); t \geq 0\}$ is a homogeneous Poisson process with unit intensity being independent of Λ. The distribution of a first passage time $\tau(X)$ of $\{X(t)\}$ can be calculated by conditioning on Λ. If $\tau(N_\lambda)$ denotes the first passage time of a homogeneous Poisson process $\{N_\lambda(t); t \geq 0\}$ with intensity λ we have immediately

$$P(\tau(X) > t) = \int_0^\infty P(\tau(N_\lambda) > t) \, dP(\Lambda \leq \lambda), \qquad t \geq 0. \tag{11.23}$$

For example let us consider the Pólya-Lundberg process, a mixed Poisson process with random intensity Λ following a gamma distribution $\mathbf{Ga}(\alpha, \gamma)$, i.e., Λ has the probability density function

$$f_\Lambda(x) = \frac{\gamma^\alpha}{\Gamma(\alpha)} x^{\alpha-1} e^{-\gamma x}, \qquad x \geq 0.$$

It is well known [see Schmidt (1996, p. 96)] that the Pólya-Lundberg process is a nonhomogeneous birth process with stationary dependent increments and intensities

$$q_{n,n+1}(t) = \frac{\alpha + n}{\gamma + t}, \qquad n \in \mathbf{N}, t \geq 0.$$

The distribution of $\tau(X)$ can be computed by formula (11.23). In the upper and the two boundary case the probabilities $P(\tau(N_\lambda) > t)$ have the structure

$$P(\tau(N_\lambda) > t) = e^{-\lambda t} \sum_n \lambda^n w_n(t)$$

where the functions $w_n(t)$ do not depend on λ. Hence,

$$
\begin{aligned}
P(\tau(X) > t) &= \sum_n w_n(t) \int_0^\infty \frac{\gamma^\alpha}{\Gamma(\alpha)} \lambda^{n+\alpha-1} e^{-\lambda(t+\gamma)} \, d\lambda \\
&= \sum_n w_n(t) \frac{\gamma^\alpha}{\Gamma(\alpha)} \frac{\Gamma(\alpha+n)}{(t+\gamma)^{\alpha+n}} \\
&= \gamma^\alpha \sum_n \frac{w_n(t)}{(t+\gamma)^{\alpha+n}} \prod_{i=0}^{n-1} (\alpha + i).
\end{aligned}
$$

The lower boundary case can be treated analogously.

References

1. Baxter, G. and Donsker, M. (1957). On the distribution of the supremum functional for processes with stationary independent increments, *Transactions of the American Mathematical Society*, **85**, 73–87.

2. Borovkov, A. A. (1965). On the first passage time for one class of processes with independent increments *Theory of Probability and its Applications*, **10**, 331–334.

3. Daniels, H. E. (1963). The Poisson process with curved absorbing boundary, *Bulletin of the ISI, 34th Session, Ottawa*, 994–1008.

4. Durbin, J. (1971). Boundary-crossing probabilities for the Brownian motion and Poisson processes and techniques for computing the power of the Kolmogorov-Smirnov test, *Journal of Applied Probability*, **8**, 431–453.

5. Durbin, J. (1985). The first-passage density of a continuous Gaussian process to a general boundary, *Journal of Applied Probability*, **22**, 99–122.

6. Franz, J. (1977). Niveaudurchgangszeiten zur Charakterisierung sequentieller Schätzverfahren, *Mathematische Operationsforschung und Statistik, Series Statistics*, **8**, 499–510.

7. Gallot, S. F. L. (1966). Asymptotic absorption probabilities for a Poisson process, *Journal of Applied Probability*, **3**, 445–452.

8. Gnedenko, B. V., Belyayev, Yu. K. and Solovyev, A. D. (1969). *Mathematical Methods of Reliability Theory* New York and London: Academic Press.

9. Lisek, B. and Hochschildt, J. (1983). *Sequentielle Zuverlässigkeitsprüfung*, Leipzig: Teubner.

10. Picard, Ph. and Lefèvre, C. (1996). First crossing of basic counting processes with non-linear boundaries: A unified approach through pseudopolynomials (I), *Advances in Applied Probability*, **28**, 853–876.

11. Prabhu, N. U. (1980). *Stochastic Storage Processes: Queues, Insurance Risk and Dams*, New York: Springer-Verlag.

12. Pyke, R. (1959). The supremum and the infimum of the Poisson process, *Annals of Mathematical Statistics*, **30**, 568–576.

13. Schmidt, K.D. (1996). *Lectures on Risk Theory*, Stuttgart: Teubner.

14. Siegmund, D. (1986). Boundary crossing probabilities and statistical applications, *Annals of Statistics*, **14**, 361–404.

15. Stadje, W. (1993). Distribution of first-exit times for empirical counting and Poisson processes with moving boundaries, *Communications in Statistics–Stochastic Models*, **9**, 91–103.

16. Whittle, P. (1961). Some exact results for one-sided distribution tests of the Kolmogorov-Smirnov type, *Annals of Mathematical Statistics*, **32**, 499–505.

17. Zacks, S. (1991). Distributions of stopping times for Poisson processes with linear boundaries, *Communications in Statistics–Stochastic Models*, **7**, 233–242.

12

Optimal Sequential Estimation for Markov-Additive Processes

Ryszard Magiera

Technical University of Wrocław, Wrocław, Poland

Abstract: The problem of estimating the parameters of a Markov-additive process from data observed up to a random stopping time is considered. Markov-additive processes are a class of Markov processes which have important applications to queueing and data communication models. They have been used to model queueing-reliability systems, arrival processes in telecommunication networks, environmental data, neural impulses etc. The problem of obtaining optimal sequential estimation procedures, i.e., optimal stopping times and the corresponding estimators, in estimating functions of the unknown parameters of Markov-additive processes is considered. The parametric functions and sequential procedures which admit minimum variance unbiased estimators are characterized. In the main, the problem of finding optimal sequential procedures is considered in the case where the loss incurred is due not only to the error of estimation, but also to the cost of observing the process. Using a weighted squared error loss and assuming the cost is a function of the additive component of a Markov-additive process (for example, the cost depending on arrivals at a queueing system up to the moment of stopping), a class of minimax sequential procedures is derived for estimating the ratios between transition intensities of the embedded Markov chain and the mean value parameter of the additive part of the Markov-additive process considered.

Keywords and phrases: Sequential estimation procedure, minimax estimation, efficient estimation, stopping time, Markov-additive process

12.1   Introduction

Denote $I = \{1, 2, \ldots, m\}$. In accordance with the definition given in Barndorff-Nielsen (1978), a process $(A(t), X(t)), t \geq 0$, with the continuous time parameter t, on the state space $R \times I$ is said to be a Markov-additive process if

1. $(A(t), X(t)), t \geq 0$, is a Markov process;

2. the conditional distribution of $(A(s+t) - A(s), X(s+t))$, given $(A(s), X(s))$, depends only on $X(s)$.

The basic theory of Markov-additive processes is presented in Çinlar (1972), Ezhov and Skorohod (1969), and Prabhu (1991), where a more general state space for $X(t)$ is considered. Recall [see Pacheco and Prabhu (1995)] the most important properties of these processes. The transition probability measure of a Markov-additive process is given by

$$P(A(s + t) \in B, X(s + t) = j | A(s) = y, X(s) = i)$$
$$= P(A(s + t) - A(s) \in B - y, X(s + t) = j | X(s) = i) \qquad (12.1)$$

for $s, t > 0$, $i, j \in I$, $y \in R$ and $B \in \mathcal{B}_R$, where $\mathcal{B}_R$ is the σ-algebra of Borel subsets of R. Since $(A(t), X(t)), t \geq 0$, is Markov, it follows from (12.1) that the component $X(t)$ is Markov and that $A(t)$ has conditionally independent increments, given $X(t)$. For $0 \leq t_1 \leq t_2 \leq \ldots \leq t_n, n > 2$, the increments

$$A(t_1) - A(0), A(t_2) - A(t_1), \ldots, A(t_n) - A(t_{n-1})$$

are conditionally independent given $X(0), X(t_1), \ldots, X(t_n)$. The process $A(t)$, which is, in general, not Markovian is called the additive component of the Markov-additive process. Markov renewal processes are discrete time versions of Markov-additive processes with the additive component taking values in R_+.

A particularly important class of Markov-additive processes is the class of Markov-additive processes $(A(t), X(t))$ with the additive component $A(t)$ taking values in the set of nonnegative integers. The increments of $A(t)$ can be related to events. A typical example is that of arrivals at a queueing system. In the case when $A(t)$ has the state space $\{1, 2, \ldots\}$, the process $(A(t), X(t))$ is called a Markov-additive process of arrivals. Markov-additive processes of arrivals are studied in Pacheco and Prabhu (1995). An important special case of the Markov-additive process of arrivals is the Markov-modulated Poisson process in which the rate λ of occurrence of Poisson events changes instantaneously with the change of state in $X(t)$.

Markov-additive processes of arrivals are of great importance in applied probability. They appear in modelling more complex systems, especially in queueing and data communication models. A survey of possible applications of Markov-additive processes of arrivals one can find in Pacheco and Prabhu (1995). For example, these processes have been used to model overflow from trunk groups, superpositioning of packeted voice streams, and input to ATM (Asynchronous Transfer Mode) networks, which will be used in high-speed communication networks. Using Markov-additive processes, some theoretical queueing results were established in Neuts (1992).

Although there is a vast literature on Markov-additive processes and their use in modelling many stochastic phenomena, the statistical issues for these processes have not been the subject of so much study. Nonparametric and Bayesian estimation for Markov renewal processes was considered in Gill (1980), Phelan (1990b) and Phelan (1990b). Some problems of parameter estimation for Markov-modulated Poisson processes were treated in Rydén (1994). Asymptotic normality of sequential and nonsequential maximum likelihood estimators for Markov renewal and Markov-additive processes was established in Stefanov (1995). A sequential estimation scheme based on "small" samples was considered in Fygenson (1991), where efficient (optimal in the sense of the Cramér-Rao-Wolfowitz lower bound) sequential procedures were investigated for Markov renewal processes.

We consider the problem of obtaining optimal sequential estimation procedures, i.e., optimal stopping times and the corresponding estimators, in estimating functions of the unknown parameters of Markov-additive processes. We characterize the parametric functions and sequential procedures which admit minimum variance unbiased estimators. In the main, the problem of finding optimal sequential procedures is considered in the case where the loss incurred is due not only to the error of estimation, but also to the cost of observing the process (for example, to the cost depending on arrivals at a queueing system up to the moment of stopping). Under a weighted squared error loss, a class of minimax sequential procedures is derived for estimating the values of certain functions of the unknown parameters. The procedures obtained are optimal in the class of all sequential procedures having finite risk. It is shown that for each underlying class of functions values of which are to be estimated, the minimax sequential procedures are obtainable explicitly under certain conditions on the distributions of the additive component of the Markov-additive process considered. In particular, the results presented are applicable for the Markov-additive processes of arrivals most frequently involved in the literature, i.e., for the Markov-modulated Poisson processes.

12.2   The Model and Sampling Times

Let $\lambda = (\lambda_{ij})_{i,j=1}^m$ be the transition intensity matrix of the embedded m-state Markov chain $X(t)$. We assume that the conditional distribution of $A(t) - A(s)$, given $X(u) = i$ for all $u \in [s, t]$, is given by the density

$$\exp[v_i x - f_i(v_i)(t-s)] \tag{12.2}$$

with respect to a σ-finite measure which may depend on the state i in general, and v_i is a real parameter, $v_i \in \Upsilon_i \subseteq R$. Υ_i is assumed to be the interior of the natural parameter space of the respective exponential family given by (12.2). It is assumed that $X(0) = 1$ with probability 1.

The likelihood function corresponding to the observation of the process up to time t has the following form [see Stefanov (1995)]:

$$L(t, \lambda, v) = \prod_{\substack{i,j=1 \\ i \neq j}}^m \lambda_{ij}^{N_{ij}(t)} \exp[-S_i(t)\lambda_{ij}] \prod_{i=1}^m \exp[A_i(t)v_i - S_i(t)f_i(v_i)]$$

$$= \prod_{\substack{i,j=1 \\ i \neq j}}^m \lambda_{ij}^{N_{ij}(t)} \prod_{i=1}^m \exp\left\{ A_i(t)v_i - S_i(t)[\lambda_{ii} + f_i(v_i)] \right\}, \tag{12.3}$$

where $v = (v_1, \ldots, v_m)$, $N_{ij}(t)$ is the number of the transitions from state i to state j of the Markov process $X(s)$ in the time interval $[0, t]$, $S_i(t)$ is the sojourn time in state i of the process $X(s)$, and

$$A_i(t) = \sum_{n=1}^\infty \left\{ [A(\eta_n^*(i)) - A(\eta_n(i))] \mathbf{1}_{[0,\infty)}[t - \eta_n^*(i)] \right.$$

$$\left. + [A(t) - A(\eta_n(i))] \mathbf{1}_{(\eta_n(i), \eta_n^*(i))}(t) \right\},$$

where $\eta_n(i)$ is the n-th consecutive time of first entrance of $X(s)$ to state i, and $\eta_n^*(i)$ is the time of exit from state i after $\eta_n(i)$.

By the fundamental identity of sequential analysis, for any finite stopping time τ the sequential version of (12.3) has the following form:

$$L(\tau, \lambda, v) = \prod_{\substack{i,j=1 \\ i \neq j}}^m \lambda_{ij}^{N_{ij}(\tau)} \prod_{i=1}^m \exp\left\{ A_i(\tau)v_i - S_i(\tau)[\lambda_{ii} + f_i(v_i)] \right\} \tag{12.4}$$

$$= \exp\left\{ \sum_{\substack{i,j=1 \\ i \neq j}}^m N_{ij}(\tau) \log \lambda_{ij} + \sum_{i=1}^m A_i(\tau)v_i - \sum_{i=1}^m S_i(\tau)[\lambda_{ii} + f_i(v_i)] \right\}.$$

From a theoretical point of view, namely, in view of among others the possibility of exploiting the very useful tools associated with exponential families, as well as in view of possible applicability, the most relevant stopping times which can be taken into account in searching for optimal sequential estimation procedures are the following ones defined by

$$
\begin{aligned}
\tau^i_{J,s} &= \inf\{t : \sum_{j\in J} N_{ji}(t) = s\}, \ s = 1,2,\ldots; \\
\tau^i_{i,s} &= \inf\{t : S_i(t) = s\}, \ s > 0; \\
\tilde{\tau}^i_{i,s} &= \inf\{t : A_i(t) = s\}; \\
\tau^i_{iJ,s} &= \inf\{t : A_i(t) + \sum_{j\in J} N_{ji}(t) = s\}, \ s = 1,2,\ldots,
\end{aligned}
$$

for each $i \in I$ and each $J \subseteq I$ such that $\sum_{j\in J} \lambda_{ji} > 0$. It is additionally assumed that $P_{\lambda,v}(\tilde{\tau}^i_{i,s} < \infty) = 1$ and $P_{\lambda,v}(\tau^i_{iJ,s} < \infty) = 1$ for each (λ, v). This class of stopping times was described by Stefanov (1995). By his Proposition 5.1, for each stopping time from this class the curved (in general) exponential family of (12.4) becomes a noncurved exponential one of order equal to the dimension of the parameter (λ, v). The condition $P_{\lambda,v}(\tilde{\tau}^i_{i,s} < \infty) = 1$ is satisfied, for example, if the process $A_i(t)$ is nonnegative with continuous trajectories and $A_i(t) \to \infty$ as $t \to \infty$, or if the process $A_i(t)$ is a Poisson process; in the latter case s must be a natural number [see Remark 5.1 in Stefanov (1995)]. The condition $P_{\lambda,v}(\tau^i_{iJ,s} < \infty) = 1$ is satisfied, for example, in the case of Markov-modulated Poisson processes. The stopping times $\tilde{\tau}^i_{i,s}$ and $\tau^i_{iJ,s}$ are new in comparison to those relevant in sequential estimation problems for finite-state Markov processes. Stopping times of such a type will be considered in the next sections.

12.3 Efficient Sequential Procedures

To prove optimality in the sense of the Cramér-Rao-Wolfowitz lower bound of the stopping times presented in the previous section one can use the methods and results established in Stefanov (1986) and Magiera and Stefanov (1989). Therefore, only some basic facts in the method of deriving efficient sequential procedures will be presented and the proofs will be omitted.

Let $\delta = (\tau, d(\tau))$ be a sequential procedure satisfying standard regularity conditions and let $h(\vartheta)$ be a real valued function of the unknown parameter $\vartheta \in \Theta \subseteq R^n$ of the process observed. Then as a special case of information inequalities in general statistical structures [see, for example, the information inequalities considered in Franz and Magiera (1997)], for any unbiased estimator $d(\tau)$ of the function $h(\vartheta)$ the following information inequality holds:

$$
\mathrm{Var}_\vartheta(d(\tau)) \geq [\nabla h(\vartheta)]^T \mathcal{I}_\tau^{-1}(\vartheta) \nabla h(\vartheta),
$$

where $\nabla h(\vartheta) = (\partial h(\vartheta)/\partial \vartheta_1, \ldots, \partial h(\vartheta)/\partial \vartheta_n)^T = (h'_1(\vartheta), \ldots, h'_n(\vartheta))^T$ (T denotes transposition) and

$$\mathcal{I}_\tau(\vartheta) = E_\vartheta \left(\frac{\partial \log L(\tau, \vartheta)}{\partial \vartheta_i} \frac{\partial \log L(\tau, \vartheta)}{\partial \vartheta_j} \right)^n_{i,j=1} = -E_\vartheta \left(\frac{\partial^2 \log L(\tau, \vartheta)}{\partial \vartheta_i \partial \vartheta_j} \right)^n_{i,j=1}$$

is the Fisher information matrix on ϑ at the stopping time τ.

For the Markov-additive process the unknown parameter is $\vartheta = (\lambda, \upsilon)$, where $\lambda = (\lambda_{12}, \ldots, \lambda_{1m}, \ldots, \lambda_{m1}, \ldots, \lambda_{m,m-1})$, $\upsilon = (\upsilon_1, \ldots, \upsilon_m)$. Since

$$\frac{\partial^2 \log L(\tau, \lambda, \upsilon)}{\partial \lambda_{ij} \lambda_{kl}} = \begin{cases} -N_{ij}(\tau)/\lambda_{ij}^2 \,, & k = i, l = j, \\ 0 \,, & k \neq i \text{ or } l \neq j, \end{cases}$$

$$\frac{\partial^2 \log L(\tau, \lambda, \upsilon)}{\partial \upsilon_i \upsilon_j} = \begin{cases} -f''_i(\upsilon_i) S_i(\tau) \,, & j = i, \\ 0 \,, & j \neq i, \end{cases}$$

the Fisher information matrix is diagonal and the information inequality can be easily shown to take the form

$$\mathrm{Var}_\vartheta(d(\tau)) \geq \sum_{\substack{i,j=1 \\ i \neq j}}^m \frac{\lambda_{ij}^2 (h'_{ij})^2}{E_{\lambda,\upsilon} N_{ij}(\tau)} + \sum_{i=1}^m \frac{(h'_i)^2}{f''_i(\upsilon_i) E_{\lambda,\upsilon} S_i(\tau)}$$

$$= \sum_{i=1}^m \frac{1}{E_{\lambda,\upsilon} S_i(\tau)} \Big[\sum_{j=1, j\neq i}^m \lambda_{ij}(h'_{ij})^2 + \frac{(h'_i)^2}{f''_i(\upsilon_i)} \Big], \qquad (12.5)$$

where $h'_{ij} = \partial h/\partial \lambda_{ij}$ and $h'_i = \partial h/\partial \upsilon_i$. For example, for any unbiased sequential estimator $d(\tau)$ of the ratio $\lambda_{ij}/f'_i(\upsilon_i)$ we have $\mathrm{Var}_\vartheta(d(\tau)) \geq \lambda_{ij}[(f'_i(\upsilon_i))^2 + \lambda_{ij} f'_i(\upsilon_i)] / (f'_i(\upsilon_i))^4 E_{\lambda,\upsilon} S_i(\tau)$.

Inequality (12.5) becomes an equality at a particular value of (λ, υ) if and only if

$$d(\tau) = \sum_{i=1}^m \frac{1}{E_{\lambda,\upsilon} S_i(\tau)} \Big\{ \sum_{j=1, j\neq i}^m h'_{ij} [N_{ij}(\tau) - \lambda_{ij} S_i(\tau)]$$

$$+ \frac{h'_i}{f''_i(\upsilon_i)} [A_i(\tau) - f'_i(\upsilon_i) S_i(\tau)] \Big\} + h(\lambda, \upsilon)$$

a.s. Hence efficient sequential estimators are characterized as follows: if under the sampling rule τ a nonconstant estimator $d(\tau)$ is efficient for $E_{\lambda,\upsilon} d(\tau) = h(\lambda, \upsilon)$ at (λ, υ), then there exist constants a_{ij}, b_k, c_k, $i, j, k = 1, \ldots, m$; $i \neq j$, not all zero, and a constant c such that

$$d(\tau) = \sum_{\substack{i,j=1 \\ i\neq j}}^m a_{ij} N_{ij}(\tau) + \sum_{k=1}^m [b_k S_k(\tau) + c_k A_k(\tau)] + c$$

a.s.

Suppose that h is a function only of $\bar{\lambda}_i = (\lambda_i, v_i)$ (i fixed), where $\lambda_i = (\lambda_{i1}, \ldots, \lambda_{i,i-1}, \lambda_{i,i+1}, \ldots, \lambda_{im})$. If under the sampling rule τ there exists a nonconstant estimator $d(\tau)$ which is efficient for the function $h(\bar{\lambda}_i)$, then there exist constants $\alpha_{i1}, \ldots, \alpha_{i,i-1}, \alpha_{i,i+1}, \ldots, \alpha_{im}, \beta_i, \gamma_i$, not all zero, and $\kappa \neq 0$ such that

$$\sum_{j=1,j\neq i}^{m} \alpha_{ij} N_{ij}(\tau) + \beta_i S_i(\tau) + \gamma_i A_i(\tau) = \kappa$$

a.s. All efficiently estimable functions are of the form

$$E_{\bar{\lambda}_i}\Big[\sum_{j=1,j\neq i}^{m} a_{ij} N_{ij}(\tau) + b_i S_i(\tau) + c_i A_i(\tau) + c \Big]$$

$$= \kappa \frac{\sum_{j=1,j\neq i}^{m} a_{ij}\lambda_{ij} + b_i + c_i f_i'(v_i)}{\sum_{j=1,j\neq i}^{m} \alpha_{ij}\lambda_{ij} + \beta_i + \gamma_i f_i'(v_i)} + c.$$

In particular, for the sequential procedure determined by the stopping time

$$\tau_{a_i} = \inf\{t : A_i(t) = a_i\}$$

(which is assumed to be finite for each $\bar{\lambda}_i$), the efficiently estimable function is of the form

$$h(\bar{\lambda}_i) = \frac{a_i}{f_i'(v_i)}\Big(\sum_{j=1,j\neq i}^{m} a_{ij}\lambda_{ij} + b_i \Big) + \bar{c}_i,$$

and

$$d(\tau_{a_i}) = \sum_{j=1,j\neq i}^{m} a_{ij} N_{ij}(\tau_{a_i}) + b_i S_i(\tau_{a_i}) + \bar{c}_i$$

is its only efficient estimator. In the next section we shall show that the stopping time τ_{a_i} determines a minimax procedure.

Analogously, for the sequential procedure defined by

$$\bar{\tau}_{l_i} = \inf\Big\{ t : A_i(t) + \sum_{j=1,j\neq i}^{k_i} N_{i,\sigma_i(j)}(t) = l_i \Big\}, \quad l_i = 1, 2, \ldots,$$

where $(\sigma_i(1), \ldots, \sigma_i(i-1), \sigma_i(i+1), \ldots, \sigma_i(m))$ is a permutation of $(1, \ldots, i-1, i+1, \ldots, m)$ and k_i is an integer $2 \leq k_i \leq m$ (we recall that the finiteness of $\bar{\tau}_{l_i}$ is assumed), we obtain that

$$h(\bar{\lambda}_i) = l_i \frac{\sum_{j=1,j\neq i}^{k_i} a_{ij}\lambda_{i,\sigma_i(j)} + b_i + c_i f_i'(v_i)}{\sum_{j=1,j\neq i}^{k_i} \lambda_{i,\sigma_i(j)}} + c$$

is efficiently estimable, and

$$d(\bar{\tau}_{l_i}) = \sum_{j=1, j\neq i}^{k_i} a_{ij} N_{i,\sigma_i(j)}(\bar{\tau}_{l_i}) + b_i S_i(\bar{\tau}_{l_i}) + c_i A_i(\bar{\tau}_{l_i}) + c$$

is its only efficient estimator.

The remaining two sequential procedures (a special case of the first two stopping times from the class presented in the previous section), which together with the stopping times τ_{a_i} and $\bar{\tau}_{l_i}$ exhaust the class of all sequential procedures in estimating a function of a single row of the transition matrix, are the same as those obtained in Trybula (1982) for finite state Markov processes.

12.4   Minimax Sequential Procedures

The situation becomes more complex in investigations concerning the optimality of the stopping times from the class of Section 12.2 subject to minimax criterion when a cost of observation must be taken into account. Considering the problem of estimation for some special functions of the unknown parameter (λ, v) we exhibit the idea and tools for proving the minimaxity of sequential procedures determined by the stopping times from this class.

Suppose that the unknown parameter vector is $\bar{\lambda}_i = (\lambda_i, v_i)$ (i fixed), where $\lambda_i = (\lambda_{i1}, \ldots, \lambda_{i,i-1}, \lambda_{i,i+1}, \ldots, \lambda_{im})$. Then the likelihood function is

$$L_i(\tau, \bar{\lambda}_i) = \prod_{j=1, j\neq i}^{m} \lambda_{ij}^{N_{ij}(\tau)} \exp\left\{ A_i(\tau)v_i - S_i(\tau)[\lambda_{ii} + f_i(v_i)] \right\}. \qquad (12.6)$$

It will be shown that if the ratios of λ_{ij} ($j = 1, \ldots, m; j \neq i$) to $f_i'(v_i)$ are of interest, then to estimate them one can use a sequential procedure defined by the following stopping time

$$\tau_{a_i} = \inf\{t : A_i(t) = a_i\}, \quad a_i > 0, \qquad (12.7)$$

for which it is assumed that $P_{\bar{\lambda}_i}(\tau_{a_i} < \infty) = 1$ for every $\bar{\lambda}_i \in \Lambda_i \times \Upsilon_i, \Lambda_i = (0, \infty)^{m-1}$. Assume that $f_i'(v_i) > 0$ for each $v_i \in \Upsilon_i$. Since formula (12.2) implies that the derivative $f_i'(v_i)$ is the mean value parameter of the increments for $A(t)$, this assumption corresponds to the natural case of positive increments (arrivals) of the additive component $A(t)$. We will then show that the stopping time of (12.7) determines a minimax sequential procedure under a weighted squared error loss and the cost depending on the value of the process $A_i(t)$ at the moment of stopping. In particular, for a Markov-additive process of arrivals this cost will be a function of arrivals at a queueing system. For the stopping time τ_{a_i} the sequential likelihood function of (12.6) takes the following form

$$L_i(\tau_{a_i}, \bar{\lambda}_i) = \prod_{j=1, j\neq i}^{m} \lambda_{ij}^{N_{ij}(\tau_{a_i})} \exp\{a_i v_i - S_i(\tau_{a_i})[\lambda_{ii} + f_i(v_i)]\}. \tag{12.8}$$

The family of (12.8) is an exponential family in which the dimension of the sufficient statistic equals the dimension of the unknown parameter. It then follows from the well-known analytical properties of the non-curved exponential families [see Barndorff-Nielsen (1978) or Brown (1986)] that the regularity conditions which allow to differentiate twice under the integral sign with respect to the parameter $\bar{\lambda}_i$ in the identity $\int L_i(\tau_{a_i}, \bar{\lambda}_i) d\mu_{\tau_{a_i}} = 1$ are satisfied. Thus, for the stopping time τ_{a_i} the following Wald identities can be obtained:

$$E_{\bar{\lambda}_i} S_i(\tau_{a_i}) = \frac{a_i}{f_i'(v_i)}; \tag{12.9}$$

$$E_{\bar{\lambda}_i} N_{ij}(\tau_{a_i}) = \lambda_{ij} E_{\bar{\lambda}_i} S_i(\tau_{a_i}) = \frac{\lambda_{ij} a_i}{f_i'(v_i)}, \quad j \neq i; \tag{12.10}$$

$$E_{\bar{\lambda}_i}\left[S_i(\tau_{a_i}) - \frac{a_i}{f_i'(v_i)}\right]^2 = \frac{f_i''(v_i)}{(f_i'(v_i))^2} E_{\bar{\lambda}_i} S_i(\tau_{a_i}) = \frac{a_i f_i''(v_i)}{(f_i'(v_i))^3}; \tag{12.11}$$

$$E_{\bar{\lambda}_i}\left\{\left[N_{ij}(\tau_{a_i}) - \frac{\lambda_{ij} a_i}{f_i'(v_i)}\right]\left[S_i(\tau_{a_i}) - \frac{a_i}{f_i'(v_i)}\right]\right\}$$
$$= \lambda_{ij} E_{\bar{\lambda}_i}\left[S_i(\tau_{a_i}) - \frac{a_i}{f_i'(v_i)}\right]^2 = \frac{\lambda_{ij} a_i f_i''(v_i)}{(f_i'(v_i))^3}, \quad j \neq i; \tag{12.12}$$

$$E_{\bar{\lambda}_i}\left[N_{ij}(\tau_{a_i}) - \frac{\lambda_{ij} a_i}{f_i'(v_i)}\right]^2 = \frac{a_i \lambda_{ij}}{f_i'(v_i)}\left\{1 + \frac{\lambda_{ij} f_i''(v_i)}{(f_i'(v_i))^2}\right\}, \quad j \neq i. \tag{12.13}$$

In the Bayesian approach to the problem of finding optimal estimation procedures, it is relevant to give certain characterization of the prior distributions on $\Lambda_i \times \Upsilon_i$, which should be conjugate to the family of (12.6). Suppose that there exists a number $\rho_{0,i} > 0$ such that the following conditions

$(i1)$ $\quad \int_{\Upsilon_i} f_i''(v_i)(f_i'(v_i))^2 \exp[\rho_i v_i - \beta_i f_i(v_i)] dv_i < \infty,$

$(i2)$ $\quad \int_{\Upsilon_i} (f_i'(v_i))^{-1} \exp[\rho_i v_i - \beta_i f_i(v_i)] dv_i < \infty,$

$(i3)$ $\quad \int_{\Upsilon_i} \frac{d}{dv_i}\left\{[f_i''(v_i) + f_i'(v_i)(\rho_i - \beta_i f_i'(v_i))] \exp[\rho_i v_i - \beta_i f_i(v_i)]\right\} dv_i = 0,$

are satisfied for every $\rho_i > \rho_{0,i}$ and $\beta_i > 0$.

The natural prior distribution density of the parameter $\bar{\lambda}_i$ is proportional to

$$g(\bar{\lambda}_i; r_i, \alpha_i) := \prod_{j=1, j \neq i}^{m} \lambda_{ij}^{r_{ij}-1} \exp(-r_{ii}\lambda_{ij}) f_i'(v_i) \exp\left[\alpha_i v_i - r_{ii} f_i(v_i)\right], \quad (12.14)$$

$r_i = (r_{i1}, \ldots, r_{im})$, and it is proper for all $r_{ij} > 0, j = 1, \ldots, m$ and each $\alpha_i > 0$ [the integral of (12.14) with respect to dv_i equals α_i/r_{ii}; for results relative to conjugate priors for exponential families of processes see Magiera and Wilczyński (1991)]. Modified priors of the parameter $\bar{\lambda}_i$ will be considered which are defined according to the following density

$$C(r_i, \alpha_i) w(\bar{\lambda}_i) g(\bar{\lambda}_i; r_i, \alpha_i), \quad (12.15)$$

where

$$w(\bar{\lambda}_i) := \frac{f_i''(v_i) \left(\sum_{j=1, j \neq i}^{m} \lambda_{ij}^2 + 1\right) + (f_i'(v_i))^2 \lambda_{ii}}{(f_i'(v_i))^3}.$$

It is easy to see that under condition $(i1)$ there exists the norming constant $C(r_i, \alpha_i)$ such that formula (12.15) represents a probability distribution for all $r_{ij} > 0, j = 1, \ldots, m$ and each $\alpha_i > \rho_{0,i}$. Conditions $(i2)$ and $(i3)$ are needed to derive finite posterior expected loss under the weighted squared error defined below. Let us remark that the function $w(\bar{\lambda}_i)$ equals the sum of variances $\sum_{j=1, j \neq i}^{m} \text{Var}_{\bar{\lambda}_i} N_{ij}(\tau_{a_i}) + \text{Var}_{\bar{\lambda}_i} S_i(\tau_{a_i})$ renormalized relative to the observed value of the process $A_i(t)$ at the moment of stopping. Moreover, suppose that

$$(i4) \qquad \sup_{\bar{\lambda}_i \in \Lambda_i \times \Upsilon_i} \frac{f_i'(v_i) \left(\sum_{j=1, j \neq i}^{m} \lambda_{ij}^2 + 1\right)}{f_i''(v_i) \left(\sum_{j=1, j \neq i}^{m} \lambda_{ij}^2 + 1\right) + (f_i'(v_i))^2 \lambda_{ii}} = \frac{1}{\rho_{0,i}}.$$

Remark that conditions $(i1) - (i4)$ determine a class of the additive components $A(t)$ of the Markov-additive process.

Let the loss function be defined by

$$\mathcal{L}(\bar{\lambda}_i, d_i) = \frac{1}{w(\bar{\lambda}_i)} \left\{ \sum_{\substack{j=1 \\ j \neq i}}^{m} \left[d_{ij} - \frac{\lambda_{ij}}{f_i'(v_i)}\right]^2 + \left[d_{ii} - \frac{1}{f_i'(v_i)}\right]^2 \right\}, \quad (12.16)$$

$d_i = (d_{i1}, \ldots, d_{im})$, and let the cost function $c(\cdot)$ depend only on the value of the process $A_i(t)$ at the moment of stopping.

Proposition 12.4.1 *Suppose that conditions $(i1) - (i4)$ are satisfied. If there exists a_i^* such that*

$$\frac{1}{\rho_{0,i} + a_i^*} + c(a_i^*) = \min_{a_i} \left[\frac{1}{\rho_{0,i} + a_i} + c(a_i)\right],$$

then the sequential procedure $\delta_{a_i^*} = (\tau_{a_i^*}, d_i^0(\tau_{a_i^*}))$ with $\tau_{a_i^*}$ defined by (12.7) and

$$d_i^0(\tau_{a_i^*})$$
$$= \frac{1}{\rho_{0,i} + a_i^*} \left(N_{i,1}(\tau_{a_i^*}), \ldots, N_{i,i-1}(\tau_{a_i^*}), S_i(\tau_{a_i^*}), N_{i,i+1}(\tau_{a_i^*}), \ldots, N_{i,m}(\tau_{a_i^*}) \right)$$

is minimax.

PROOF. In the proof, the method of Theorem 1 of Wilczyński (1985) for finding minimax sequential estimation procedures will be exploited. Let $\pi_{\alpha_i}^* = \pi^*(\bar{\lambda}_i; r_i, \alpha_i)$ be the family of modified priors on $\Lambda_i \times \Upsilon_i$ defined by (12.15). Denote

$$Z_i(t) = (N_{i1}(t), \ldots, N_{i,i-1}(t), S_i(t), N_{i,i+1}(t), \ldots, N_{im}(t)).$$

Let τ be any finite stopping time with respect to $\mathcal{F}_t = \sigma\{(Z_i(s), A_i(s)), s \leq t\}, t \geq 0$. It is easy to see that the posterior probability distribution of the parameter $\bar{\lambda}_i$, given $Z_i(\tau) = z_i = (n_{i1}, \ldots, n_{i,i-1}, s_i, n_{i,i+1}, \ldots, n_{im},)$ and $A_i(\tau) = a_i$, is determined by $\pi^*(\bar{\lambda}_i; \tilde{r}_i, \tilde{\alpha}_i)$, where $\tilde{r}_i = r_i + z_i$ and $\tilde{\alpha}_i = \alpha_i + a_i$. The posterior risk is

$$\tilde{\mathcal{R}}(\pi_{\alpha_i}^*(\cdot \mid Z_i(\tau) = z_i, A_i(\tau) = a_i), d_i(z_i, a_i))$$
$$= C(\tilde{r}_i, \tilde{\alpha}_i) \int_{\Lambda_i} \int_{\Upsilon_i} \left\{ \sum_{\substack{j=1 \\ j \neq i}}^{m} \left[d_{ij} - \frac{\lambda_{ij}}{f_i'(v_i)} \right]^2 + \left[d_{ii} - \frac{1}{f_i'(v_i)} \right]^2 \right\} g(\bar{\lambda}_i; \tilde{r}_i, \tilde{\alpha}_i) d\lambda_i dv_i.$$

This risk attains its minimum if

$$d_{ij} = \frac{\int_{\Lambda_i} \int_{\Upsilon_i} (\lambda_{ij}/f_i'(v_i)) g(\bar{\lambda}_i; \tilde{r}_i, \tilde{\alpha}_i) d\lambda_i dv_i}{\int_{\Lambda_i} \int_{\Upsilon_i} g(\bar{\lambda}_i; \tilde{r}_i, \tilde{\alpha}_i) d\lambda_i dv_i}$$
$$= \frac{\tilde{r}_{ij}}{\tilde{\alpha}_i} = \frac{r_{ij} + n_{ij}}{\alpha_i + a_i} := d_{r_i, \alpha_i, ij}^*(z_i, a_i), \quad j \neq i,$$

and

$$d_{ii} = \frac{\int_{\Lambda_i} \int_{\Upsilon_i} (1/f_i'(v_i)) g(\bar{\lambda}_i; \tilde{r}_i, \tilde{\alpha}_i) d\lambda_i dv_i}{\int_{\Lambda_i} \int_{\Upsilon_i} g(\bar{\lambda}_i; \tilde{r}_i, \tilde{\alpha}_i) d\lambda_i dv_i}$$
$$= \frac{\tilde{r}_{ii}}{\tilde{\alpha}_i} = \frac{r_{ii} + s_i}{\alpha_i + a_i} := d_{r_i, \alpha_i, ii}^*(z_i, a_i).$$

The posterior risk corresponding to the estimator $d_{r_i, \alpha_i, i}^* = (d_{r_i, \alpha_i, i1}^*, \ldots, d_{r_i, \alpha_i, im}^*)$ is

$$\tilde{\mathcal{R}}(\pi_{\alpha_i}^*(\cdot \mid Z_i(\tau) = z_i, A_i(\tau) = a_i), d_{r_i, \alpha_i, i}^*(z_i, a_i))$$
$$= \frac{C(\tilde{r}_i, \tilde{\alpha}_i)}{\tilde{\alpha}_i^2} \int_{\Lambda_i} \int_{\Upsilon_i} \left\{ \sum_{\substack{j=1 \\ j \neq i}}^{m} \left[\tilde{r}_{ij} - \tilde{\alpha}_i \frac{\lambda_{ij}}{f_i'(v_i)} \right]^2 \right.$$
$$\left. + \left[\tilde{r}_{ii} - \tilde{\alpha}_i \frac{1}{f_i'(v_i)} \right]^2 \right\} g(\bar{\lambda}_i; \tilde{r}_i, \tilde{\alpha}_i) d\lambda_i dv_i.$$

Taking into account the identities

$$
\int_{\Lambda_i} \int_{\Upsilon_i} \Big[\tilde{r}_{ij} - \tilde{\alpha}_i \frac{\lambda_{ij}}{f_i'(v_i)} \Big]^2 g(\bar{\lambda}_i; \tilde{r}_i, \tilde{\alpha}_i) d\lambda_i dv_i
$$

$$
= \tilde{\alpha}_i \int_{\Lambda_i} \int_{\Upsilon_i} \frac{\lambda_{ij}^2 f_i''(v_i) + \lambda_{ij}(f_i'(v_i))^2}{(f_i'(v_i))^3} g(\bar{\lambda}_i; \tilde{r}_i, \tilde{\alpha}_i) d\lambda_i dv_i, \ j \neq i;
$$

$$
\int_{\Lambda_i} \int_{\Upsilon_i} \Big[\tilde{r}_{ii} - \tilde{\alpha}_i \frac{1}{f_i'(v_i)} \Big]^2 g(\bar{\lambda}_i; \tilde{r}_i, \tilde{\alpha}_i) d\lambda_i dv_i
$$

$$
= \tilde{\alpha}_i \int_{\Lambda_i} \int_{\Upsilon_i} \frac{f_i''(v_i)}{(f_i'(v_i))^3} g(\bar{\lambda}_i; \tilde{r}_i, \tilde{\alpha}_i) d\lambda_i dv_i,
$$

obtained in view of $(i3)$ by differentiating by parts, gives

$$
\tilde{\mathcal{R}}(\pi_{\tilde{\alpha}_i}^*(\cdot \mid Z_i(\tau) = z_i, A_i(\tau) = a_i), d_{\tilde{r}_i, \tilde{\alpha}_i, i}^*(z_i, a_i))
$$

$$
= \frac{1}{\tilde{\alpha}_i} C(\tilde{r}_i, \tilde{\alpha}_i) \int_{\Lambda_i} \int_{\Upsilon_i} w(\bar{\lambda}_i) g(\bar{\lambda}_i; \tilde{r}_i, \tilde{\alpha}_i) d\lambda_i dv_i = \frac{1}{\alpha_i + a_i}.
$$

Consider now the sequential procedure $\delta_{a_i} = (\tau_{a_i}, d_i^0(\tau_{a_i}))$ with τ_{a_i} defined by (12.7) and

$$
d_i^0(\tau_{a_i}) = \frac{1}{\rho_{0,i} + a_i} Z_i(\tau_{a_i}).
$$

The risk corresponding to the estimator $d_i^0(\tau_{a_i})$ is

$$
\mathcal{R}_0(\bar{\lambda}_i, d_i^0(\tau_{a_i}))
$$

$$
= \frac{1}{w(\bar{\lambda}_i)} \Big\{ \sum_{\substack{j=1 \\ j \neq i}}^m E_{\bar{\lambda}_i} \Big[\frac{N_{ij}(\tau_{a_i})}{\rho_{0,i} + a_i} - \frac{\lambda_{ij}}{f_i'(v_i)} \Big]^2 + E_{\bar{\lambda}_i} \Big[\frac{S_i(\tau_{a_i})}{\rho_{0,i} + a_i} - \frac{1}{f_i'(v_i)} \Big]^2 \Big\}
$$

$$
= \frac{1}{(\rho_{0,i} + a_i)^2 w(\bar{\lambda}_i)} \Big\{ \sum_{\substack{j=1 \\ j \neq i}}^m E_{\bar{\lambda}_i} \Big[N_{ij}(\tau_{a_i}) - (\rho_{0,i} + a_i) \frac{\lambda_{ij}}{f_i'(v_i)} \Big]^2
$$

$$
+ E_{\bar{\lambda}_i} \Big[S_i(\tau_{a_i}) - (\rho_{0,i} + a_i) \frac{1}{f_i'(v_i)} \Big]^2 \Big\}.
$$

Using identities $(12.9) - (12.13)$ yields

$$
\mathcal{R}_0(\bar{\lambda}_i, d_i^0(\tau_{a_i})) = \frac{1}{(\rho_{0,i} + a_i)^2 w(\bar{\lambda}_i)}
$$

$$
\cdot \Big\{ a_i \frac{f_i''(v_i) \left(\sum_{j=1, j \neq i}^m \lambda_{ij}^2 + 1 \right) + (f_i'(v_i))^2 \lambda_{ii}}{(f_i'(v_i))^3} + \rho_{0,i}^2 \frac{\sum_{j=1, j \neq i}^m \lambda_{ij}^2 + 1}{(f_i'(v)_i)^2} \Big\}
$$

$$
= \frac{1}{(\rho_{0,i} + a_i)^2} \Big\{ a_i + \rho_{0,i}^2 \frac{f_i'(v_i) \left(\sum_{j=1, j \neq i}^m \lambda_{ij}^2 + 1 \right)}{f_i''(v_i) \left(\sum_{j=1, j \neq i}^m \lambda_{ij}^2 + 1 \right) + (f_i'(v_i))^2 \lambda_{ii}} \Big\}.
$$

Thus, under condition $(i4)$,

$$
\begin{aligned}
\sup_{\bar{\lambda}_i} \mathcal{R}_0(\bar{\lambda}_i, d_i^0(\tau_{a_i})) &= \frac{1}{\rho_{0,i} + a_i} \\
&= \lim_{\alpha_i \to \rho_{0,i}} \tilde{\mathcal{R}}(\pi_{\alpha_i}^*(\cdot \mid Z_i(\tau) = z_i, A_i(\tau) = a_i), d_{r_i,\alpha_i,i}^*(z_i, a_i)),
\end{aligned}
$$

which implies the desired result. ∎

Let us note that the sequential procedure $\delta_{a_i^*} = (\tau_{a_i^*}, d_i^0(\tau_{a_i^*}))$ is the unique minimax procedure under the loss function given by (12.16) and the assumed cost function.

In particular, conditions $(i1) - (i4)$ are satisfied in the case when the conditional density of $A(t) - A(s)$ given $X(u) = i, u \in [s, t]$ is the Poisson density with intensity μ_i, i.e., if $f_i(v_i) = \exp(v_i)$, where $v_i = \log \mu_i$. In this case we have $\rho_{0,i} = 1$, and condition $(i4)$ is satisfied because $f_i''(v_i) = f_i'(v_i)$ for each $v_i \in \Upsilon_i = (-\infty, \infty)$.

As it will be shown below, to estimate the intensities λ_{ij} $(j = 1, \ldots, m; j \neq i)$ and the mean value parameter $f_i'(v_i)$ one can apply a sequential procedure defined by

$$
\tau_{s_i} = \inf\{t : S_i(t) = s_i\}, \quad s_i > 0. \tag{12.17}
$$

For this stopping time the sequential likelihood function of (12.6) takes the following form

$$
L_i(\tau_{s_i}, \bar{\lambda}_i) = \prod_{j=1, j \neq i}^{m} \lambda_{ij}^{N_{ij}(\tau_{s_i})} \exp\left\{ A_i(\tau_{s_i})v_i - s_i[\lambda_{ii} + f_i(v_i)] \right\}.
$$

Suppose that for any $\rho_i > 0$ and $\beta_i > 0$ the following conditions are satisfied:

$(s1)$ $\displaystyle\int_{\Upsilon_i} f_i''(v_i) \exp[\rho_i v_i - \beta_i f_i(v_i)] dv_i < \infty,$

$(s2)$ $\displaystyle\int_{\Upsilon_i} (f_i'(v_i))^2 \exp[\rho_i v_i - \beta_i f_i(v_i)] dv_i < \infty,$

$(s3)$ $\displaystyle\int_{\Upsilon_i} \frac{d}{dv_i}\Big\{ [\rho_i - \beta_i f_i'(v_i)] \exp[\rho_i v_i - \beta_i f_i(v_i)] \Big\} dv_i = 0.$

The natural prior distribution of the parameter $\bar{\lambda}_i$ is proportional to

$$
\prod_{j=1, j \neq i}^{m} \lambda_{ij}^{r_{ij}-1} \exp(-\alpha_i \lambda_{ij}) \exp\left[r_{ii} v_i - \alpha_i f_i(v_i) \right], \tag{12.18}
$$

and it is proper for all $r_{ij} > 0, j = 1, \ldots, m$ and each $\alpha_i > 0$ (the integral of (12.18) with respect to dv_i is finite for each $r_{ii} > 0$ and $\alpha_i > 0$). The modified prior, i.e., that of (12.18) multiplied by the sum $\sum_{j=1, j \neq i}^{m} \lambda_{ij} + f_i''(v_i)$

$= \lambda_{ii} + f_i''(v_i)$ is proper under condition $(s1)$ for all $r_{ij} > 0, j = 1, \ldots, m$, and each $\alpha_i > 0$.

Let the loss function be defined by

$$\mathcal{L}(\bar{\lambda}_i, d_i) = [\lambda_{ii} + f_i''(v_i)]^{-1} \Big[\sum_{j=1, j\neq i}^{m} (d_{ij} - \lambda_{ij})^2 + (d_{ii} - f_i'(v_i))^2 \Big]$$

and let the cost function $c(\cdot)$ depend only on the value of the process $S_i(t)$ at the moment of stopping. In an analogous way as in Proposition 12.4.1 we obtain the following result.

Proposition 12.4.2 *Suppose that conditions $(s1)$ – $(s3)$ are satisfied. If there exists s_i^* such that*

$$\frac{1}{s_i^*} + c(s_i^*) = \min_{s_i} \Big[\frac{1}{s_i} + c(s_i) \Big],$$

then the sequential procedure $\delta_{s_i^*} = (\tau_{s_i^*}, d_i^0(\tau_{s_i^*}))$ with $\tau_{s_i^*}$ defined by (12.17) and

$$d_i^0(\tau_{s_i^*}) = \frac{1}{s_i^*} \Big(N_{i,1}(\tau_{s_i^*}), \ldots, N_{i,i-1}(\tau_{s_i^*}), A_i(\tau_{s_i^*}), N_{i,i+1}(\tau_{s_i^*}), \ldots, N_{i,m}(\tau_{s_i^*}) \Big)$$

is minimax.

For example, conditions $(s1)$ – $(s3)$ are satisfied for the Poisson density with intensity μ_i, i.e., if $f_i(v_i) = \exp(v_i)$, $v_i = \log \mu_i$.

References

1. Barndorff-Nielsen, O. (1978). *Information and Exponential Families*, New York: John Wiley & Sons.

2. Brown, L. (1986). *Fundamentals of Statistical Exponential Families*, Hayward, CA: IMS.

3. Çinlar, E. (1972). Markov additive processes. I. *Z. Wahrschein. verw. Gebiete*, **24**, 85–93.

4. Ezhov, I. and Skorohod, A. (1969). Markov processes with homogeneous second component. I. *Teor. Verojatn. Primen.*, **14**, 3–14 (in Russian).

5. Franz, J. and Magiera, R. (1997). On information inequalities in sequential estimation for stochastic processes, *Mathematical Methods of Operations Research*, **46**, 1–27.

6. Fygenson, M. (1991). Optimal sequential estimation for semi-Markov and Markov renewal processes, *Communications in Statistics—Theory and Methods*, **20**, 1427–1444.

7. Gill, R. (1980). Nonparametric estimation based on censored observation of a Markov renewal process, *Z. Wahrschein. verw. Gebiete*, **53**, 97–116.

8. Karr, A. (1991). *Point Processes and their Statistical Inference*, Second edition, New York: Marcel Dekker.

9. Magiera, R. and Stefanov, V. T. (1989). Sequential estimation in exponential-type processes under random initial conditions, *Sequential Analysis*, **8**, 147–167.

10. Magiera, R. and Wilczyński, M. (1991). Conjugate priors for exponential-type processes, *Statistics and Probability Letters*, **12**, 379–384.

11. Neuts, M. (1992). Models based on the Markovian arrival process, *IEICE Transactions and Communications*, **E75-B**, 1255–1265.

12. Pacheco, A. and Prabhu, N. (1995). Markov-additive processes of arrivals, In *Advances in Queueing: Theory, Methods, and Open Problems* (Ed., J. H. Dshalalow), pp. 167–194, Boca Raton: CRC Press.

13. Phelan, M. (1990a). Bayes estimation from a Markov renewal process, *Annals in Statistics*, **18**, 603–616.

14. Phelan, M. (1990b). Estimating the transition probabilities from censored Markov renewal processes, *Statistics & Probability Letters*, **10**, 43–47.

15. Prabhu, N. (1991). Markov renewal and Markov-additive processes, *Technical Report 984*, College of Engineering, Cornell University.

16. Rydén, T. (1994). Parameter estimation for Markov modulated Poisson processes, *Communications in Statistics—Stochastic Models*, **10**, 795–829.

17. Stefanov, V. (1986). Efficient sequential estimation in exponential-type processes, *Annals in Statistics*, **14**, 1606–1611.

18. Stefanov, V. (1995). Explicit limit results for minimal sufficient statistics and maximum likelihood estimators in some Markov processes: exponential families approach, *Annals in Statistics*, **23**, 1073–1101.

19. Trybuła, S. (1982). Sequential estimation in finite-state Markov processes, *Zastos. Matem.*, **17**, 227–248.

20. Wilczyński, M. (1985). Minimax sequential estimation for the multinomial and gamma processes, *Zastos. Matem.*, **18**, 577–595.

13

Some Models Describing Damage Processes and Resulting First Passage Times

Heide Wendt

Otto-von-Guericke-Universität Madgeburg, Magdeburg, Germany

Abstract: The system reliability for a shock model with accumulating damages will be described by marked point processes Φ. The likelihood function is determined for different information about Φ. Usually, it depends on unknown parameters. The several resulting maximum–likelihood–estimates are derived for a mixed Poisson process.

Keywords and phrases: Point process, filtration, history, predictable process, martingale, compensator, stochastic intensity, doubly stochastic and mixed Poisson process, likelihood function, maximum–likelihood–estimate, independent right–censoring

13.1 Introduction

In connection with the investigation of the reliability of technical systems it is mostly necessary to consider damage processes which are able to occur at these systems. The failure behaviour of technical systems is often influenced by shocks. We examine a shock model with accumulating damages. Here the system is suffered from shocks where the shocks happen at time points $(T_n)_{n \geq 1}$ and bring about a gradual damage of the system. Each damage from a shock at random time T_n is described by a scalar non–negative r.v. X_n ($n = 1, 2 \ldots$). The damages accumulate so that at any time t the whole damage of the system is given by an r.v. Z_t with $Z_t = \sum_{n=1}^{\infty} I_{(T_n \leq t)} \cdot X_n$. Let us suppose the existence of a certain wear level h where the system is intact as long as Z_t doesn't exceed the value h (see Figure 13.1).

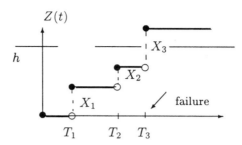

Figure 13.1: Realization of the MPP Φ

The model will be described by marked point processes $\Phi = ((T_n, X_n))_{n \geq 1}$. We will use the case of independent marking.

13.2   Basic Definitions

Let be given a fixed probability space $[\Omega, \mathcal{F}, P]$. A point process is a sequence $(T_n)_{n \geq 1}$ of positive random numbers T_n. We often have additional information on the points T_n. Then we can define random elements X_n which are called the marks of T_n. Assume that $(\mathbf{X}, \mathcal{X})$ is a measurable space and $\mathbf{X}_\infty := \mathbf{X} \cup \{x_\infty\}$ where $x_\infty \notin \mathbf{X}$. A marked point process MPP is a sequence $\Phi = ((T_n, X_n))_{n \geq 1}$ with

$$T_n < T_{n+1} \quad , \quad X_n \in \mathbf{X} \quad \text{if } T_n < \infty \,,$$
$$T_n = T_{n+1} = \infty \quad , \quad X_n = x_\infty \quad \text{if } T_n = \infty \,.$$

An unique correspondence between the description of an MPP as a sequence and as a counting measure $\Phi(A) = card\{n : (T_n, X_n) \in A\}$ with $A \in \mathcal{B}^+ \otimes \mathcal{X}$ is provided by Last and Brandt (1995). We use the same symbol Φ to denote the sequence or the corresponding counting measure, respectively.

A family $\{\mathcal{F}_t : t \geq 0\} = \{\mathcal{F}_t\}$ of sub–σ–fields of $\mathcal{F}$ is called a filtration if $\mathcal{F}_s \subseteq \mathcal{F}_t$, $s \leq t$. We denote $\{\mathcal{F}_t\}$ as a history of a stochastic process (Y_t) if Y_t is $\mathcal{F}_t$–measurable for all $t \in \mathbf{R}^+$. The smallest history of (Y_t) is called the internal history $\{\mathcal{F}_t^Y\}$ of (Y_t), i.e. $\mathcal{F}_t^Y := \sigma(Y_s : s \leq t)$. Different information about an MPP Φ can be described by several histories of Φ. Let $\{\mathcal{F}_t\}$ be a history of the MPP Φ and let $E[card\{n \geq 1 : T_n \leq t\}] < \infty$. An $\{\mathcal{F}_t\}$–predictable random measure ν on $\mathbf{R}^+ \times \mathbf{X}$ is referred to as an $(P, \mathcal{F}_t)$–compensator of Φ if

$$\Phi((0, t] \times B) - \nu((0, t] \times B) \,, \qquad t \geq 0 \,,$$

is an $\{\mathcal{F}_t\}$–martingale for all $B \in \mathcal{X}$ (see [3]). Just as the hazard measure determines the distribution of a random point, the compensator determines the distribution of an MPP. Therefore it is enough to consider the $(P, \mathcal{F}_t)$–compensator for a given MPP.

13.3 System Failure Time in the Case of Independent Marking

Now we can understand the shock model as MPP $\Phi = ((T_n, X_n))_{n \geq 1}$ where X_n is a damage increment at time T_n. For instance we identify the damage with an increase of a split spreading. Usual the space $\mathbf{X}$ will be a subspace of the real or natural numbers. That means in the second case we can only observe jump units of splits.

Let $(X_n^\star)$ be a sequence of i.i.d. $\mathbf{X}$–valued random elements with distribution G independent of an arbitrary point process (T_n). $\Phi = ((T_n, X_n))_{n \geq 1}$ is called an independent G–marking of (T_n) if $X_n = X_n^\star$ in the case $T_n < \infty$ and else $X_n = x_\infty$. Then the $(P, \mathcal{F}_t)$–compensator of Φ is given as follows

$$\nu(\omega, t, B) := \nu(\omega, (0, t] \times B) = \nu(\omega, t, \mathbf{X}) \cdot \int_B G(dx) \qquad B \in \mathcal{X}, \qquad (13.1)$$

so that the distribution of an independent G–marking is uniquely determined by G and the distribution of (T_n).

First we will only consider the sequence (T_n) of jump times. The corresponding counting process we denote by N, i.e. $N_t := \Phi((0, t] \times \mathbf{X}) = card\{n \geq 1 : T_n \leq t\}$. In what follows we want examine the distribution of system failure time where several distributions of marks (X_n) are supposed. We assume the existence of the $(P, \mathcal{F}_t)$–stochastic intensity $\tilde{\lambda}$ of the point process (T_n), i.e. $\nu(\omega, t, \mathbf{X}) = \int_0^t \tilde{\lambda}(\omega, s) ds$ where $\tilde{\lambda}$ is an $\{\mathcal{F}_t\}$–predictable function. A heuristic equation is given as (see [3])

$$P(\Phi(dt \times \mathbf{X}) > 0 \mid \mathcal{F}_{t-}) = \tilde{\lambda}(\omega, t) dt.$$

Let $\mathcal{F}_t = \mathcal{F}_t^N \vee \sigma(\Lambda)$ be a history of (T_n) where $\mathcal{F}_t^N = \sigma(\Phi((0, t] \times \mathbf{X}))$ is the internal history of (T_n) and Λ – generating $\mathcal{F}_0$ – is a non–negative random element of any measurable space. The $(P, \mathcal{F}_t)$–stochastic intensity $\tilde{\lambda}(\omega, t)$ is assumed to be given by

$$\tilde{\lambda}(\omega, t) = \Lambda(\omega) \cdot a(t). \qquad (13.2)$$

There $a(t)$ denotes a deterministic function. If $a(t)$ is left–continuous with right–hand limits then $\tilde{\lambda}$ will be $\{\mathcal{F}_t\}$–predictable. Usually Λ and $a(t)$ will depend on unknown parameters. We use reference probabilities (referring to standard Poisson processes) and determine likelihood ratios in form of Radon–Nikodym derivatives (see [3]). Then we obtain for $N_t = card\{n \geq 1 : T_n \leq t\}$

and $j \in \mathbf{Z}^+$ (with $0^0 := 1$)

$$P(N_t = j) = E\left[\frac{(\Lambda \int_0^t a(s)ds)^j}{j!}\exp(-\Lambda \int_0^t a(s)ds)\right] \quad (13.3)$$

where we form the expectation referring to Λ. Let us mention that (T_n) is called doubly stochastic Poisson process. In case $a(t) = const.$, the sequence is denoted as mixed Poisson process. We will consider several distributions of Λ and alternatives of function $a(t)$.

Example 13.3.1 We assume that $\tilde{\lambda}(\omega, t) = \Lambda(\omega) \cdot (m+1) \cdot t^m$ where $m > -1$ and Λ is subject to Gamma distribution with parameters c and b, i.e. the distribution function of Λ has the density

$$I_{\{\lambda \geq 0\}}\frac{c^b}{\Gamma(b)}\lambda^{b-1}\exp(-c\lambda) .$$

When we consider a fixed value of Λ then we can interpret the sequence (T_n) as process with decreasing failure rate $(m < 0)$, constant failure rate $(m = 0)$ or increasing failure rate $(m > 0)$.

Using the definition of the Gamma distribution with parameters $c + t^{m+1}$, $j + b$ we obtain for $j \in \mathbf{Z}^+$

$$P(N_t = j) = \int_0^\infty \frac{(\lambda t^{m+1})^j}{j!}e^{-\lambda t^{m+1}}\frac{c^b}{\Gamma(b)}\lambda^{b-1}e^{-c\lambda}d\lambda$$

$$= \frac{\Gamma(j+b)}{\Gamma(b)\Gamma(j+1)}\frac{c^b \cdot (t^{m+1})^j}{(c+t^{m+1})^{b+j}}\int_0^\infty \frac{(c+t^{m+1})^{b+j}}{\Gamma(j+b)}\lambda^{j+b-1}e^{-\lambda[c+t^{m+1}]}d\lambda$$

$$= \frac{\Gamma(j+b)}{\Gamma(b)\Gamma(j+1)} \cdot \frac{c^b}{(c+t^{m+1})^b} \cdot \frac{(t^{m+1})^j}{(c+t^{m+1})^j} \quad (13.4)$$

If $b \in \mathbf{N}$ we have a negative binomial distribution.

Example 13.3.2 Let $\tilde{\lambda}$ be given as $\tilde{\lambda}(\omega, t) = \Lambda(\omega)$ and suppose that the distribution function of Λ possesses the density $I_{\{\lambda \in [a,b]\}} \cdot \frac{1}{b-a}$ with $[a, b] \in \mathcal{B}^+$. It follows that

$$P(N_t = j) = \int_a^b \frac{(\lambda t)^j}{j!} \cdot e^{-\lambda t} \cdot \frac{1}{b-a}d\lambda$$

and after repeatedly partial integration $(t > 0)$

$$P(N_t = j) = \frac{1}{t(b-a)}\sum_{i=0}^j \left(\frac{(at)^i}{i!} \cdot e^{-at} - \frac{(bt)^i}{i!} \cdot e^{-bt}\right) . \quad (13.5)$$

Now we want to return to the MPP $\Phi = ((T_n, X_n))_{n \geq 1}$ and consider the system failure time. It is wise to regard a (possibly unknown) state of first X_0 at time $T_0 := 0$. Then the system failure time $Z_h^\star$ can be written as

$$Z_h^\star := \inf\{t : \sum_{n=0}^{\infty} I_{(T_n \leq t)} \cdot X_n \geq h\} . \tag{13.6}$$

We are interested in the intact probability of the system, i.e. in the survival probability of $Z_h^\star$. Using the formula of total probability we obtain

$$P(Z_h^\star > t) = \sum_{n=0}^{\infty} P(N_t = n) \cdot P(X_0 + X_1 + \ldots + X_n < h | T_n \leq t < T_{n+1}) .$$

In the special case of independent marking we have got

$$P(Z_h^\star > t) = \sum_{n=0}^{\infty} P(N_t = n) \cdot P(X_0 + X_1 + \ldots + X_n < h) .$$

We will give some examples for the distributions of $Z_h^\star$ by independent marking. First we consider discrete random variables X_n where the event $\{X_n = 0\}$ doesn't occur $(n \geq 1)$.

Example 13.3.3 Let the sequence (Y_n) be i.i.d. with binomial distribution and $X_n := Y_n + 1$ where the parameters are $k - 1$ and p. Then it follows for $j \in \mathbf{Z}^+$

$$P(\sum_{m=1}^{n} X_m = j) = I_{(n \leq j \leq n \cdot k)} \binom{n(k-1)}{j-n} p^{j-n}(1-p)^{nk-j} .$$

In the case $X_0 = x_0 = const.$ we find

$$P(Z_h^\star > t) = P(N_t = 0)I_{(x_0 < h)}+$$

$$+ \sum_{n=1}^{\lfloor h-x_0-1 \rfloor} \sum_{j=n}^{\lfloor h-x_0-1 \rfloor} I_{(j \leq nk)} P(N_t = n) \binom{n(k-1)}{j-n} \left(\frac{p}{1-p}\right)^j \left(\frac{(1-p)^k}{p}\right)^n$$

where $\lfloor \cdot \rfloor$ denotes the greatest integer part of a number.

Example 13.3.4 We assume that $P(X_n = j) = \frac{1}{e^\mu - 1} \frac{\mu^j}{j!}$ where $n, j \geq 1$. This is the truncated Poisson distribution. Now we want to determine the distribution function of $\sum_{m=1}^{n} X_m$ of i.i.d. X_m. We will inductively show that for $j \in \mathbf{Z}^+$

$$P(\sum_{m=1}^{n} X_m = j) = \frac{\mu^j}{(e^\mu - 1)^n} \cdot \frac{n!}{j!} \cdot S(j, n) \tag{13.7}$$

where $S(j,n)$ denotes the Stirling number second kind, i.e. the number of partitions of j elements in exact n non–empty disjoint subsets. Equation (13.7) is valid in the case $n = 1$. Furthermore we compute for $j \geq n+1$

$$
\begin{aligned}
P(\sum_{m=1}^{n+1} X_m = j) &= \sum_{l=1}^{j-n} P(X_{n+1} = l) \cdot P(\sum_{m=1}^{n} X_m = j - l) \\
&= \sum_{l=1}^{j-n} \frac{1}{e^\mu - 1} \frac{\mu^l}{l!} \cdot \frac{\mu^{j-l}}{(e^\mu - 1)^n} \frac{n!}{(j-l)!} S(j-l, n) \\
&= \frac{\mu^j}{(e^\mu - 1)^{n+1}} \cdot \frac{n!}{j!} \sum_{l=1}^{j} \binom{j}{l} S(j-l, n) .
\end{aligned}
$$

The last step follows from $S(j - l, n) = 0$ for $j - l < n$. We use that

$$
\sum_{l=1}^{j} \binom{j}{l} S(j-l, n) = S(j+1, n+1) - S(j, n) = (n+1) \cdot S(j, n+1) .
$$

Finally we obtain

$$
P(\sum_{m=1}^{n+1} X_m = j) = \frac{\mu^j}{(e^\mu - 1)^{n+1}} \cdot \frac{(n+1)!}{j!} S(j, n+1) .
$$

Consequently equation (13.7) is valid for all $n \in \mathbf{N}$ and $j \geq n+1$. It's easily to see that this equation is also true in the case $j \leq n$. Therefore, in the case $X_0 = x_0 = const.$ the survival distribution of $Z_h^\star$ is given by

$$
P(Z_h^\star > t) = P(N_t = 0) I_{(x_0 < h)} + \sum_{j=1}^{\lfloor h - x_0 - 1 \rfloor} \sum_{n=1}^{j} P(N_t = n) \frac{\mu^j}{(e^\mu - 1)^n} \cdot \frac{n!}{j!} \cdot S(j, n)
$$

Example 13.3.5 Let be $X_n := Y_n + 1$ where Y_n is subject to geometric distribution, i.e. $P(Y_n = j) = (1 - p) \cdot p^j$ $(j = 0, 1, \ldots)$. First we will prove that the following equation for the sum of i.i.d. Y_m is valid

$$
P(\sum_{m=1}^{n} Y_m = j) = (1 - p)^n \cdot p^j \cdot \binom{j + n - 1}{n - 1} \qquad j = 0, 1, \ldots \quad .
$$

In the case $n = 1$ the statement is true. Further we inductively find out

$$
\begin{aligned}
P(\sum_{m=1}^{n+1} Y_m = j) &= \sum_{i=0}^{j} P(Y_{m+1} = i) \cdot P(\sum_{m=1}^{n} Y_m = j - i) \\
&= \sum_{i=0}^{j} (1 - p) \, p^i \cdot (1 - p)^n \, p^{j-i} \binom{j - i + n - 1}{n - 1} \\
&= \sum_{i=0}^{j} (1 - p)^{n+1} \, p^j \left(\binom{j - i + n}{n} - \binom{j - i + n - 1}{n} \right) \\
&= (1 - p)^{n+1} \cdot p^j \cdot \binom{j + n}{n}
\end{aligned}
$$

With that it follows for $j = n, n+1, \ldots$

$$P(\sum_{m=1}^{n} X_m = j) = (1-p)^n \cdot p^{j-n} \cdot \binom{j-1}{n-1} \tag{13.8}$$

and for $X_0 = x_0 = const. < h$ we get

$$P(Z_h^\star > t) = P(N_t = 0) + \sum_{j=1}^{\lfloor h-x_0-1 \rfloor} \sum_{n=1}^{j} P(N_t = n) \cdot (1-p)^n p^{j-n} \binom{j-1}{n-1}.$$

Additionally let us assume that (T_n) is given as mixed Poisson process with $\tilde{\lambda}(\omega, t) = \Lambda \sim Exp(c)$, i.e. $P(N_t = n) = \frac{c}{c+t} \cdot \left(\frac{t}{c+t}\right)^n$ (see Example 13.3.1 with $b = 1$ and $m = 0$). Then we determine the survival distribution of $Z_h^\star$ as follows

$$
\begin{aligned}
P(Z_h^\star > t) &= \frac{c}{c+t} + \sum_{j=1}^{\lfloor h-x_0-1 \rfloor} \sum_{n=1}^{j} \frac{c}{c+t} \left[\frac{t}{c+t}(1-p)\right]^n \cdot p^{j-n} \binom{j-1}{n-1} \\
&= \frac{c}{c+t} + \sum_{j=1}^{\lfloor h-x_0-1 \rfloor} \frac{c}{c+t} \left[\frac{t}{c+t}(1-p)\right] \cdot \left(\frac{t}{c+t}(1-p)+p\right)^{j-1} \\
&= \frac{c}{c+t} + \frac{c}{c+t}\frac{t}{c+t}(1-p) \cdot \frac{1 - \left[\frac{t+c\cdot p}{t+c}\right]^{\lfloor h-x_0-1 \rfloor}}{1 - \frac{t+c\cdot p}{t+c}} \\
&= 1 - \frac{t}{c+t} \cdot \left[\frac{t+c\cdot p}{c+t}\right]^{\lfloor h-x_0-1 \rfloor}. \tag{13.9}
\end{aligned}
$$

We obtain for the density $f_{Z_h^\star}$ of the r.v. $Z_h^\star$ that

$$f_{Z_h^\star}(t) = \frac{c \cdot (t+cp)^{\lfloor h-x_0 \rfloor-2}}{(c+t)^{\lfloor h-x_0 \rfloor+1}} \cdot \left(\lfloor h-x_0 \rfloor t \cdot (1-p) + (c+t) \cdot p\right) \quad (t \geq 0).$$

Concluding we will give an example where the non–negative random variables X_n are continuous.

Example 13.3.6 Let X_n have an exponential distribution with parameter p. It is well–known that (in the case of independent marking) the sum $X_1 + \ldots + X_n$ has got an Erlang distribution. As in Example 13.3.5 we assume that $P(N_t = n) = \frac{c}{c+t} \cdot \left(\frac{t}{c+t}\right)^n$ and $X_0 = x_0 = const. < h$.

We easily prove the uniform convergence so that we can write

$$
\begin{aligned}
P(Z_h^\star > t) &= P(N_t = 0) + \sum_{n=1}^{\infty} \int_0^{h-x_0} p^n \frac{x^{n-1}}{(n-1)!} e^{-p\cdot x} dx \cdot \frac{c}{c+t} \left(\frac{t}{c+t}\right)^n \\
&= \frac{c}{c+t} + \frac{c}{c+t} \int_0^{h-x_0} e^{-p\cdot x} \left[\sum_{n=1}^{\infty} p^n \left(\frac{t}{c+t}\right)^n x^{n-1} \frac{1}{(n-1)!}\right] dx
\end{aligned}
$$

$$= \frac{c}{c+t} + \frac{c}{c+t} \cdot \frac{pt}{c+t} \int_0^{h-x_0} \exp\left(-x\,[p - \frac{pt}{c+t}]\right) dx$$

$$= 1 - \frac{t}{c+t} \cdot \exp\left(-(h-x_0)\,p \cdot \frac{c}{c+t}\right). \tag{13.10}$$

Here the density $f_{Z_h^\star}$ of $Z_h^\star$ is given, for $t \geq 0$, by

$$f_{Z_h^\star}(t) = \exp\left(-(h-x_0)\,p \cdot \frac{c}{c+t}\right) \cdot \left(\frac{c}{(c+t)^2} + (h-x_0)\frac{p \cdot c \cdot t}{(c+t)^3}\right)$$

13.3.1   ML–Estimates for parameters in the distribution of the system failure time

As we have seen from Example 13.3.3 to 13.3.6 the survival distribution of $Z_h^\star$ can depend on unknown parameters. Now we will demonstrate a parameter estimation on the basis of observations (T_n, X_n). Let us assume that (T_n) is a mixed Poisson process with stochastic intensity $\tilde{\lambda}(\omega, t) = \Lambda \sim Exp(c)$ and $X_n = Y_n + 1$ where the i.i.d. sequence (Y_n) has a binomial distribution $Bi(k-1, p)$ (see Example 13.3.3).

At the beginning we consider observations without censoring, i.e. the threshold h is sufficiently large so that a system failure doesn't occur.

- In the first case let the history of the MPP $\Phi = ((T_n, X_n))_{n \geq 1}$ be given by $\mathcal{F}_t = \mathcal{F}_t^\Phi \vee \sigma(\Lambda)$, $t \geq 0$, and the T_n–past is generated as follows $\mathcal{F}_{T_n} = \sigma(\Lambda, T_1, X_1, \ldots, T_n, X_n)$. Using reference probability we obtain the following likelihood function at time T (see [3])

$$L(T; c, p) = c\,e^{-c \cdot \Lambda}\Lambda^{N_T}\left(\prod_{n:T_n \leq T} \binom{k-1}{X_n-1}p^{X_n-1}(1-p)^{k-X_n}\right) \cdot e^{kT - \Lambda T}.$$

Generally the r.v. T is an $\{\mathcal{F}_t\}$–stopping time, but here let T be a constant. Now we consider M independent realizations of Φ where Λ_i, $X_{i,n}$ respectively $N_i(T)$ are the corresponding observations of the i-th realization $(i = 1, \ldots, M)$. Using maximum–likelihood principles we immediately find that

$$\hat{c} = \frac{M}{\sum_{i=1}^M \Lambda_i} \quad \text{and} \quad \hat{p} = \sum_{i=1}^M \sum_{n=1}^{N_i(T)} \frac{X_{i,n} - 1}{N_\bullet(T) \cdot (k-1)},$$

where $N_\bullet(T) = \sum_{j=1}^M N_j(T)$ is the total number of jumps in M realizations of Φ up to time T.

- In the second case let be $\mathcal{F}_t = \mathcal{F}_t^\Phi$, $t \geq 0$; the T_n–past $\mathcal{F}_{T_n}$ is generated by $\sigma(T_1, X_1, \ldots, T_n, X_n)$ without observation of Λ. We are interested in the estimate of the parameter c. Under the assumption of independent marking and noncensoring observations it's enough to use the likelihood function for the sequence (T_n). Let Q denote a probability measure referring to a standard Poisson process. And let L_t be the $(P, \mathcal{F}_t^N \vee \sigma(\Lambda))-$ likelihood function in more general case so that

$$L_t = L_0(\prod_{n:T_n \leq t} \tilde{\lambda}(T_n)) \exp(\int_0^t [1 - \tilde{\lambda}(s)] ds)$$

where L_0 is a non–negative random variable with expectation one. Then we determine the $(P, \mathcal{F}_t^N)$–likelihood function at time t as [see Bremaud (1981)]

$$\hat{L}_t = E_Q[L_t|\mathcal{F}_t^N] = (\prod_{n:T_n \leq t} \tilde{\lambda}^\star(T_n)) \exp(\int_0^t [1 - \tilde{\lambda}^\star(s)] ds) . \qquad (13.11)$$

Here $\tilde{\lambda}^\star$ is the $(P, \mathcal{F}_t^N)$–stochastic intensity of (T_n). Using the innovation theorem we find for the mixed Poisson process [see Anderson et al. (1993)]

$$\tilde{\lambda}^\star(t) = E_P \left[\tilde{\lambda}(t)|\mathcal{F}_{t-}^N \right] = \lim_{s \uparrow\uparrow t} \frac{E_Q[\Lambda L_s|\mathcal{F}_s^N]}{E_Q[L_s|\mathcal{F}_s^N]} = \frac{\int_0^\infty \lambda^{N_{t-}+1} e^{-\lambda t} d\,F(\lambda)}{\int_0^\infty \lambda^{N_{t-}} e^{-\lambda t} dF(\lambda)}$$

where F denotes the distribution function of Λ. If $dF(\lambda) = c \cdot e^{-c\lambda} d\lambda$ we compute

$$\tilde{\lambda}^\star(t) = \frac{(N_{t-} + 1)! \, c}{(c+t)^{N_{t-}+2}} : \frac{(N_{t-})! \, c}{(c+t)^{N_{t-}+1}} = \frac{N_{t-} + 1}{c+t} .$$

Then the $(P, \mathcal{F}_t^N)$–compensator of (T_n) is given by $\quad (T_0 = 0)$

$$\int_0^t \tilde{\lambda}^\star(s) ds = \sum_{j=1}^{N_t} \int_{T_{j-1}}^{T_j} \frac{j}{c+s} ds + \int_{T_{N_t}}^t \frac{N_t + 1}{c+s} ds$$

$$= -\ln(c) - \sum_{j=1}^{N_t} \ln(c + T_j) + (N_t + 1) \ln(c + t) \quad (13.12)$$

Finally we obtain for the log–likelihood–function of (T_n) at time T by M independent realizations

$$\ln \hat{L}(T; c) = \sum_{i=1}^M \left\{ \sum_{j=1}^{N_i(T)} \ln(j) + T + \ln(c) - (N_i(T) + 1) \ln(c + T) \right\} .$$

With that the following maximum–likelihood–estimate of c can be written in the case $N_\bullet(T) > 0$

$$\hat{c} = \frac{T \cdot M}{\sum_{i=1}^M N_i(T)} = \frac{T \cdot M}{N_\bullet(T)} .$$

Remark 13.3.1 The first estimate of c has a positive bias (i.e. $E[\hat{c}] - c > 0$) and doesn't take on the lower variance bound $\frac{M}{(M-1)^2} \cdot c^2$ because we can show that

$$E\left(\frac{M}{\sum_{i=1}^{M} \Lambda_i}\right) = \frac{M}{M-1} \cdot c \quad \text{and} \quad D^2\left(\frac{M}{\sum_{i=1}^{M} \Lambda_i}\right) = \frac{M^2}{(M-1)^2} \cdot c^2 \cdot \frac{1}{M-2} .$$

By Jensen's inequality $E[g(Z)] \geq g(E[Z])$ for the convex function $g(z) = z^{-1}$ we also obtain a positive bias for the second estimate. We find out at fixed time T that a lower variance bound is given as

$$D^2\left(\frac{TM}{\sum_{i=1}^{M} N_i(T)}\right) \geq \frac{c^3}{M \cdot T} + \frac{c^2}{M} = \left(E\left[-\frac{\partial^2 \ln \hat{L}_T}{\partial c^2}\right]\right)^{-1} .$$

Remark 13.3.2 If we estimate $\theta = 1/c$ we obtain in both cases unbiased estimates with minimal variance. There the variances are given as follows

$$D^2\left(\sum_{i=1}^{M} \frac{\Lambda_i}{M}\right) = \frac{\theta^2}{M} \quad \text{and} \quad D^2\left(\sum_{i=1}^{M} \frac{N_i(T)}{TM}\right) = \frac{\theta}{M \cdot T} + \frac{\theta^2}{M} .$$

Now we want to formulate a parameter estimate for censored processes (without an observation of the r.v. Λ). That means we are only observing the sequence $(T_n, X_n)_{n \geq 1}$ up to random time $T \wedge Z_h^\star = \min(T, Z_h^\star)$. Let N^c denote the right–censored counting process where

$$N^c(t) = \Phi(\,(0, t \wedge Z_h^\star] \times \mathbf{X}\,) = N(t \wedge Z_h^\star) = \int_0^t I_{(s \leq Z_h^\star)} \, dN(s) .$$

The right–censoring process $I_{(t \leq Z_h^\star)}$ is $\{\mathcal{F}_t\}$–predictable (with $\mathcal{F}_t = \mathcal{F}_t^\Phi$). Let us mention that the censoring is often predictable referring to an enlarged filtration $\{\mathcal{S}_t\} \supseteq \{\mathcal{F}_t\}$. The relation $\{\mathcal{S}_t\} = \{\mathcal{F}_t\}$ is a special case of independent right–censoring so that the $(P, \mathcal{F}_t)$–compensator of N^c is given as follows [see Anderson et al. (1993)]

$$\bar{\nu}^{cen}(t) = \nu^{cen}(\,(0, t] \times \mathbf{X}\,) = \int_0^t I_{(s \leq Z_h^\star)} \cdot \tilde{\lambda}^\star(s) \, ds$$

$$= \sum_{j=1}^{N^c(t)} \int_{T_{j-1}}^{T_j} \frac{j}{c+s} ds + I_{(t < Z_h^\star)} \cdot \int_{T_{N(t)}}^t \frac{N(t)+1}{c+s} ds$$

$$= -\ln(c) - \sum_{j=1}^{N^c(t)-1} \ln(c + T_j) -$$

$$-I_{(t < Z_h^\star)} \cdot \ln(c + T_{N(t)}) + (N^c(t) + 1) \cdot \ln(c + t \wedge Z_h^\star) .$$

Because we have supposed an independent marking we obtain for the $(P, \mathcal{F}_t)$–compensator of the censored k–variate point process $(N^{1,c}, \dots, N^{k,c})$ by equation (13.1)

$$\nu^{i,cen}(t) = \nu^{cen}((0,t] \times \{i\}) = P(X = i) \cdot \int_0^t I_{(s \leq Z_h^\star)} \tilde{\lambda}^\star(s) \, ds \quad (i = 1, \dots, k) .$$

Now we can describe the observations at time t as

$$\mathcal{F}_t^{cen} = \sigma\left(N^{i,c}(u), \, \nu^{i,cen}(u) \, ; \, 0 \leq u \leq t, \, i = 1, \dots, k \right) .$$

By the innovation theorem the family $(\nu^{i,cen}(u), 0 \leq u \leq t, i = 1, \dots, k)$ is also the family of $(P, \mathcal{F}_t^{cen})$–compensators for the right–censored counting process $(N^{1,c}, \dots, N^{k,c})$.

We determine ML–estimates on the basis of partial likelihoods. Because of independent right–censoring we can use a partial likelihood $L_t^\star$ with the same form as the likelihood function based on the uncensored process [see Anderson et al. (1993)]:

$$L^\star(t) = \left(\prod_{n:T_n \leq t \wedge Z_h^\star} \binom{k-1}{X_n - 1} p^{X_n - 1} (1-p)^{k-X_n} \cdot \frac{n}{c + T_n} \right) \cdot \exp(-\bar{\nu}^{cen}(t)) .$$

This way we obtain in the case of M independent realizations at fixed time T the following ML–equation for the parameter c respectively ML–estimate for the parameter p

$$\sum_{i=1}^M \left\{ \frac{1}{\hat{c}} - \frac{N_i^c(T) + 1}{\hat{c} + (T \wedge Z_h^\star(i))} \right\} = 0 \quad \text{and} \quad \hat{p} = \frac{1}{N_\bullet^c(T)} \sum_{i=1}^M \sum_{n=1}^{N_i^c(T)} \frac{X_n^i - 1}{k - 1} .$$

Remark 13.3.3 It is followed by Anderson et al. (1993) that the asymptotical properties of the estimates $\hat{c}$ and $\hat{p}$ are the same as these for uncensored data based on the full likelihood (because the martingale properties stay the same). $\hat{\theta} = (\hat{c}, \hat{p})$ is asymptotically multinormally distributed around the true value θ. The asymptotic covariance matrix can be estimated by $\left[-\frac{\partial^2 \ln L^\star(T;\theta)}{\partial \theta^2} \big|_{\theta = \hat{\theta}} \right]^{-1}$. We obtain asymptotically independent estimates $\hat{c}$ and $\hat{p}$. A lower variance bound of $\hat{c}$ is given as follows

$$D^2(\hat{c}) \geq \frac{c^2(c+T)}{M \cdot T - c^2 \sum_{i=1}^M E_\theta \left[I_{(T \geq Z_h^\star(i))} \frac{(T - Z_h^\star(i)) \cdot (N_i^c(T) + 1)}{(c + Z_h^\star(i))^2} \right]} \quad (13.13)$$

Concluding we will give a simulation example. Let us consider a mixed Poisson process where the sequence (T_n) has the stochastic intensity $\tilde{\lambda}(\omega, t) = \Lambda \sim EXP(2.5)$ and the shocks X_n satisfy $(X_n - 1) \sim Bi(50, 0.4)$. We suppose a threshold $h = 1000$. For several numbers M of independent realizations of Φ and for several observation times T we determine the estimates of the parameter $c = 2.5$ in cases of different information.

We can see in Table 13.1 the obtained empirical expectations and variances by 1000 simulation repetitions.

Table 13.1: Empirical moments in the simulation example

| M | T | $\mathcal{F}_t = \sigma(\Lambda) \vee \mathcal{F}_t^{\Phi}$ | | $\mathcal{F}_t = \mathcal{F}_t^{\Phi}$ | | $\mathcal{F}_t = \mathcal{F}_t^{cen}$ | |
		$E[\hat{c}]$	$D^2(\hat{c})$	$E[\hat{c}]$	$D^2(\hat{c})$	$E[\hat{c}]$	$D^2(\hat{c})$
50	50	2.5425	0.1353	2.5458	0.1442	2.5441	0.1454
	100	2.5671	0.1363	2.5670	0.1399	2.5680	0.1430
	200	2.5404	0.1270	2.5390	0.1298	2.5364	0.1334
	300	2.5442	0.1324	2.5414	0.1328	2.5428	0.1372
100	50	2.5397	0.0618	2.5397	0.0640	2.5393	0.0647
	100	2.5396	0.0636	2.5376	0.0653	2.5376	0.0665
	200	2.5261	0.0625	2.5234	0.0621	2.5215	0.0634
	300	2.5419	0.0691	2.5394	0.0691	2.5394	0.0724

The estimate of c in case of complete information is independent of the value T (i.e. $\mathcal{F}_t = \sigma(\Lambda) \vee \mathcal{F}_t^{\Phi}$). We have got different estimates by fixed M because the generated random sequences were not the same. The bias and the variance of all estimates decrease if we consider fixed time T and increasing number M.

As we have expected from the variance bounds in Remark 13.3.1, the variance differences of both uncensored estimates are decreasing by fixed number M of realizations and increasing observation time T. That means, we obtain a knowledge of the value of Λ at time $T_0 = 0$ when T is sufficiently large.

As we can see from equation (13.13) the variance of the censored estimate will be increasing if the probability of the event $\{T \geq Z_h^\star\}$ will greater. If the probability of $\{T \geq Z_h^\star\}$ is zero we obtain the same lower bound as in the uncensored case where $\mathcal{F}_t = \mathcal{F}_t^{\Phi}$ (see Remark 13.3.1). We can see from the estimates based on incomplete information that the variance differences will greater by fixed M and increasing value T, because the event $\{T \geq Z_h^\star\}$ will highly probable.

References

1. Anderson, P., Borgan, Ø., Gill, R. and Keiding, N. (1993). *Statistical Models Based on Counting Processes*, New York: Springer-Verlag.

2. Bremaud, P. (1981). *Point Processes and Queues*, New York, Berlin, Heidelberg: Springer-Verlag.

3. Last, G. and Brandt, A. (1995). *Marked Point Processes on the Real Line*, New York, Berlin, Heidelberg: Springer-Verlag.

14

Absorption Probabilities of a Brownian Motion in a Triangular Domain

Erik Zierke

Otto-von-Guericke-Universität Madgeburg, Magdeburg, Germany

Abstract: The absorption probabilities of a two-dimensional Brownian motion with independent components in a triangular domain are evaluated for special parameter cases. They are obtained from a known random walk result.

Keywords and phrases: Brownian motion, absorption probabilities, functional limit theorem

14.1   Introduction

Brownian motion is an often used model in various fields (e.g. neurophysiology, insurance mathematics, sequential analysis). One point of interest is the random times and/or the probabilities of absorption in one or more absorbing barriers. Often the absorption is a formal synonym for a failure of the real system. In other problems the question is not the time point of the absorption but *if* or *where* the absorption occurs (e.g. ruin problems).

Usually absorption times and absorption probabilities of Brownian motion are calculated as solutions of differential equations. In the case of a one-dimensional Brownian motion a lot of problems were solved via this method; but it comes quickly to its limits in the case of two dimensions [see Dominé (1993, 1996)].

Another approach to evaluate absorption probabilities of Brownian motion is based on the fact that a Brownian motion can be described as a limit of random walks with decreasing step sizes and decreasing intervals between two steps. The absorption times or probabilities of the Brownian motion are calculated as limits of absorption times or probabilities of random walks, respectively.

Such an approach makes it possible to get an exact formula of the absorption

probabilities of a Brownian motion in the following special triangular domain—
a still not investigated problem. Unfortunately, the used random walk result
and the discretization of the Brownian motion itself require very hard conditions
to the parameters of the Brownian motion.

Given a two-dimensional Brownian motion $W = (W^1, W^2)$ with independent
components, drifts μ_1 and μ_2 and diffusion coefficients σ_1^2 and σ_2^2. Suppose
$\sigma_1 = \sigma_2 = \sigma$ and $\mu_1 = \pm\mu_2$. W starts a.s. in the point (a, b) inside the triangle
D constructed by $x = 0$, $y = 0$, and $x + y = m$, where $m = a + b + c$ with
$a, b, c > 0$. The sides of the triangle **D** are absorbing barriers, i.e. after reaching
a barrier point, the process remains there.

This paper gives exact results for the probabilities of absorption in the
vertical side $y = 0$, the horizontal side $x = 0$, and the diagonal side $x + y = m$,
which are denoted by $\overline{\alpha}$, $\overline{\beta}$, and $\overline{\gamma}$, respectively.

In Section 14.2 some known random walk results and a necessary limit
theorem are stated. In Sections 14.3 and 14.4 the results for the two possible
cases are calculated. In Section 14.5 we discuss the quality of the given results.

14.2   A Random Walk Result and Some Used Limit Theorems

Given a two-dimensional random walk on the points with integer valued co-
ordinates. In each step the process jumps to one of the four neighbor points:
with probability p to the right, probability q to the left, probability r upward,
and probability s downward $(p+q+r+s = 1)$. The random walk is restricted
by the absorbing barriers $x = 0$, $x = N$, $y = 0$, $y = N$ (N integer) and starts
a.s. in (A, B) $(0 < A < N, 0 < B < N, A$ and B integers). Let $\pi^{A,B}(x, y)$
denote the "average times of leaving" (ATL) the point (x, y) by the random
walk $(0 < x < N, 0 < y < N)$. Then one can find the following result in
Barnett (1963):

Define the parameters λ, ϱ, Θ and Φ by

$$p = \lambda e^{\Theta}, q = \lambda e^{-\Theta}, r = \varrho e^{\Phi}, s = \varrho e^{-\Phi} \qquad (14.1)$$

and $\{\delta_j\}_{j=1}^{N-1}$ by

$$\lambda \cos \frac{j\pi}{N} + \varrho \cosh \delta_j = \frac{1}{2}. \qquad (14.2)$$

Then

$$\pi^{A,B}(x, y) = \begin{cases} \pi_1^{A,B}(x, y) & \text{if } y \leq B \\ \pi_2^{A,B}(x, y) & \text{if } y > B, \end{cases}$$

where

$$
\left.\begin{aligned}
\pi_1(x,y) &= \frac{2e^{(x-A)\Theta}e^{(y-B)\Phi}}{\varrho N} \sum_{j=1}^{N-1} \sin\frac{j\pi x}{N} \sin\frac{j\pi A}{N} \frac{\sinh(N-B)\delta_j \sinh y\delta_j}{\sinh N\delta_j \sinh \delta_j} \\
\pi_2(x,y) &= \frac{2e^{(x-A)\Theta}e^{(y-B)\Phi}}{\varrho N} \sum_{j=1}^{N-1} \sin\frac{j\pi x}{N} \sin\frac{j\pi A}{N} \frac{\sinh(N-y)\delta_j \sinh B\delta_j}{\sinh N\delta_j \sinh \delta_j}
\end{aligned}\right\}.
$$

$$(14.3)$$

Now we consider the case of the triangle. Let $N := A+B+C$ $(C>0)$, and the upper and right absorbing barriers above are replaced by the new absorbing barrier $x+y=N$. Moreover, we suppose $pq = rs$. The starting point remains (A,B). Let $\mathrm{E}_{A,B}(x,y)$ be the ATL for the inner point (x,y) of the triangle in this new problem. Then it holds [see Barnett (1964)]

$$
\mathrm{E}_{A,B}(x,y) = \pi^{A,B}(x,y) - \left(\frac{p}{s}\right)^{x+y-N} \pi^{A,B}(N-y, N-x). \qquad (14.4)
$$

If β is the probability of total absorption in the horizontal barrier $y=0$ then

$$
\beta = \sum_{x=1}^{N-2} s\,\mathrm{E}_{A,B}(x,1). \qquad (14.5)
$$

To get Brownian motion results from this it is necessary to have a limit theorem:

Theorem 14.2.1 *Let* $\mathrm{W} = (\mathrm{W}^1, \mathrm{W}^2)$ *be a two-dimensional Brownian motion with independent components,* W^1 *has the parameters* μ_1 *and* σ_1^2, W^2 *has the parameters* μ_2 *and* σ_2^2, *and* $\mathrm{W}_0 = (a,b)$ *a.s.*

Further, let $\{Z_{m,n}\}_{m,n}$ $(m = 0,1,2,\dots\,; n = 1,2,\dots)$ *be a two-dimensional array of two-dimensional real-valued random variables.* $\{Z_{m,n}\}_{m=1}^{\infty}$ *is a sequence of independent and identically distributed (iid) random variables for fixed* n *with*

$$
P(Z_{m,n} = (+\Delta_{1,n},0)) = \frac{1}{4}\left(1 + \frac{\mu_1\sqrt{2}}{\sigma_1\sqrt{n}}\right),
$$

$$
P(Z_{m,n} = (-\Delta_{1,n},0)) = \frac{1}{4}\left(1 - \frac{\mu_1\sqrt{2}}{\sigma_1\sqrt{n}}\right),
$$

$$
P(Z_{m,n} = (0, +\Delta_{2,n})) = \frac{1}{4}\left(1 + \frac{\mu_2\sqrt{2}}{\sigma_2\sqrt{n}}\right),
$$

$$
P(Z_{m,n} = (0, -\Delta_{2,n})) = \frac{1}{4}\left(1 - \frac{\mu_2\sqrt{2}}{\sigma_2\sqrt{n}}\right),
$$

where $\Delta_{1,n} := \sigma_1\sqrt{2}/\sqrt{n}$ and $\Delta_{2,n} := \sigma_2\sqrt{2}/\sqrt{n}$. *Moreover,* $Z_{0,n} = ([a/\Delta_{1,n}]$ $\Delta_{1,n}, [b/\Delta_{2,n}]\Delta_{2,n})$ *a.s. The two-dimensional random walk* $\mathrm{S}^{(n)}$ *is constructed by* $\mathrm{S}_m^{(n)} := \sum_{i=0}^m Z_{i,n}$. *Define the continuous process* $\mathrm{X}^{(n)}$ *by*

$$\mathrm{X}^{(n)}(t) := ([nt] + 1 - nt)\mathrm{S}_{[nt]}^{(n)} + (nt - [nt])\mathrm{S}_{[nt]+1}^{(n)} \quad (t \in [0, \infty)).$$

Then $\mathrm{X}^{(n)} \xrightarrow{\mathcal{D}} \mathrm{W}$ *for* $n \to \infty$, *where* [] *denotes the integer part and* $\xrightarrow{\mathcal{D}}$ *denotes weak convergence of probability measures.*

PROOF. This theorem is a direct consequence of Theorem 11.2.3 in Stroock and Varadhan (1979). ∎

Given now the situation of the introduction where the two-dimensional Brownian motion $\mathrm{W} = (\mathrm{W}^1, \mathrm{W}^2)$ starts in (a, b) a.s. and it is absorbed in the sides of the triangle $\mathbf{D}$. Denote the corresponding absorption probabilities of the process $\mathrm{X}^{(n)}$ in the sides of the triangle $x = 0$, $y = 0$ and $x + y = M$ by α_n, β_n and γ_n, respectively.

Theorem 14.2.2 *Under the given conditions it holds* $\alpha_n \to \overline{\alpha}$, $\beta_n \to \overline{\beta}$ *and* $\gamma_n \to \overline{\gamma}$ *for* $n \to \infty$.

PROOF. Let $\mathrm{C}[0,\infty)$ be the space of continuous functions $f : [0, \infty) \to (-\infty, \infty)$ with its uniform topology. As one of the numerous metrics we take

$$\varrho(x, y) = \sum_{k=1}^\infty 2^{-k}(\sup_{0 \le t \le k} |x(t) - y(t)|)/(1 + \sup_{0 \le t \le k} |x(t) - y(t)|).$$

We define the mapping h from $\mathrm{C}[0,\infty)$ onto $\{0, 1\}$ where $h(f) = 1$ iff f starts in the interior of $\mathbf{D}$ *and* leaves it first via the horizontal side exclusively its two end points.

Furthermore, we denote by $\mathcal{A}_z$ (z positive integer) the set of functions starting in the interior of the new triangle $x = 0$, $y = 1/z$, $x + y = m$ and leaving it first by *crossing* the horizontal side exclusively the two end points. Regarding the metric above it is easy to see that the sets $\mathcal{A}_z$ are open. Moreover $h^{-1}(\{1\}) = \bigcup_{j=z_0}^\infty \bigcap_{z=j}^\infty \mathcal{A}_z$ and $h^{-1}(\{0\}) = \overline{h^{-1}(\{1\})}$, where we choose z_0 larger than $1/m$. Thus h is measurable.

Let $\mathrm{W}_t(\omega)$ be a realization of a Brownian motion starting in (a, b). In order to be a discontinuity point of h, the realization $\mathrm{W}_t(\omega)$ has to fulfill one of the following four conditions:

(i) The second component of $\mathrm{W}_t(\omega)$ only tangent $y = 0$.

(ii) $\mathrm{W}_t(\omega)$ reaches the point $(0,0)$.

(iii) $\mathrm{W}_t(\omega)$ reaches the point $(m,0)$.

(iv) $W_t(\omega)$ does not leave the triangle, i.e. it remains in a bounded area forever.

The probabilities of all four events are zero [e.g. Bhattacharia and Waymire (1990, Theoretical Complements to Section I.9 and Chapter V.14)].

Now, by the continuous mapping theorem [Billingsley (1968, Theorem 5.1)], $h(X^{(n)}) \xrightarrow{\mathcal{D}} h(W)$ and hence $\beta_n = P(h(X^{(n)}) = 1) \to P(h(W) = 1) = \overline{\beta}$ as $n \to \infty$.

The other two relations $\alpha_n \to \overline{\alpha}$ and $\gamma_n \to \overline{\gamma}$ can be shown analogously. ■

Obviously the absorption probabilities of $S^{(n)}$ and $X^{(n)}$ are equal. After discretization of Brownian motion according to Theorem 14.2.1 and replacing $X^{(n)}$ by $S^{(n)}$ we get for each n a random walk with step sizes nonequal to 1. But after multiplying the two components of the state space with $\Delta_{1,n}^{-1}$, and $\Delta_{2,n}^{-1}$ respectively, we can use the above formulae. The "formal discretization" leads to the following.

Continuous problem (W)		**Discretized problem after axis transformation ($S^{(n)}$)**
μ_1, μ_2, σ^2	$\to$	$p_n := \dfrac{1}{4}\left(1 + \dfrac{\mu_1\sqrt{2}}{\sigma\sqrt{n}}\right),$
		$q_n := \dfrac{1}{4}\left(1 - \dfrac{\mu_1\sqrt{2}}{\sigma\sqrt{n}}\right),$
		$r_n := \dfrac{1}{4}\left(1 + \dfrac{\mu_2\sqrt{2}}{\sigma\sqrt{n}}\right),$
		$s_n := \dfrac{1}{4}\left(1 - \dfrac{\mu_2\sqrt{2}}{\sigma\sqrt{n}}\right),$
a	$\to$	$A = \left[\dfrac{a\sqrt{n}}{\sigma\sqrt{2}}\right]$
b	$\to$	$B = \left[\dfrac{b\sqrt{n}}{\sigma\sqrt{2}}\right]$
c	$\to$	$C = \begin{cases} \left[\dfrac{c\sqrt{n}}{\sigma\sqrt{2}}\right] & \text{if } \dfrac{c\sqrt{n}}{\sigma\sqrt{2}} = \left[\dfrac{c\sqrt{n}}{\sigma\sqrt{2}}\right] \\ \left[\dfrac{c\sqrt{n}}{\sigma\sqrt{2}}\right] + 1 & \text{else.} \end{cases}$

$$(14.6)$$

Remark 14.2.1 The demand of the independence of the Brownian motion components is necessary because of the functional limit theorem[1]. The condition $\mu_1 = \pm\mu_2$ ensures $p_n q_n = r_n s_n$ for all n. But only if $\sigma_1 = \sigma_2$, there is no distortion of the triangle form after the axis transformation.

---

[1] A sequence of random walks with steps only parallel to the axis cannot lead to a Brownian motion with dependent components.

14.3   The Case of Equal Drifts

Suppose $\mu_1 = \mu_2 =: \mu$. From (14.1) and (14.6) we conclude $p_n = r_n$, $q_n = s_n$ and therefore

$$e^\Theta = e^\Phi = \sqrt{\tfrac{p_n}{q_n}} = \sqrt{\tfrac{p_n}{s_n}} \tag{14.7}$$

and

$$\lambda = \varrho = \sqrt{p_n q_n} = \tfrac{1}{4}\sqrt{1 - \tfrac{2\mu^2}{\sigma^2 n}}. \tag{14.8}$$

Since $\pi^{A,B}(x, y) = \pi^{B,A}(y, x)$ for all x and y between 0 and N we obtain with (14.4) and (14.5)

$$
\begin{aligned}
\overline{\beta} &= \lim_{n\to\infty}\left\{ s_n \sum_{x=1}^{N-2} \pi^{A,B}(x, 1) - s_n \sum_{x=1}^{N-2} \left(\frac{p_n}{s_n}\right)^{1-(N-x)} \pi^{B,A}(N - x, N - 1)\right\} \\
&= \lim_{n\to\infty}\left\{ s_n \sum_{x=1}^{N-2} \pi_1^{A,B}(x, 1) - s_n \sum_{x=2}^{N-1} e^{2\Theta(1-x)} \pi_2^{B,A}(x, N - 1)\right\},
\end{aligned} \tag{14.9}
$$

where in the second sum we made an index transformation. Taking (14.3) and setting $w_n := \left(1 - \tfrac{2\mu^2}{\sigma^2 n}\right)^{-1/2}$ we have

$$
\begin{aligned}
\overline{\beta} = 2 \lim_{n\to\infty} \Bigg\{ &\left(1 - \frac{\mu\sqrt{2}}{\sigma\sqrt{n}}\right) w_n \\
&\cdot \Bigg(\sum_{x=1}^{N-2} \frac{e^{\Theta(x+1-A-B)}}{N} \sum_{j=1}^{N-1} \sin\frac{j\pi x}{N} \sin\frac{j\pi A}{N} \frac{\sinh(A + C)\delta_{j,n}}{\sinh N\delta_{j,n}} \\
&- \sum_{x=2}^{N-1} \frac{e^{\Theta(1-x+C)}}{N} \sum_{j=1}^{N-1} \sin\frac{j\pi x}{N} \sin\frac{j\pi B}{N} \frac{\sinh A\delta_{j,n}}{\sinh N\delta_{j,n}} \Bigg) \Bigg\},
\end{aligned} \tag{14.10}
$$

where $\delta_{j,n}$ is defined by

$$\cos\frac{j\pi}{N} + \cosh\delta_{j,n} = 2w_n. \tag{14.11}$$

Note that from now on we choose only the positive solution of (14.11). Now we find as $n \to \infty$

$$\left(1 - \frac{\mu\sqrt{2}}{\sigma\sqrt{n}}\right) w_n = 1 + o(1), \tag{14.12}$$

$$
\left.
\begin{aligned}
A &= \frac{a\sqrt{n}}{\sigma\sqrt{2}}(1 + o(1)), \quad B = \frac{b\sqrt{n}}{\sigma\sqrt{2}}(1 + o(1)), \\
C &= \frac{c\sqrt{n}}{\sigma\sqrt{2}}(1 + o(1)), \quad N = \frac{m\sqrt{n}}{\sigma\sqrt{2}}(1 + o(1)),
\end{aligned}
\right\} \tag{14.13}
$$

and

$$e^{\Theta} = \left(\frac{1 + \frac{\mu\sqrt{2}}{\sigma\sqrt{n}}}{1 - \frac{\mu\sqrt{2}}{\sigma\sqrt{n}}}\right)^{\frac{1}{2}} = \left(1 + \frac{\mu\sqrt{2}}{\sigma\sqrt{n}}\right)(1 + o(1)). \tag{14.14}$$

Hence

$$e^{\Theta(1-A-B)} = \exp\left(-\frac{\mu(a+b)}{\sigma^2}\right)(1 + o(1)) \tag{14.15}$$

and

$$e^{\Theta(1+C)} = \exp\left(\frac{\mu c}{\sigma^2}\right)(1 + o(1)). \tag{14.16}$$

Now the relation (14.10) reduces to

$$\overline{\beta} = 2\exp\left(-\frac{\mu(a+b)}{\sigma^2}\right)\lim_{n\to\infty}\sum_{j=1}^{N-1}\sin\frac{j\pi A}{N}\cdot\frac{\sinh(A+C)\delta_{j,n}}{\sinh N\delta_{j,n}}\cdot\frac{1}{N}\sum_{x=1}^{N-2}e^{\Theta x}\sin\frac{j\pi x}{N}$$

$$-2\exp\left(\frac{\mu c}{\sigma^2}\right)\lim_{n\to\infty}\sum_{j=1}^{N-1}\sin\frac{j\pi B}{N}\cdot\frac{\sinh A\delta_{j,n}}{\sinh N\delta_{j,n}}\cdot\frac{1}{N}\sum_{x=2}^{N-1}e^{-\Theta x}\sin\frac{j\pi x}{N}. \tag{14.17}$$

Next we investigate the limits of the six factors in (14.17).

With (14.13) we get

$$\lim_{n\to\infty}\sin\frac{j\pi A}{N} = \sin\frac{j\pi a}{m}, \quad \lim_{n\to\infty}\sin\frac{j\pi B}{N} = \sin\frac{j\pi b}{m}. \tag{14.18}$$

Using now

$$e^{\Theta x} = \left(\frac{p_n}{q_n}\right)^{\frac{x}{2}} = e^{\frac{\mu m}{\sigma^2}(1+o(1))\frac{x}{N}} = \left(e^{\frac{\mu m}{\sigma^2}}(1 + o(1))\right)^{\frac{x}{N}} \tag{14.19}$$

as $n \to \infty$, e.g. $N \to \infty$, Lemma 14.5.2 in the Appendix proves the relations

$$\lim_{n\to\infty}\frac{1}{N}\sum_{x=1}^{N-2}e^{\Theta x}\sin\frac{j\pi x}{N} = \frac{\sigma^4 j\pi}{\mu^2 m^2 + \sigma^4 j^2 \pi^2}\left(\exp\left(\frac{\mu m}{\sigma^2}\right)(-1)^{j+1} + 1\right), \tag{14.20}$$

$$\lim_{n\to\infty}\frac{1}{N}\sum_{x=2}^{N-1}e^{-\Theta x}\sin\frac{j\pi x}{N} = \frac{\sigma^4 j\pi}{\mu^2 m^2 + \sigma^4 j^2 \pi^2}\left(\exp\left(-\frac{\mu m}{\sigma^2}\right)(-1)^{j+1} + 1\right). \tag{14.21}$$

Further we consider $\lim_{n\to\infty} e^{N\delta_{j,n}}$. With (14.13), $w_n = 1 + \frac{\mu^2}{\sigma^2 n}(1 + o(1))$ and $\cos\frac{j\pi}{N} = 1 - \frac{j^2\pi^2}{2N^2}(1 + o(1))$ as $n \to \infty$ (Taylor expansions), from (14.11) it follows that

$$\cosh\delta_{j,n} = \frac{1}{2}\left(e^{\delta_{j,n}} + e^{-\delta_{j,n}}\right) = 1 + \frac{1}{n}\left(\frac{2\mu^2}{\sigma^2} + \frac{j\pi\sigma^2}{m^2}\right)(1 + o(1)). \tag{14.22}$$

We chose $\delta_{j,n} > 0$, hence

$$e^{\delta_{j,n}} = 1 + \frac{2}{\sqrt{n}}\sqrt{\frac{2\mu^2}{\sigma^2} + \frac{j\pi\sigma^2}{m^2}} \tag{14.23}$$

and therefore

$$\lim_{n\to\infty} e^{N\delta_{j,n}} = \exp\sqrt{j^2\pi^2 + \frac{2\mu^2 m^2}{\sigma^4}}. \tag{14.24}$$

With (14.13) and (14.24) we find

$$\lim_{n\to\infty} \frac{\sinh(A+C)\delta_{j,n}}{\sinh N\delta_{j,n}} = \frac{\sinh\left(\frac{a+c}{m}\sqrt{j^2\pi^2 + \frac{2\mu^2 m^2}{\sigma^4}}\right)}{\sinh\sqrt{j^2\pi^2 + \frac{2\mu^2 m^2}{\sigma^4}}} \tag{14.25}$$

$$\text{and}\quad \lim_{n\to\infty} \frac{\sinh A\delta_{j,n}}{\sinh N\delta_{j,n}} = \frac{\sinh\left(\frac{a}{m}\sqrt{j^2\pi^2 + \frac{2\mu^2 m^2}{\sigma^4}}\right)}{\sinh\sqrt{j^2\pi^2 + \frac{2\mu^2 m^2}{\sigma^4}}}. \tag{14.26}$$

Now we can establish the result.

Theorem 14.3.1 *Let* $W = (W^1, W^2)$ *be a two-dimensional Brownian motion with independent components which starts in* (a,b) *a.s. Suppose* $\mu_1 = \mu_2 = \mu$ *for the drift parameters and* $\sigma_1 = \sigma_2 = \sigma$ *for the diffusion coefficients. Then*

(i) $\overline{\beta} = \overline{\beta}(a,b,c)$

$$\begin{aligned}
= \quad & \exp\left(-\frac{\mu(a+b)}{\sigma^2}\right) \sum_{j=1}^{\infty} \frac{2\sigma^4 j\pi}{\mu^2(a+b+c)^2 + \sigma^4 j^2\pi^2} \sin\left(\frac{j\pi a}{a+b+c}\right) \\
\times \quad & \left(\exp\left(\frac{\mu(a+b+c)}{\sigma^2}\right)(-1)^{j+1} + 1\right) \\
\times \quad & \frac{\sinh\left(\frac{a+c}{a+b+c}\sqrt{j^2\pi^2 + \frac{2\mu^2(a+b+c)^2}{\sigma^4}}\right)}{\sinh\sqrt{j^2\pi^2 + \frac{2\mu^2(a+b+c)^2}{\sigma^4}}} \\
- \quad & \exp\left(\frac{\mu c}{\sigma^2}\right) \sum_{j=1}^{\infty} \frac{2\sigma^4 j\pi}{\mu^2(a+b+c)^2 + \sigma^4 j^2\pi^2} \sin\left(\frac{j\pi b}{a+b+c}\right) \\
\times \quad & \left(\exp\left(-\frac{\mu(a+b+c)}{\sigma^2}\right)(-1)^{j+1} + 1\right) \\
\times \quad & \frac{\sinh\left(\frac{a}{a+b+c}\sqrt{j^2\pi^2 + \frac{2\mu^2(a+b+c)^2}{\sigma^4}}\right)}{\sinh\sqrt{j^2\pi^2 + \frac{2\mu^2(a+b+c)^2}{\sigma^4}}}, \tag{14.27}
\end{aligned}$$

(ii) $\overline{\alpha} = \overline{\beta}(b,a,c)$,

(iii) $\overline{\gamma} = 1 - \overline{\alpha} - \overline{\beta}$.

PROOF.

(i) By Corollary 14.5.1 to Theorem 14.5.1 (Appendix) we can change sum and limes in both summands of (14.17), so setting (14.18), (14.20), (14.21), (14.25) and (14.26), into (14.17) leads to (14.27).

(ii) This we get by reflecting the triangle problem on $y = x$.

(iii) The last formula holds because the probability of a Brownian motion not leaving a bounded domain forever is zero. ■

14.4   The Case of Opposite Drifts

Suppose $\mu_1 = \mu_2 =: -\mu$. From (14.1) and (14.6) we conclude $p_n = s_n$, $q_n = r_n$ and therefore

$$e^{\Theta} = e^{-\Phi} = \sqrt{\frac{p_n}{q_n}} \tag{14.28}$$

and

$$\lambda = \varrho = \sqrt{p_n q_n} = \tfrac{1}{4}\sqrt{1 - \frac{2\mu^2}{\sigma^2 n}}. \tag{14.29}$$

Like in the previous section we transform the second summand of formula (14.5) by reflecting the square problem on $y = x$. This we shortly describe by $\pi^{A,B,\mu}(x, y) = \pi^{B,A,-\mu}(y, x)$ for all x and y between 0 and N, so we obtain [see (14.9)]

$$\overline{\beta} = \lim_{n \to \infty} \left\{ s \sum_{x=1}^{N-2} \pi_1^{A,B,\mu}(x, 1) - s \sum_{x=2}^{N-1} \pi_2^{B,A,-\mu}(x, N-1) \right\}. \tag{14.30}$$

Taking (14.3) and setting again $w_n := \left(1 - \frac{2\mu^2}{\sigma^2 n}\right)^{-1/2}$ we have

$$
\begin{aligned}
\overline{\beta} = 2 \lim_{n \to \infty} &\left\{ \left(1 + \frac{\mu\sqrt{2}}{\sigma\sqrt{n}}\right) w_n \right. \\
&\cdot \left(\sum_{x=1}^{N-2} \frac{e^{\Theta(x-A+B-1)}}{N} \sum_{j=1}^{N-1} \sin\frac{j\pi x}{N} \sin\frac{j\pi A}{N} \frac{\sinh(A+C)\delta_{j,n}}{\sinh N\delta_{j,n}} \right. \\
&\left.\left. - \sum_{x=2}^{N-1} \frac{e^{\Theta(2B+C-1-x)}}{N} \sum_{j=1}^{N-1} \sin\frac{j\pi x}{N} \sin\frac{j\pi B}{N} \frac{\sinh A\delta_{j,n}}{\sinh N\delta_{j,n}} \right) \right\},
\end{aligned}
\tag{14.31}
$$

where $\delta_{j,n}$ is defined by (14.11). All terms of (14.31) are very similar or identical to those in (14.10), so we get analogously

Theorem 14.4.1 *Let* $W = (W^1, W^2)$ *be a two-dimensional Brownian motion with independent components which starts in* (a, b) *a.s. Suppose* $\mu_1 = -\mu_2 = \mu$ *for the drift parameters and* $\sigma_1 = \sigma_2 = \sigma$ *for the diffusion coefficients. Then*

(i) $\overline{\beta} = \overline{\beta}(a, b, c, \mu)$

$$
= \exp\left(\frac{\mu(-a+b)}{\sigma^2}\right) \sum_{j=1}^{\infty} \frac{2\sigma^4 j\pi}{\mu^2(a+b+c)^2 + \sigma^4 j^2 \pi^2} \sin\left(\frac{j\pi a}{a+b+c}\right)
$$

$$
\times \left(\exp\left(\frac{\mu(a+b+c)}{\sigma^2}\right)(-1)^{j+1} + 1\right)
$$

$$
\times \frac{\sinh\left(\frac{a+c}{a+b+c}\sqrt{j^2\pi^2 + \frac{2\mu^2(a+b+c)^2}{\sigma^4}}\right)}{\sinh\sqrt{j^2\pi^2 + \frac{2\mu^2(a+b+c)^2}{\sigma^4}}}
$$

$$
- \exp\left(\frac{\mu(2b+c)}{\sigma^2}\right) \sum_{j=1}^{\infty} \frac{2\sigma^4 j\pi}{\mu^2(a+b+c)^2 + \sigma^4 j^2 \pi^2} \sin\left(\frac{j\pi b}{a+b+c}\right)
$$

$$
\times \left(\exp\left(-\frac{\mu(a+b+c)}{\sigma^2}\right)(-1)^{j+1} + 1\right)
$$

$$
\times \frac{\sinh\left(\frac{a}{a+b+c}\sqrt{j^2\pi^2 + \frac{2\mu^2(a+b+c)^2}{\sigma^4}}\right)}{\sinh\sqrt{j^2\pi^2 + \frac{2\mu^2(a+b+c)^2}{\sigma^4}}}, \tag{14.32}
$$

(ii) $\overline{\alpha} = \overline{\beta}(b, a, c, -\mu)$,

(iii) $\overline{\gamma} = 1 - \overline{\alpha} - \overline{\beta}$.

14.5   Discussion of the Results

In both cases we get infinite series in the results. Since the fractions with the hyperbolicus terms have the exponential decreases

$$
\exp\left\{-\frac{u}{a+b+c}\sqrt{j^2\pi^2 + \frac{2\mu^2(a+b+c)^2}{\sigma^4}}\right\},
$$

where $u = b$ or $u = b + c$, for large j, these sums can be easily calculated by computers. First computations confirmed two expected simple properties: The nearer the starting point to a triangle side or the smaller the drift away from this side, the higher is the absorption probability in this side. A more detailed analysis must be adapted to the respective practical problem.

Note that for $\mu_1 = \mu_2 = 0$ formulae (14.27) and (14.32) are identical. Moreover, they are independent of σ then. Hence, in the case of zero drifts the absorption probabilities only depends on the starting point.

Appendix

Lemma 14.5.1

$$\exists\, K > 0 \;\forall\, 0 < x < \pi:$$

$$g(x) := \left(2 - \cos x + \sqrt{3 - 4\cos x + \cos^2 x}\right)^{\frac{1}{x}} \geq e^K > 1.$$

PROOF.

(a) Let be $f(u) := 2 - u + \sqrt{3 - 4u + u^2}$, $-1 < u < 1$. Of course $f(u) - 1$ is positive, and (regarding the first derivation) $f(u)$ is decreasing, i.e. $f(\cos x)$ is increasing on $(0, \pi)$.

(b) $\lim_{x \to 0} g(x) = \lim_{x \to 0} \exp\left\{ \dfrac{\ln\left(2 - \cos x + \sqrt{3 - 4\cos x + \cos^2 x}\right)}{x} \right\}$

$\overset{\text{L'Hospital}}{=} \exp\left\{ \lim_{x \to 0} \dfrac{\sin x + \dfrac{2\sin x - \sin x \cos x}{\sqrt{3 - 4\cos x + \cos^2 x}}}{2 - \cos x + \sqrt{3 - 4\cos x + \cos^2 x}} \right\}$

$= \exp\left\{ \lim_{x \to 0} \dfrac{\sin x \left(2 - \cos x + \sqrt{3 - 4\cos x + \cos^2 x}\right)}{\sqrt{3 - 4\cos x + \cos^2 x}\left(2 - \cos x + \sqrt{3 - 4\cos x + \cos^2 x}\right)} \right\}$

$= \exp\left\{ \lim_{x \to 0} \dfrac{\sin x}{x} \dfrac{1}{\sqrt{(3 - \cos x)\frac{1 - \cos x}{x^2}}} \right\}$

$\overset{\text{L'Hospital}}{=} \exp\left\{ \left(2 \lim_{x \to 0} \dfrac{\sin x}{2x}\right)^{-\frac{1}{2}} \right\}$

$= e.$

Setting $\eta := 1$, there exists a (sufficiently small) $\varepsilon > 0$ such that $0 < x < \varepsilon$ $\Rightarrow |g(x) - e| < \eta$, i.e. $g(x) > e - 1 > 1$ if $0 < x < \varepsilon$. And because of (a): $g(x) \geq \left(2 - \cos\varepsilon + \sqrt{3 - 4\cos\varepsilon + \cos^2\varepsilon}\right)^{\frac{1}{\pi}} > 1$ if $\varepsilon \leq x < \pi$. Choosing

$$e^K := \min\left\{ e - 1, \left(2 - \cos\varepsilon + \sqrt{3 - 4\cos\varepsilon + \cos^2\varepsilon}\right)^{\frac{1}{\pi}} \right\}$$

completes the proof. ■

Theorem 14.5.1 *Suppose that for all integers $n \geq n_0$ the function $f : \mathbb{R}^2 \to \mathbb{R}$ fulfills the following three conditions:*

(i) $\sum\limits_{j=1}^{\infty} f(j,n)$ *exists.*

(ii) *There exists a sequence $\{b(j)\}_{j=1}^{\infty}$ such that $|f(j,n)| \leq b(j)$ and $\sum\limits_{j=1}^{\infty} b(j) < \infty$.*

(iii) $\lim\limits_{n\to\infty} f(j,n)$ *exists for all $j \geq 1$.*

Then it holds $\lim\limits_{n\to\infty} \sum\limits_{j=1}^{\infty} f(j,n) = \sum\limits_{j=1}^{\infty} \lim\limits_{n\to\infty} f(j,n)$.

This theorem can be found e.g. in Ferschl (1970).

Corollary 14.5.1 *In both summands of (14.17) sum and lines can be changed.*

PROOF.

(i) Because the sums in (14.17) are finite, this is trivially fulfilled.

(ii) If we take first the second summand of (14.17), with (14.14) and for sufficiently large n we can estimate

$$\left| \sin \frac{j\pi B}{N} \frac{e^{\delta_{j,n}\cdot A} - e^{-\delta_{j,n}\cdot A}}{e^{\delta_{j,n}\cdot N} - e^{-\delta_{j,n}\cdot N}} \cdot \frac{1}{N} \sum_{x=2}^{N-1} \left(e^{-\frac{\mu m}{\sigma^2}}(1+o(1)) \right)^{\frac{x}{N}} \sin \frac{j\pi x}{N} \right|$$

$$< \quad 1 \cdot \frac{1}{e^{\delta_{j,n}\cdot(N-A)} - e^{-\delta_{j,n}\cdot(N+A)}} \cdot 2\max\{1, e^{-\frac{\mu m}{\sigma^2}}\}$$

$$< \quad \frac{2\max\{1, e^{-\frac{\mu m}{\sigma^2}}\}}{\left(e^{\delta_{j,N}}\right)^{N\left(1-\frac{a}{m}\right)(1+o(1))} - 1}$$

$$< \quad \frac{2\max\{1, e^{-\frac{\mu m}{\sigma^2}}\}}{\left(2w_n - \cos\frac{\pi j}{N} + \sqrt{4w_n\left(w_n - \cos\frac{\pi j}{N}\right) + \cos^2\frac{\pi j}{N} - 1}\right)^{N\frac{b+c}{2m}} - 1}$$

$$\underset{w_n \gtrless 1}{<} \quad \frac{2\max\{1, e^{-\frac{\mu m}{\sigma^2}}\}}{\left(2 - \cos\frac{\pi j}{N} + \sqrt{4\left(1 - \cos\frac{\pi j}{N}\right) + \cos^2\frac{\pi j}{N} - 1}\right)^{\left(\frac{\pi j}{N}\right)^{-1}\pi j\frac{b+c}{2m}} - 1}$$

$$\underset{\text{Lemma 14.5.1}}{<} \quad \frac{2\max\{1, e^{-\frac{\mu m}{\sigma^2}}\}}{e^{\frac{K\pi(b+c)}{2m}j} - 1}. \tag{14.33}$$

Adding up (14.33) from $j = 1$ to infinity we get a convergent sum. For the first sum of (14.17) there exists a similar estimation in the same manner.

(iii) See (14.18), (14.20), (14.21), (14.25), and (14.26).   ∎

Lemma 14.5.2 *Let* $\varepsilon_N \to 0$ *as* $N \to \infty$ *and let* L, c *and* j *be some real numbers, then*

$$\lim_{N \to \infty} \frac{1}{N} \sum_{x=1}^{N} e^{L\frac{x}{N}} (1 + \varepsilon_N)^{\frac{x}{N}} \sin c\frac{x}{N} = \int_0^1 e^{Lu} \sin cu \, du.$$

$$= \frac{e^L}{c^2 + L^2} (L \sin c - c \cos c) + \frac{c}{c^2 + L^2}.$$

PROOF. Consider

$$\mathbf{I}_N := \sum_{x=1}^{N} \int_{\frac{x-1}{N}}^{\frac{x}{N}} \left(e^{L\frac{x}{N}} (1 + \varepsilon_N)^{\frac{x}{N}} \sin c\frac{x}{N} - e^{Lu} \sin cu \right) du.$$

If $0 \le \frac{x-1}{N} \le u \le \frac{x}{N} \le 1$, then we find

$$|e^{L\frac{x}{N}} - e^{Lu}| \le e^L \frac{L}{N},$$
$$|\sin c\frac{x}{N} - \sin cu| \le \frac{c}{N},$$
$$|(1 + \varepsilon_N)^{\frac{x}{N}} - 1| \le |\varepsilon_N|.$$

Some obvious calculus leads to $\mathbf{I}_N \to 0$ as $N \to \infty$. ∎

References

1. Barnett, V. D. (1963). Some explicit results for an asymmetric two-dimensional random walk, *Proceedings of Cambridge Philosophical Society*, **59**, 451–462.

2. Barnett, V. D. (1964). A three-player extension of the Gambler's ruin problem, *Journal of Applied Probability*, **1**, 321–334.

3. Bhattacharia, R. N. and Waymire, E. C. (1990). *Stochastic Processes with Applications*, New York: John Wiley & Sons.

4. Billingsley, P. (1968). *Convergence of Probability Measures*, New York: John Wiley & Sons.

5. Dominé, M. (1993). Erstpassagenprobleme für ausgewählte Diffusionsprozesse, *Ph.D. Dissertation*, Otto-von-Guericke-University Magdeburg.

6. Dominé, M. (1996). First passage time distribution of a Wiener process with drift concerning two elastic barriers, *Journal of Applied Probability*, **33**, 164–175.

7. Ferschl, F. (1970). *Markoffketten*, Berlin: Springer-Verlag.

8. Stroock, D. W. and Varadhan, S. R. S. (1979). *Multidimensional Diffusion Processes*, Berlin: Springer-Verlag.

PART III
NETWORK ANALYSIS

15

A Simple Algorithm for Calculating Approximately the Reliability of Almost Arbitrary Large Networks

---

Elart von Collani

Volkswirtschaftliches Institut, Würzburg, Germany

Abstract: Complex systems, whose components are subject to failure, are generally modelled by undirected and connected graphs where an edge e connecting two nodes is available only with a certain probability $p_e < 1$. Generally network reliability is measured by the so-called "all-terminal reliability", which is the probability that all nodes of the network are connected with one another. Besides, the so-called K-terminal reliability is used in reliability analysis of systems, where K is a subset of the set of nodes, and the K-terminal reliability is the probability that all the nodes of K are connected with one another. The problem is to determine these reliability measures for a general system, which turns out to be possible only for rather small (w.r.t. the number of nodes and edges) or specially structured (parallel or serial) systems. Thus simple approximations are needed to enable to compute the reliability for general systems.

Keywords and phrases: Large network, K-terminal reliability, approximate reliability

---

15.1 Introduction

The determination of network reliability for large networks is of great significance, but also of great difficulty.

- "Significant", because number, complexity and significance of networks are worldwide increasing and the effects of failures become more and more serious.

- "Difficult", because the determination of network reliability is a so-called NP-hard problem, thus leading to computationally intensive algorithms even for small and middle size networks.

The here proposed algorithm is based on the assumption that in complex systems, failure to reach a node is due to some local failures with high probability. For example if in a large and complex telephone network there is no reply from subscriber then there will be caused by a local failure almost with certainty. Thus the probability of failures of edges which don't lie within a vicinity of the node in question can be neglected. The resulting algorithm is extremely simple and requires even for very large and complex networks only an electronic pocket calculator to determine approximately the network reliability. The examples show that the approximation is very good in the case of high system reliability and decreases with decreasing system reliability. For system reliability of $p \geq 0.8$ the approximation has proved to be rather accurate for sufficiently complex networks, for $0.7 \leq p \leq 0.8$ the accuracy is satisfactory, and for $0.6 \leq p \leq 0.7$ it is still sufficient. Only if system reliability drops lower than 0.6 the algorithm should not be used in the general case.

15.2   Notations

The following notations are used to describe the networks as undirected graphs:

- V: set of nodes

- E: set of edges

- $G(V, E)$: undirected graph with set of nodes V and set of edges E

- n: number of nodes with $n = |V|$

- m: number of edges with $m = |E|$

- $E_i \subset E$ set of edges connected with node $v_i \in V$

- $m_i = |E_i|$ number of edges connected with node v_i

- $K = \{v_i^K\}_{i=1,\cdots,|K|}$: set of terminal nodes with $K \subset V$

Aim of the reliability analysis is to determine the probability that each node of K is able to communicate with each other node of K which is called K-terminal reliability. The extreme cases are $|K| = n$ and $|K| = 2$ called *all-terminal reliability* and *two-terminal reliability*, respectively.

For evaluating system reliability we use the following notations:

- p_e: availability of edge $e \in E$

- $R(G, K)$: K-terminal reliability of G

- $\underline{R}(G, K)$: lower bound for $R(G, K)$

- $\overline{R}(G, K)$: upper bound for $R(G, K)$

- $R(\widehat{G, K})$: approximation for $R(G, K)$

The relative, absolute deviation of the approximation from the exact value is denoted by r and used for measuring its quality:

-
$$r = \frac{\left| R(G,K) - R(\widehat{G,K}) \right|}{R(G,K)} \, .$$

15.3 The Approximation

With respect to the graph $G(V, E)$ we distinguish two cases:

- The system to be represented by $G(V, E)$ consists of several subsystems $G_i(V, E)$, which are connected by a few edges to an overall system. We call the corresponding system a *compound network*.

- The system to be represented by $G(V, E)$ does not consist of several subsystems. We call the corresponding system a *simple network*.

Remark 15.3.1 The above distinction between simple and compound networks does not mean that one should look for a possibility to break a given large network into several smaller networks, e.g. by using so-called decomposition techniques, but it shall take into account the fact that in real world in ever increasing cases smaller networks are linked together by a few connections thus forming a larger network.

15.3.1 Simple network

In this chapter we derive an approximation for simple networks, which by straightforward modification will be later extended to compound networks.

Further notations

Let

- $c(G) = \frac{m}{\binom{n}{2}}$, where $c(G)$ is called the complexity of $G(V, E)$, with $\frac{n-1}{\binom{n}{2}} \leq c(G) \leq 1$

- $e_j = \begin{cases} 1 & \text{if} & e_j \text{ is operating} \\ 0 & \text{if} & e_j \text{ has failed} \end{cases}$

- $K = \{v_i^K\}_{i=1,\cdots|K|} \subset V$

- $X(v_i^K) = \begin{cases} 1 & \text{if} & v_i^K \text{ connected with each node in } K \\ 0 & \text{if} & v_i^K \text{ disconnected with at least one node } v_j^K \in K \end{cases}$

- $E_i^K = \{e_{i_k}^K\}_{k=1,\cdots,m_i^K}$ set of edges connected with $v_i^K \in K$

Derivation

The problem is to compute the K-terminal reliability

$$R(G, K) = \mathbf{P}\Big(\text{all nodes of } K \text{ are connected with one another}\Big)$$

where $K \subset V$ with $2 \leq |K| \leq n$.

Define the following events

$\mathcal{R} =$ all nodes of K are connected with each other
$\mathcal{S} =$ complement of $\mathcal{R}$, i.e. there is at least one node of K which is not connected with each of the other nodes of K

then

$$R(G, K) = \mathbf{P}(\mathcal{R}) = 1 - \mathbf{P}(\mathcal{S}).$$

Moreover let

$$\mathcal{T} = \bigcup_{i=1}^{|K|} \left\{ \sum_{k=1}^{m_i^K} e_{i_k}^K = 0, X\left(v_j^{K\setminus\{v_i^K\}}\right) = 1, j \neq i \right\}$$

and

$$\mathcal{U} = \mathcal{S} \setminus \mathcal{T}$$

then

$$R(G, K) = 1 - \mathbf{P}(\mathcal{S}) = 1 - \mathbf{P}(\mathcal{T}) - \mathbf{P}(\mathcal{U})$$

$\mathcal{T}$ is the event that exactly one node of K cannot be reached from the rest of the terminal nodes, which are connected with each other. Hence, $\mathcal{U}$ describes the event that K is decomposed into two or more subsets which cannot communicate with each other where either the number of subsets is larger that 2 or if it is 2 than the number of nodes in each of the subsets is larger than 1.

Our aim is to develop an algorithm for calculating the K-terminal reliability of *large* and *complex* systems. Moreover, it is assumed here that the edge availability and therefore also $R(G, K)$ is sufficiently large too.

Assumption A: $\qquad c(G)$ and p_e for $e \in E$ are large.

From **A** we conclude:

$$\mathbf{P}\big(\mathcal{T}\big) \approx \sum_{i=1}^{|K|} \mathbf{P}\Big(\sum_{k=1}^{m_i^K} e_{i_k}^K = 0\Big) \tag{15.1}$$

$$\mathbf{P}\big(\mathcal{U}\big) \approx 0. \tag{15.2}$$

Inserting (15.1) and (15.2) into (15.3.1) yields

$$
\begin{aligned}
R(G, K) \quad &\approx \quad 1 - \sum_{i=1}^{|K|} \mathbf{P}\Big(\sum_{k=1}^{m_i^K} e_{i_k}^K = 0\Big) \\
&= \quad 1 - \sum_{i=1}^{|K|} \prod_{k=1}^{m_i^K} P\big(e_{i_k}^K = 0\big) \\
&= \quad 1 - \sum_{i=1}^{|K|} \prod_{k=1}^{m_i^K} (1 - p_{i_k}^K).
\end{aligned} \tag{15.3}
$$

Because

$$\mathbf{P}\big(\mathcal{T}\big) < \mathbf{P}\Big(\cup_{i=1}^{|K|} \{\sum_{k=1}^{m_i^K} e_{i_k}^K = 0\}\Big) < \sum_{k=1}^{|K|} \mathbf{P}\Big(\sum_{k=1}^{|K|} e_{i_k}^K = 0\Big)$$

and

$$\mathbf{P}\big(\mathcal{U}\big) > 0$$

the approximations (15.1) and (15.2) are countermoving and therefore compensate one another to a certain extent.

15.4 Algorithms

Based on the approximation (15.3), three algorithms are proposed. The first one for simple networks and at least three terminal nodes (i.e. $|K| > 2$), the second one for simple networks and two terminal nodes, and finally an algorithm for compound networks.

Algorithm 1

For the case that the network does not consists of several partial networks, the K-terminal reliability for $|K| > 2$ is calculated approximately in three steps:

- **Step 1.1:**
  Determine for each node $v_i^K \in K$ the set of edges E_i^K and compute

$$\prod_{k=1}^{m_i^K} (1 - p_{i_k}^K).$$

- **Step 1.2:**
  Compute

$$\sum_{i=1}^{|K|} \prod_{k=1}^{m_1^K} (1 - p_{i_k}^K).$$

- **Step 1.3:**
  Set

$$\hat{R}(G, K) = 1 - \sum_{i=1}^{|K|} \prod_{k=1}^{m_i^K} (1 - p_{i_k}^K).$$

Example. For illustration consider the following example network ARTI[1] with $n = |V| = 11$, $m = |E| = 24$ and with identical edge availabilities $p_e = 0.8$ for any $e \in E$. Let K be given by $K = \{1, 3, 6, 8, 11\}$.

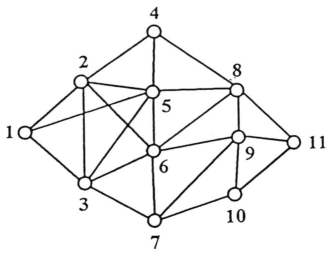

Figure 15.1: Network ARTI, n = 11, m = 24

---

[1]compare Blechschmidt (1993)

Because of the identical edge availabilities, it is sufficient to know the cardinality of each set E_i^K:

$$|E_1^K| = 3 \qquad |E_3^K| = 5 \qquad |E_6^K| = 6 \qquad |E_8^K| = 5 \qquad |E_{11}^K| = 3$$

hence

$$\hat{R}(ARTI, K) = 1 - \left(2 \cdot 0.2^3 + 2 \cdot 0.2^5 + 0.2^6\right) = 0.9832.$$

Algorithm 2

If a 2-terminal reliability is to be calculated for a simple network, then Algorithm 1 can easily be refined by taking additionally into account that simultaneously one of the terminal nodes together with one of the adjacent nodes are not reachable.

Let

- $K = \{v_i, v_\ell\}$

- $E_i = \{e_{i_j}\}_{j=1,\cdots,m_i}$ the set of edges connected with v_i

- $E_\ell = \{e_{\ell_j}\}_{j=1,\cdots,m_\ell}$ the set of edges connected with v_ℓ

- $v_{i,i_j} = (v_i, e_{i_j}, v_{i_j})$, $j = 1, \cdots, m_i$, the m_i subsystems consisting out of the terminal node v_i and one adjacent node v_{i_j} and the connecting edge $e_{i_j} \in E_i$

- $v_{\ell,\ell_j} = (v_\ell, e_{\ell_j}, v_{\ell_j})$, $j = 1, \cdots, m_\ell$, the m_ℓ subsystems consisting out of the terminal node v_ℓ and one adjacent node $v_{\ell_j} \in E_\ell$ and the connecting edge e_{ℓ_j}

Then the 2-terminal reliability $\hat{R}(G, \{v_i, v_\ell\})$ for a simple network is approximately determined in four steps:

- **Step 2.1:**
  Determine for the two nodes v_i and v_ℓ of K:

  a) $E_i = \{e_{i_j}\}_{j=1,\cdots,m_i}$ and $E_\ell = \{e_{\ell_j}\}_{j=1,\cdots,m_\ell}$ and the corresponding edge availabilities $\{p_{i_j}\}_{j=1,\cdots,m_i}$ and $\{p_{\ell_j}\}_{j=1,\cdots,m_\ell}$, respectively.

  b) $\prod_{j=1}^{m_i}(1 - p_{i_j})$ and $\prod_{j=1}^{m_\ell}(1 - p_{\ell_j})$

  Determine for each subsystem $v_{i,i_j} = (v_i, e_{i_j}, v_{i_j})$, $j = 1, \cdots, m_i$

  c) $E_{i,i_j} = \{e_{i,i_j}^{(k)}\}_{k=1,\cdots,m_i+m_{i_j}-2}$ and the corresponding edge availabilities
  $\{p_{i,i_j}^{(k)}\}_{k=1,\cdots,m_i+m_{i_j}-2}$

  d) $\prod_{k=1}^{m_i+m_{i_j}-2}(1 - p_{i,i_j}^{(k)})$

and for each subsystem $v_{\ell,\ell_j} = (v_\ell, e_{\ell_j}, v_{\ell_j})$, $j = 1, \cdots, m_\ell$

e) $E_{\ell,\ell_j} = \{e_{\ell,\ell_j}^{(k)}\}_{k=1,\cdots,m_\ell+m_{\ell_j}-2}$ and the corresponding edge availabilities
$\{p_{\ell,\ell_j}^{(k)}\}_{k=1,\cdots,m_\ell+m_{\ell_j}-2}$

f) $\prod_{k=1}^{m_\ell+m_{\ell_j}-2}(1 - p_{\ell,\ell_j}^{(k)})$

- **Step 2.2:**
  Compute
  $$\sum_{j=1}^{m_i} \prod_{k=1}^{m_i+m_{i_j}-2} (1 - p_{i,i_j}^{(k)})$$

  and

  $$\sum_{j=1}^{m_\ell}{}' \prod_{k=1}^{m_\ell+m_{\ell_j}-2} (1 - p_{\ell,\ell_j}^{(k)}).$$

Remark 15.4.1 $\sum'$ means that if v_i and v_ℓ are connected by edge e_{ℓ_j}, then the corresponding summand is omitted.

- **Step 2.3:**
  Set

  $$\hat{R}(G, \{v_i, v_\ell\}) = 1 - \left\{\sum_{j=1}^{m_i} \prod_{k=1}^{m_i+m_{i_j}-2} (1 - p_{i,i_j}^{(k)}) + \sum_{j=1}^{m_\ell} \prod_{k=1}^{m_\ell+m_{\ell_j}-2} (1 - p_{\ell,\ell_j}^{(k)})\right\}.$$

Example. For illustration consider again the network ARTI with $n = |V| = 11$, $m = |E| = 24$ with identical edge availabilities $p_e = 0.8$ for any $e \in E$. Let K be given by $K = \{4, 7\}$.

Because of the identical edge availability, it suffices to know the cardinalities of the sets E_4 and E_7:
$$|E_4| = 3 \qquad |E_7| = 4.$$

There are three subsystems generated by the node $4 \in K$ and four subsystems generated by the node $7 \in K$, the cardinality of these sets is given by:
$$|E_{4,2}| = 6 \qquad |E_{4,5}| = 7 \qquad |E_{4,8}| = 6$$
$$|E_{7,3}| = 7 \qquad |E_{7,6}| = 8 \qquad |E_{7,9}| = 7 \qquad |E_{7,10}| = 5$$

hence

$$\hat{R}(ARTI, \{4, 7\}) = 1 - \left(0.2^3 + 0.2^4 + 0.2^5 + 2 \times 0,2^6 + 3 \times 0.2^7 + 0.2^8\right) = 0.9899$$

15.4.1 Compound system

Very often there are several partial networks, which are connected by few edges to an overall network. In this case the above given algorithm is modified in order to take into account the special structure of the overall network.

Assume that the overall network $G(V, E)$ consists of h partial networks $G_\nu(V^{(\nu)}, E^{(\nu)})$, $\nu = 1, \cdots, h$ with

$$\cup_{\nu=1}^h V^{(\nu)} = V \tag{15.4}$$

and

$$\cup_{\nu=1}^h E^{(\nu)} \cap E = E_c \neq \emptyset \tag{15.5}$$

where E_c is the set of edges connecting the partial networks. Moreover let $E_{c,\nu} \subset E_c$ be the set of edges connected with some nodes of $V^{(\nu)}$, $\nu = 1, \cdots, h$, and $\{p_{\nu,m}^{(c)}\}_{m=1,\cdots,|E_{c,\nu}|}$ be the corresponding edge availabilities.

Moreover, for given set of terminals K, let

$$W_K = \left\{ \nu | K \cap V^{(\nu)} \neq \emptyset \right\}$$

be the partial systems containing at least one of the terminal nodes of K.

Algorithm 3

- **Step 3.1:**
  Determine for each node $v_i^K \in K$

  $$E_i^K = \{e_{i_j}^K\}_{j=1,\cdots,m_i^K} \text{ and the corresponding}$$

  edge availabilities $\{p_{i_j}^K\}_{j=1,\cdots,m_i^K}$

  $$\prod_{j=1}^{m_i^K}(1 - p_{i_j}^K)$$

  and for each partial Graph $G_\nu(V^{(\nu)}, E^{(\nu)})$ with $\nu \in W_K$

  $$\prod_{r=1}^{|E_{c,\nu}|}(1 - p_{\nu,r}^{(c)}).$$

- **Step 3.2:**
  Compute

  $$\sum_{i=1}^{|K|}\prod_{j=1}^{m_i^K}(1 - p_{i_j}^K) + \sum_{\nu \in W_K}\prod_{r=1}^{|E_{c,\nu}|}(1 - p_{\nu,r}^{(c)}).$$

- **Step 3.3:**
  Set

  $$\hat{R}(G, K) = 1 - \sum_{i=1}^{|K|}\prod_{j=1}^{m_i^K}(1 - p_{i_j}^K) - \sum_{\nu \in W_K}\prod_{r=1}^{|E_{c,\nu}|}(1 - p_{\nu,r}^{(c)}).$$

Example. For illustration we use the network TECL [Blechschmidt (1993)] with $n = |V| = 19$ and $m = |E| = 47$ and $h = 3$ with identical edge availabilities $p_e = 0.8$ for any $e \in E$. Let K be given by $K = \{2, 4, 5, 10, 15, 16, 18\}$. It is $|W_K| = 3$, i.e. each of the three subsystems is represented with at least one node in K.

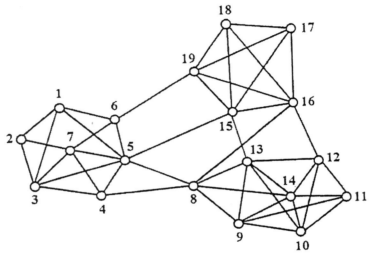

Figure 15.2: TECL, n = 19, m = 47

Again because of the identical edge availabilities ($p_{i_j} = 0.8$) we only need the cardinalities of the sets $E_i, i \in K$, and the cardinality of the sets $E_{c,\nu}, \nu = 1, 2, 3$. From the figure we immediately obtain:

$$|E_2| = 3 \qquad |E_4| = 4 \qquad |E_5| = 7 \qquad |E_{10}| = 5$$

$$|E_{15}| = 6 \qquad |E_{16}| = 6 \qquad |E_{18}| = 4$$

and for the three subsystems

$$|E_{c,1}| = 4 \qquad |E_{c,2}| = 5 \qquad |E_{c,3}| = 5$$

hence

$$\hat{R}(TECL, K) = 1 - \left(0.2^3 + 3 \times 0.2^4 + 3 \times 0.2^5 + 2 \times 0.2^6 + 0.2^7\right) = 0.9861.$$

15.5   Accuracy

As already noted the accuracy of the algorithms shall be examined by numerical examples. These examples will be taken from the Dr.-dissertation of Andrea Blechschmidt [Blechschmidt (1993)], where the software package **RELNET** is described in detail. Furthermore, the capability of **RELNET** is tested by means of a number of network examples. The exact all-terminal reliability as well as the K-terminal reliability for certain sets K are calculated and listed in the dissertation. The edge reliability is assumed to be the same for all the edges of the investigated graphs. The range of the edge reliability is $0.1 - 0.9$.

The necessary time for determination of the exact K-terminal reliabilities by means of **RELNET** was between some seconds for the smaller networks, some minutes for the medium sized networks, and exceeded one hour (in one case even considerably) for the more complex examples.

The approximation algorithm developed above is applied to the examples given in Blechschmidt (1993) for illustration. The calculations were made by a simple pocket calculator, and the necessary time was a few minutes for each graph.

15.5.1   Example 1: Network ARTI

For the Network ARTI (see Figure 15.1) the following values for the exact and approximate K-terminal reliability are obtained:

Table 15.1: Exact and approximate all-terminal reliability of ARTI

K	p_e	$R(G,K)$	$\hat{R}(G,K)$	r
V	0.9	0.9957452	0.99586	0.012%
	0.8	0.9630958	0.96512	0.21%
	0.7	0.8655019	0.87272	0.83%
	0.6	0.6719208	0.66924	0.40%

Table 15.2: Exact and approximate $\{1, 11\}$-terminal reliability of ARTI

K	p_e	$R(G,K)$	$\hat{R}(G,K)$	r
$\{1,11\}$	0.9	0.9978832	0.9979	0.0017%
	0.8	0.9816715	0.9821	0.044%
	0.7	0.9314981	0.9348	0.35%
	0.6	0.8203521	0.8283	0.98%

15.5.2   Example 2: Network K6

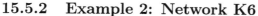

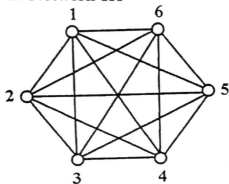

Figure 15.3: Network K6, n = 6, m =15

Table 15.3: Exact and approximate all-terminal reliability of K6

K	p_e	$R(G,K)$	$\hat{R}(G,K)$	r
V	0.9	0.9999399	0.99994	0.00001%
	0.8	0.9980538	0.99800	0.0054%
	0.7	0.9849667	0.98642	0.046%
	0.6	0.9365213	0.93856	0.22%

Table 15.4: Exact and approximate $\{1,2\}$-terminal reliability of K6

K	p_e	$R(G,K)$	$\hat{R}(G,K)$	r
$\{1,2\}$	0.9	0.99998	0.999980	0.00001%
	0.8	0.9993418	0.99934	0.00018%
	0.7	0.9947327	0.9945	0.018%
	0.6	0.9762498	0.9736	0.27%

Remark 15.5.1 K6 has a maximum of complexity for a network with seven nodes, in the sense that all possible connections between the nodes are realized. It turns out that in such a case the performance of the method is excellent even for very small edge availabilities. Thus, the accuracy improves with the ratio $\frac{|E|}{|V|}$, i.e. the more difficult an exact determination of the system reliability becomes, the better is the performance of the here proposed approximation.

15.5.3 Example 3: Network K7

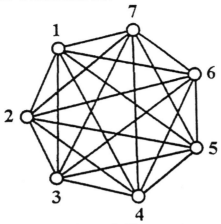

Figure 15.4: Network K7, n = 7, m = 21

Table 15.5: Exact and approximate all-terminal reliability of K7

K	p_e	$R(G, K)$	$\hat{R}(G, K)$	r
V	0.9	0.9999930	0.999993	0.0%
	0.8	0.9995506	0.999552	0.00014%
	0.7	0.9948374	0.993897	0.0060%
	0.6	0.9707218	0.971328	0.062%

Table 15.6: Exact and approximate {1, 2, 3}-terminal reliability of K7

K	p_e	$R(G, K)$	$\hat{R}(G, K)$	r
{1, 2, 3}	0.9	0.9999970	0.999970	0.0%
	0.8	0.9998067	0.999808	0.00013%
	0.7	0.9977458	0.997813	0.0067%
	0.6	0.9866756	0.987712	0.11%

Remark 15.5.2 K7 also has maximum complexity, but the number of nodes (and hence of edges) is larger than for K6. An comparison of the relative deviation indicates that the approximation gets better for increasing number of nodes and complexity.

15.5.4   Example 4: Network ALG

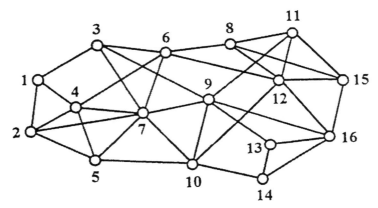

Figure 15.5: Network ALG, n = 16, m = 36

Table 15.7: Exact and approximate all-terminal reliability of ALG

K	p_e	$R(G,K)$	$\hat{R}(G,K)$	r
V	0.9	0.9962433	0.9963579	0.12%
	0.8	0.9530021	0.9649792	0.21%
	0.7	0.8520372	0.8590033	0.82%
	0.6	0.6200604	0.6036095	2.65%

Table 15.8: Exact and approximate {1, 16}-terminal reliability of ALG

K	p_e	$R(G,K)$	$\hat{R}(G,K)$	r
{1, 16}	0.9	0.9989555	0.99897	0.0015%
	0.8	0.9903893	0.99083	0.044%
	0.7	0.9591122	0.96326	0.43%
	0.6	0.8724132	0.0.89083	2.1%

15.5.5 Example 5: LGR

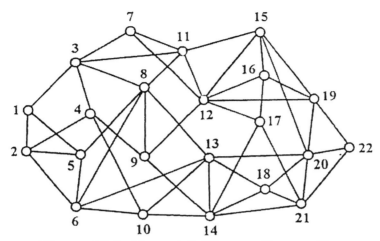

Figure 15.6: Network LGR, n = 22, m = 51

Table 15.9: Exact and approximate all-terminal reliability of LGR

K	p_e	$R(G, K)$	$\hat{R}(G, K)$	r
V	0.9	0.9961057	0.9961359	0.0030%
	0.8	0.9603600	0.9610112	0.068%
	0.7	0.8377879	0.8364853	0.16%
	0.6	0.5793283	0.5401216	6.8%

Table 15.10: Bounds and approximation for the $\{10, 16\}$-terminal reliability of LGR

K	p_e	$\underline{R}(G, K)$	$\overline{R}(G, K)$	$\hat{R}(G, K)$
$\{10, 16\}$	0.9	0.9997978	0.9997978	0.999798
	0.8	0.9966282	0.9966317	0.996628
	0.7	0.9810016	0.9811663	0.98153
	0.6	0.9261192	0.9283042	0.93412

15.5.6   Example 6: Network EVA

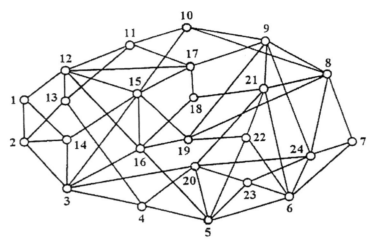

Figure 15.7: Graph EVA with $n = 24$ and $m = 59$

Table 15.11: Exact and approximate all-terminal reliability of EVA

K	p_e	$R(G, K)$	$\hat{R}(G, K)$	r
V	0.9	0.9961441	0.996170	0.0026%
	0.8	0.9614954	0.961907	0.043%
	0.7	0.8451044	0.841831	0.39%
	0.6	0.6000342	0.540122	10.0%

Table 15.12: Bounds and approximation for the $\{16, 21\}$-terminal reliability of EVA

K	p_e	$\underline{R}(G, K)$	$\overline{R}(G, K)$	$\hat{R}(G, K)$
$\{16, 21\}$	0.9	0.9999977	0.9999978	0.999998
	0.8	0.9998384	0.9998492	0.999843
	0.7	0.9979014	0.9980829	0.99798
	0.6	0.9852741	0.9867560	0.98683

15.5.7   Example 7: Network DGN

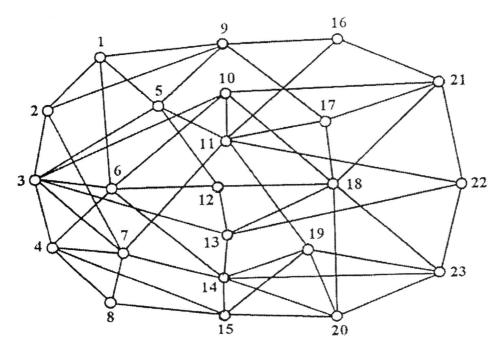

Figure 15.8: Network DGN with $n = 23$ and $m = 59$

Table 15.13: Exact and approximate all-terminal reliability of DGN

K	p_e	$R(G, K)$	$\hat{R}(G, K)$	r
V	0.9	0.9974937	0.9973976	0.0096%
	0.8	0.9740930	0.9726208	0.15%
	0.7	0.8881299	0.8788672	1.0%
	0.6	0.6830516	0.6268544	8.2%

Table 15.14: Exact and approximate $\{4, 9, 10, 11, 12, 13, 14, 15, 21\}$-terminal reliability of DGN

K	p_e	$R(G, K)$	$\hat{R}(G, K)$	r
$\{4, 9, 10,$	0.9	0.9998555	0.9998399	0.0016%
$11, 12, 13$	0.8	0.9961322	0.996454	0.032%
$14, 15, 21\}$	0.7	0.9728639	0.976883	0.41%
	0.6	0.8868939	0.909683	2.6%

Remark 15.5.3 The computation time for the exact $\{4, 9, 10, 11, 12, 13, 14, 15, 21\}$-terminal reliability of DGN for the edge availability ranging from 0.1 to 0.9 took 4 hours and 16 minutes with RELNET.

15.5.8   Example 8: FNW

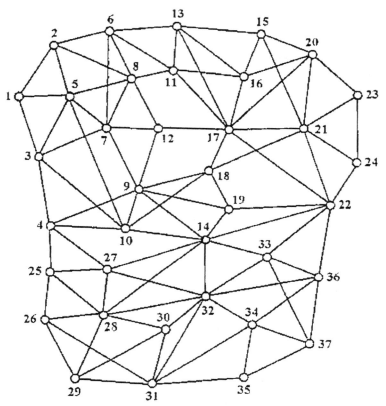

Figure 15.9: Network FNW with $n = 37$ and $m = 95$

Table 15.15: Exact and approximate all-terminal reliability of FNW

K	p_e	$R(G, K)$	$\hat{R}(G, K)$	r
V	0.9	0.9948807	0.994974	0.0094%
	0.8	0.9486070	0.949381	0.082%
	0.7	0.7882278	0.785223	0.38%

15.5.9 Example 9: Network TECL

Network TECL (see Figure 15.2) was already used for illustrating Algorithm 3.

Table 15.16: Exact and approximate all-terminal reliability of TECL

K	p_e	$R(G, K)$	$\hat{R}(G, K)$	r
V	0.9	0.9980889	0.9982149	0.013%
	0.8	0.9769850	0.9779074	0.094%
	0.7	0.8883409	0.8929963	0.52%
	0.6	0.6685481	0.6527616	2.4%

Table 15.17: Bounds and approximation for the $\{3, 16\}$-terminal reliability of TECL

K	p_e	$\underline{R}(G, K)$	$\overline{R}(G, K)$	$\hat{R}(G, K)$
$\{3, 16\}$	0.9	0.9998731	0.9998732	0.99988
	0.8	0.9972610	0.9972745	0.99757
	0.7	0.9809421	0.9811766	0.98489
	0.6	0.9203200	0.9219322	0.93965

Table 15.18: Exact and approximate all-terminal reliability of RCG

K	p_e	$R(G, K)$	$\hat{R}(G, K)$	r
V	0.9	0.9938925	0.994238	0.035%
	0.8	0.9412976	0.9509248	1.0%
	0.7	0.7745929	0.7915762	2.2%

Remark 15.5.4 Similar to TECL, the network RCG consists of four partial networks which are connected by a few edges.

15.5.10   Example 10: Network RCG

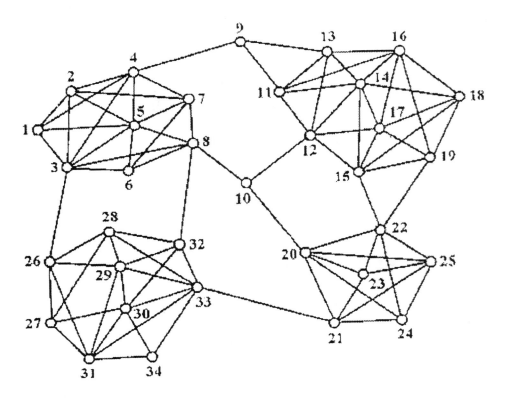

Figure 15.10: Network RCG with $n = 34$ and $m = 89$

15.6   Conclusions

The approximation developed here has the following features:

- The method is extremely simple so that the calculations can be done by means of a pocket calculator, even for large systems.

- Determination of $R(G, K)$ does not depend on the size of the set of nodes, V, or the size of the set of edges, E, but only of the number of nodes and edges contained in K.

- If the edge availabilities are identical and constant, then only the determination of the size of the sets of edges E_i belonging to nodes in K depends on the size and complexity of K, the calculation of $R(G, K)$ itself is essentially independent of the size and complexity of K.

- The method is also applicable if the edge availabilities are different.

- The method is applicable even in the case that the edge availabilities are functions of certain parameters, for instance *time* or *stress*.

- The method can also be used for

  - evaluating single nodes for detecting bottlenecks
  - assessing the network design for possible improvements and their effects on the K-terminal reliabilities.

The method may serve for determining and evaluating the reliability of complex networks which cannot be treated with traditional methods because of their size and complexity.

References

1. Beichelt, F. (1993). Decomposition and reduction techniques, In *New Trends in System Reliability Evaluation* (Ed., K. B. Misra), pp. 75–116, Elsevier.

2. Bienstock, D. (1986). An algorithm for the reliability analysis of planar graphs, *Networks*, **16**, 411–422.

3. Blechschmidt, A. (1993). Rechnergestützte Zuverlässigkeitsanalyse stochastischer Netzstrukturen auf der Basis von Dekompositions- und Reduktionstechniken, *Ph.D. Dissertation*, Ingenieurhochschule Mittweida, Mittweida.

4. Colbourn, Ch. J. (1987). *The Combinatorics of Network Reliability*, New York: Oxford University Press.

5. Kohlas, J. (1987). *Zuverlässigkeit und Verfügbarkeit*, Stuttgart: Teubner-Verlag.

6. Politof, T. and Santyanarayana, A. (1986). Efficient algorithms for reliability analysis of planar networks—A survey, *IEEE Transactions on Reliability*, **35**, 384–393.

7. Tittmann, P. and Blechschmidt, A. (1991). Reliability bounds based on networks splitting, *J. Infor. Process. Cybern.*, **27**, 317–326.

16

Reliability Analysis in Flow Networks

Roland Jentsch

University of Technology and Economics, Mittweida, Germany

Abstract: The estimation of data transfer rates in networks can be traced back to the problem of finding a flow of maximal value in undirected graphs. The probability that a certain requested flow is realizable within a network environment with unreliable links – the so called flow probability – can be considered as a reliability measure. This paper presents algorithms for the computation of the flow probability.

Keywords and phrases: Flow probability, network reliability, graph theory, algorithm

16.1 Introduction

An important question in design and analysis of telecommunication, transportation or computer networks is the evaluation of the network performance. A measure of the network performance is the maximum flow. However, this value is in general not constant but depends on the capacities and the state of the network components. Under the conditions of possible link failures or pre-occupied capacities it is desirable to know whether the network can be expected to handle a given flow. We will use an undirected graph as the mathematical model of a network to computate the flow probability, i.e. the probability that a requested flow is realizable within the network. The edges of the graph represent the links of the network, the vertices its nodes. Capacities and reliabilities are assigned to the edges. Furthermore, it will be assumed that failures are statistically independent, links have either zero (failure) or full (operational state) capacity, and nodes do not fail. The latter is no restriction since node failures can be incorporated into our model easily.

16.2   List of Used Symbols

$V = \{v_1, v_2, ..., v_n\}$ .... Set of nodes
$E = \{e_1, e_2, ..., v_m\}$ .... Set of edges
$G = (V, E)$ .... Graph
$n = |V|$ .... Number of nodes
$m = |E|$ .... Number of edges
$c_e \in R^+$ .... Capacity of edge e
$p_e \in [0, 1]$ .... Reliability of edge e (edge working probability)
$q_e = 1 - p_e$ .... Unreliability of edge e (edge failure probability)
$s, t \in E$ .... Source, Terminal node
$f_e \in R^+$ .... Flow on edge e
$F_{s,t}(G)$ .... Maximum flow from node s to node t
$R_{s,t}(G)$ .... Two terminal reliability of the graph G
$R_{s,t}(G, F)$ .... Flow probability for the graph G with a flow of F

16.3   Definitions

Definition 16.3.1 Let $f : E \to R$ be a function assigning a real number to every directed edge $e = (v_j, v_k)$, such that $f(e_a) = -f(e_b)$ with $e_a = (v_i, v_j)$ and $e_b = (v_j, v_i)$. Then $f_{s,t}$ $(s, t \in V, s \neq t)$ is called a **flow** if

$$\sum_{e_k \in \{(v_i, u) : (v_i, u) \in E\}} f(e_k) = \begin{cases} 0 & : \quad v_i \in V - \{s, t\} \\ f_{s,t} & : \quad v_i = s \\ -f_{s,t} & : \quad v_i = t. \end{cases}$$

A flow is said to be **valid** if $f_e \leq c_e \ \forall e \in E$.

Definition 16.3.2 The largest valid flow from a source node s to a terminal node t in a graph G is called **maximum flow** $F_{s,t}(G)$.

Definition 16.3.3 The **two terminal reliability** $R_{s,t}(G)$ of a graph G is the probability that there exists at least one path of operational edges from the source node s to the terminal node t.

16.4   Flow Probability

So far we have defined two independent characteristic values of a network, the maximum flow between two nodes and the two terminal reliability. The first one describes the amount that can be transferred from a source to a terminal node in a failure free network and the second the probability that a connection from a source to a terminal node exists. The interaction of both raises two questions:

- What is the probability that a connection for a given flow exists?

- How much flow can be transferred in the average case from the source to the terminal node?

We will deal only with the first question in this paper.

Definition 16.4.1 The **flow probability** $R_{s,t}(G, F)$ if defined as the probability that in a graph G a flow of size F is transferred from the source node s to the terminal node t. The value F is called **demanded flow**.

16.5   Computation of the Flow Probability

There are several approaches to the computation of the flow probability. The most general is the following: Edge failures induce a certain subset of edgeset E consisting of all edges in an operational state. Every subset out of the 2^m possible subsets of the edgeset is induced with a certain probability and is either able to transfer the demanded flow or not. This probability will be called **occurrence probability (of the subset)** $P_o(G)$. Summing the occurrence probabilities of the subsets which can transfer the demanded flow results in the wanted flow probability:

$$R_{s,t}(G, F) = \sum_{\substack{I \subset E \\ F_{s,t}(G-I) \geq F}} P_o(G - I) = \sum_{\substack{I \subset E \\ F_{s,t}(G-I) \geq F}} \left(\prod_{e \notin I} p_e \cdot \prod_{e \in I} q_e \right).$$

Thereby $G - I$ denotes the graph G in which all edges $e \in I$ have failed (and thus could be deleted). It is obvious that this nice compact formula requires computation time increasing exponentially with the number of edges. Additionally, for every subset the maximum flow of the remaining network has to be computed.

We will use this algorithm to start with and try to improve it later.

16.5.1   The decomposition algorithm

We have already used the idea of a 'decomposition' for the computation of the survival probability of a network. It is based on the following idea: The subsets of the edgeset can be divided into two disjoint (and equally sized) classes. The first one consists of all subsets containing a certain edge $e_1 \in E$ and the second of all others. Similarly, each of these classes can be divided into two subclasses by 'splitting' the classes on $e_2 \neq e_1 \in E$. We continue this process until every edge $e_k \in E$ has been used. After having constructed all possible subsets in this recursive process, we are able to compute the wanted flow probability recursively:

$$R_{s,t}(G, F) = p_e \cdot R_{s,t}(G_e, F) + q_e \cdot R_{s,t}(G - e, F)$$

where G_e denotes the graph G in which the edge e is operational and the term $G - e$ denotes the graph G in which the edge e has failed. The recursion tree has a depth of m and during the process 2^m different graphs will be generated. The leafs of the recursion tree consist of 'reliable' graphs G^* whose edgesets are subsets of the original edgeset. The occurrence probabilities for these edgesets are known and their maximum flows can be computed and compared to the demanded flow. By appropriate summing we obtain:

$$R_{s,t}(G^*, F) = \begin{cases} 1 & : \quad F \leq F_{s,t}(G) \\ 0 & : \quad \text{otherwise} \end{cases}$$

$$R_{s,t}(G, F) = \sum_{I \subset E} \left(\prod_{e \notin I} p_e \cdot \prod_{e \in I} q_e \cdot R_{s,t}(G^*, F) \right).$$

Let us consider the complexity of the algorithm: Each computation of the maximum flow runs in time $O(n^3)$ (or better). There are 2^m different subsets, thus $O(n^3 \cdot 2^m)$ is an upper bound for the total time needed. Reductions are possible and known but they will not be discussed in this paper.

In some special cases it is possible to truncate the decomposition or to collect identical boughs of the recursion tree.

16.5.2   Special values of the demanded flow

- **The maximum flow between source and terminal node is smaller then the demanded flow.**

  If $F > F_{s,t}(G)$ the demanded flow F cannot be transmitted from s to t even if none of the edges has failed. Accordingly, none of the subgraphs G^* can fulfill the flow demand and the flow probability is zero.

  $$R_{s,t}(G, F) = 0.$$

- **The capacity of each edge is higher then the demanded flow**

  If the capacity of each edge is larger than the demanded flow then each subset of edges which connects source to terminal warrants the demanded flow. Therefore we sum over all connections from s to t which is equal to the two terminal reliability in the graph.

  $$R_{s,t}(G, F) = R_{s,t}(G).$$

  Hence, we can use the simpler algorithms of the two terminal reliability for computing the flow probability.

16.5.3 Special structures

Trees

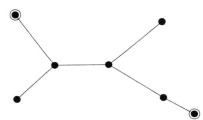

If the graph G is a tree then there is only one path $W = W_{s,t}(G)$ from the source node s to the terminal node t. This means, all the edges of the path must not fail to satisfy a demand of flow. The computation of the maximum flow is simple: it is the minimum of the capacities of all the edges on the path. $F_{s,t}(G) = \min_{e \in W}(c_e)$. For the flow probability we have:

$$R_{s,t}(G, F) = \begin{cases} \prod_{e \in W} p_e & : \quad F \leq F_{s,t}(G) \\ 0 & : \quad \text{otherwise.} \end{cases}$$

Ladder structures L_n

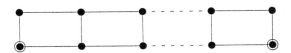

We assume that the ladder has a length of n and furthermore that all edges have the same edge reliability $p_1 = p_2 = \cdots = p$ and capacity $c_1 = c_2 = \cdots = c$. We distinguish three cases:

1. The demanded flow is smaller than or equal to c ($F \leq c$).

2. The demanded flow lies between c and $2c$ ($c < F \leq 2c$).

3. The demanded flow is larger then $2c$ ($F > 2c$).

1. If $F \leq c$ then there must be at least one path from s to t of operating edges. This is equal to the two terminal reliability of the ladder and can be computed using a recursive formula.

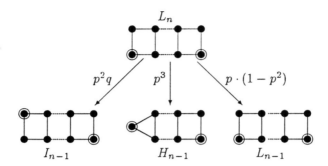

This tree can also be established for I_n and H_n. Thus:

Let $R_{s,t}(L_n, F) = L_n$, $R_{s,t}(I_n, F) = I_n$ and $R_{s,t}(H_n, F) = H_n$ then :

$$
\begin{aligned}
L_n &= (p - p^3) \cdot L_{n-1} + p^3 \cdot H_{n-1} + p^2 \cdot q \cdot I_{n-1} \\
I_n &= (p - p^3) \cdot I_{n-1} + p^3 \cdot H_{n-1} + p^2 \cdot q \cdot L_{n-1} \\
H_n &= p \cdot q \cdot [L_{n-1} + I_{n-1}] + p^2 \cdot H_{n-1}.
\end{aligned}
$$

Since the initial values are known ($L(0) = 1$, $H(0) = p + p^2 - p^3$, $L2(0) = p$) it is possible to obtain an explicit formula which is too large to be displayed in this paper.

2. For $c < F \leq 2c$, two edge disjoint paths of operating edges from s to t are necessary. Since we consider ladder structures there are only two paths simultaneously possible. This results in:

$$
R_{s,t}(L_n, F) = 2^{2n+2}.
$$

3. For $F > 2c$ three paths with disjoint edge sets are necessary. Evidently this is impossible, and therefore:

$$
R_{s,t}(L_n, F) = 0.
$$

Circles C_n

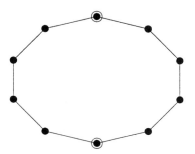

For the class $\{C_n\}$ we again assume equal values for the edge capacities c_e and the edge reliabilities p_e. In a circle there are at most two paths (one clockwise called W_1 and one counterclockwise called W_2) connecting two nodes. Let d denote the length of one of the paths. Then the length of the second is $m - d$. Again we distinguish three cases: $0 < F \leq c$, $c < F \leq 2c$ and $2c < F$. In the first case we need one path of working edges from s to t. The probability of the existence of one or more paths of working edges is $p_{W_1} + p_{W_2} - p_{W_1} \cdot p_{W_2} = p^d + p^{m-d} - p^m$. In the second case we need both paths in a working state. We obtain:

$$
R_{s.t}(C_n, F) = \begin{cases} p^d + p^{m-d} - p^m & : \quad 0 < F \leq c \\ p^m & : \quad c < F \leq 2 \cdot c \\ 0 & : \quad \text{otherwise.} \end{cases}
$$

Additionally, we obtain easily a result for the flow probability if not all of the edge capacities or edge reliabilities are equal. We simply have to replace the values of the demanded flow by c_{W_1}, c_{W_2} and $c_{W_1} + c_{W_2}$ and the occurrence probabilities of the paths by p_{W_1}, p_{W_2} and $p_{W_1} + p_{W_2}$.

The complete graph K_n

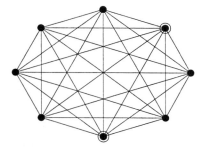

Let us assume once more equal capacities and reliabilities for the edges. The special structure particularly the symmetry of the graph implies some interesting properties: The flow probability is independent from the source and the terminal node. Furthermore there are no more than $n + 1$ different classes of demanded flows. This fact is evidently by the number of edge disjoint paths between two nodes. In a complete graph with n nodes there can be only $n-1$ of such paths and so each demanded flow F belongs to a class of $F = 0, 0 < F \leq c$, $c < F \leq 2c$, ..., $(n-2)c < F \leq (n-1)c$, $F > (n-1)c$.

If the demanded flow F is greater than 0 and lower than c then the problem can be reduced to the two terminal reliability of the complete graph (cf. 16.5.2).

A closed solution for the two terminal reliability is known, which is given by a connection to the all terminal reliability (it means the probability that all nodes belong to one connected component) of the K_n. [1]

$$R_{s,t}(K_n, F) = \sum_{j=2}^{n} \binom{n-2}{j-2} R_j q^{j(n-j)}$$

where R_j is the all terminal reliability of K_j which is known in a closed form. For all other demanded flows

$$R_{s,t}(K_n, F) = \begin{cases} C < F \leq 2 \cdot C \\ \vdots \\ (n-1)C < F \leq nC \end{cases}$$

similar results can be expected, but we will not discuss this case further.

16.5.4   Computation by a generating function

As mentioned before, the computation of the flow probability using the algorithm given above is not very fast. On the other hand it computes more information than actually needed. We obtain a maximum flow for every subset of the edgeset whereas the information whether the maximum flow is greater than or equal to the demanded flow should suffice. Remember the initial formula

$$R_{s,t}(G, F) = \sum_{\substack{I \subseteq E \\ F_{s,t}(G-I) \geq F}} \left(\prod_{e \notin I} p_e \cdot \prod_{e \in I} q_e \right).$$

For a later computation with an other demanded flow we would have to re-compute the flow probability from scratch. This unreasonable approach could be improved as follows: We collect the data of each arising maximum flow problem and select the appropriate cases afterwards. In the case of real capacities all the subsets of the edgeset could induce a different maximum flow and we have to collect, sort and compare up to 2^m different maximum flows. Fortunately, under the assumption that all capacities are small integers the number of different maximum flows is much smaller and the approach is acceptable.

We now represent the flow probability by a generating function.

---

[1] For more information about this please read Coulbourn (1987)

Definition 16.5.1 The function

$$gf_{s,t}(G) = \sum_{i=0}^{F_{s,t}(G)} P_i \cdot x^i$$

is called a **generating function of the flow probability** of G. P_i are polynomials in p and denote the probabilities that a flow of exactly i can be transferred from the source node s to the terminal node t.

Example.

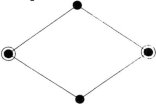

If all edges have equal reliability of p and capacity of 1 then:

$$gf_{s,t}(G) = (1 - 2p^2 + p^4) + (2p^2 - 2p^4)x + p^4 x^2.$$

$\sum_i P_i = 1$ follows immediately from the fact that every possible flow has a value between 0 and the maximum flow, thus every possible flow adds to one of the P_i and so the sum over all P_i includes all possible cases. With this observation the flow probability can be computed:

$$R_{s,t}(G, F) = \sum_{i \geq F} P_i.$$

The computational advantage of this formula is the following: Assume that the reliability of every edge is represented by a generating function $gf_e = (1 - p_e) + p_e x$. Now we can perform series and parallel reductions. In the parallel case ($G = s \, {}^{G_1}_{G_2} \, t$) the graph or part of it can be divided into two parallel subgraphs and if we have a flow of F_1 through G_1 we need a flow of $F - F_1$ through G_2. Thus $P_F = \sum_{i=0}^{F} P_i^1 \cdot P_{F-i}^2$ and the generating function is given by

$$gf(G) = gf(G_1) \cdot gf(G_2).$$

In order to get a flow F in the serial case ($G = s \, G_1 \, v \, G_2 \, t$) we need either a flow of exactly F from s to the articulation node v and a flow larger or equal to F from v to t or a flow of exactly F from v to t and a flow larger than F from s to v. So the P_i can be computed by

$$P_i = \sum_{j \geq i} P_j^1 \cdot P_i^2 + \sum_{j > i} P_i^1 \cdot P_j^2$$

where P_i^1 is the ith polynomial of the generating function of G_1 and P_i^2 is the ith polynomial of the generating function of G_2.

Let us denote this operation for all P_i by

$$gf(G) = gf(G_1) \oplus gf(G_2).$$

Using this two operations we can compute the generating function of the flow probability of each series–parallel–system in polynomial time.

Example.

(all edges have a capacity of one and a reliability of p)

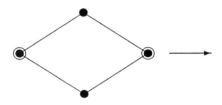

$$
\begin{aligned}
gf(G) &= gf(G_1) \cdot gf(G_2) \\
&= [gf(e) \oplus gf(e)] \cdot [gf(e) \oplus gf(e)] \\
&= \{[(1-p)+px] \oplus [(1-p)+px]\} \cdot \{[(1-p)+px] \oplus [(1-p)+px]\} \\
&= (1-2p^2+p^4) + (2p^2-2p^4)x + p^4x^2.
\end{aligned}
$$

References

1. Colbourn, C. J. (1987). *The Combinatorics of Network Reliability*, New York–Oxford: Oxford Press.

2. Sachs, H. (1988). *Einführung in die Theorie der endlichen Graphen (Teil 1)*, Teubner-Verlag, Lizenz Nr. 294/375/88/70-Es 19 B3.

3. Aggarwal, K. K., Chopra, Y. C. and Bajwa, J. S. (1982). Capacity consideration in reliability analysis of communication systems, *IEEE Transactions on Reliability*, **31**, 177–181.

4. Lee, S. H. (1980). Reliability evaluation of a flow network, *IEEE Transactions on Reliability*, **29**, 24–26.

5. Rüger, W. J. (1986). Reliability analysis of networks with capacity - constraints and failures at branches and nodes, *IEEE Transactions on Reliability*, **35**, 523–528.

6. Ahuja, R. K., Magnanti, T. L. and Orlin, J. B. *Network Flows*, New York: Prentice Hall.

17

Generalized Gram–Charlier Series A and C Approximation for Nonlinear Mechanical Systems

Carsten Sobiechowski

Otto–von–Guericke Universität Magdeburg, Magdeburg, Germany

Abstract: Generalizations of the Gram–Charlier series of type A and C are used to approximate the probability density function (p.d.f.) of stochastic nonlinear dynamical systems under parametrical or external white noise excitation. By means of a Galerkin–technique for the Fokker–Planck equation, the expansion coefficients are obtained from a system of linear equations and a system of equations with a quadratic nonlinearity respectively. The coefficients of these systems of equations are expectations that depend on the system functions. If the system functions are at least piecewise polynomials, integral formulas can be used to evaluate the expectations. The approximations of the p.d.f. are compared and critical remarks concerning applicability are made.

Keywords and phrases: Gram–Charlier series, Fokker–Planck equation, nonlinear systems

17.1 Introduction

For the approximate determination of first and second order response statistics of nonlinear systems driven by external or parametrical white noise excitations, expansions based on generalizations of the Gram–Charlier series A (GC–A) have been used for long time [Sperling (1979, 1986)]. The main advantage of this method is, that, if applied properly, it leads to a linear system of equations for the expansion coefficients.

While the results for the first and second order statistics are quite good Sperling (1986), the GC–A is in general not suited for an approximation of the p.d.f. of the system response. The approximations show an oscillating behaviour in the tail regions and even negative values for the p.d.f. are possible.

In this paper, a method for overcoming the inconveniences of the GC–A is presented. It is based on a generalization of the Gram–Charlier series C (GC–C) for the p.d.f. of the system response. However, this expansion will require a greater numerical effort than the GC–A approach.

GC–C has been applied by Hampl and Schuëller (1987) and, more recently, by Muscolino, Ricciardi and Vasta (1997) to the analysis of random dynamical systems. In Muscolino, Ricciardi and Vasta (1997), products of unidimensional Hermitian polynomials, normalized with the standard deviations obtained by Gauss closure, are used in the series expansion. Here, we adopt a more general formulation that allows to use a positive definite matrix as normalization. This offers the possibility to include information about crossvariances. Furthermore, integral formulas are used to calculate the expectations instead of expanding the system functions into a series of Hermitian polynomials.

The uniform convergence of Gram–Charlier series has been studied by Bernard (1995). The convergence of the Gram–Charlier expansion depends on the adherence of the expanded function to certain weighted Sobolev spaces; this leads to a condition for the p.d.f. of the system response that in general is difficult to verify a priori.

17.2   Formulation of the Problem

The equation of motion of an n–dimensional nonlinear dynamical system under Gaussian white noise excitation may be mathematically idealized as a Stratonovich differential equation

$$dX(t) = \xi(X, t)dt + \chi(X, t)dW(t), \ t \geq t_0, \tag{17.1}$$

where $\xi(X, t)$ is an n–dimensional deterministic function, $\chi(X, t)$ an $n \times q$–dimensional deterministic function and $W(t)$ is a q–dimensional Wiener process whose increments have the properties

$$E[dW(t)] = 0, \tag{17.2}$$
$$E[dW(t_1)dW^T(t_2)] = I\delta(t_1 - t_2)dt_1dt_2, \tag{17.3}$$

where T indicates transposition and I is the $q \times q$ intensity matrix. Equation (17.1) may be transferred to the Itô standard form

$$dX(t) = f(X, t)dt + \chi(X, t)dW(t), \tag{17.4}$$

where

$$f_i(x, t) = \xi_i(x, t) + \frac{1}{2}I_{kl}\frac{\partial \chi_{ik}(x, t)}{\partial x_j}\chi_{jl}(x, t). \tag{17.5}$$

Here and in the following, we use the convention that the multiple occurrence of the same unidimensional index in a product implies a summation within the appropriate range. For multidimensional indices, we prefer writing explicitly the sum sign.

Under known conditions, the p.d.f. of the system response satisfies the Fokker–Planck equation

$$\frac{\partial p(x,t)}{\partial t} = -\frac{\partial}{\partial x_i}(f_i(x,t)p(x,t)) + \frac{1}{2}\frac{\partial^2}{\partial x_i \partial x_j}(g_{ij}(x,t)p(x,t)), \qquad (17.6)$$

where

$$g_{ij}(x,t) = I_{kl}\chi_{ik}(x,t)\chi_{jl}(x,t), \qquad (17.7)$$

For later purposes, we also note the logarithmic version of the Fokker–Planck equation,

$$\begin{aligned}
\frac{\partial \phi}{\partial t} &= -\frac{\partial f_i}{\partial x_i} + \frac{1}{2}\frac{\partial^2 g_{ij}}{\partial x_i \partial x_j} - \frac{\partial \phi}{\partial x_i}f_i + \frac{1}{2}\frac{\partial \phi}{\partial x_i}\frac{\partial g_{ij}}{\partial x_j} + \frac{1}{2}\frac{\partial \phi}{\partial x_j}\frac{\partial g_{ij}}{\partial x_i} + \frac{1}{2}\frac{\partial^2 \phi}{\partial x_i \partial x_j}g_{ij} \\
&\quad + \frac{1}{2}\frac{\partial \phi}{\partial x_i}\frac{\partial \phi}{\partial x_j}g_{ij},
\end{aligned} \qquad (17.8)$$

where $\phi(x,t) = \ln p(x,t)$. While the Fokker–Planck equation itself is linear, the logarithmic Fokker–Planck equation has a nonlinearity of quadratic type.

17.3 Generalized Gram–Charlier–Series A Approximation

The generalized Gram–Charlier–series A expansion which is used in the following is that developed by Sperling (1986). The p.d.f. of the system response is approximated by

$$p(x,t) = \bar{p}(x,t)\left(1 + \sum_{\|s\|\geq 1}\frac{b_s(t)}{s!}H_s(x-\bar{m}(t))\right), \qquad (17.9)$$

where $\sum_{\|s\|\geq 1}$ is the sum over all multiindices s with $\|s\| := s_1 + \ldots + s_n \geq 1$ and $\bar{p}(x,t)$ the Gaussian p.d.f.

$$\frac{1}{\sqrt{(2\pi)^n \det(\bar{\mu}(t))}} \exp\left(-\frac{1}{2}(x-\bar{m}(t))^T\bar{\mu}(t)^{-1}(x-\bar{m}(t))\right). \qquad (17.10)$$

The expansion functions $H_s(x-\bar{m}(t))$ are the generalized Hermitian polynomials

$$H_s(z) = (-1)^s \exp\left(\frac{1}{2}z^T\bar{\mu}(t)^{-1}z\right)\frac{\partial^s}{\partial z_1^{s_1}\ldots \partial z_n^{s_n}}\exp\left(-\frac{1}{2}z^T\bar{\mu}(t)^{-1}z\right) \quad (17.11)$$

and the expansion coefficients are often termed generalized quasimoments. The vector function $\bar{m}(t)$ and the positive definite matrix function $\bar{\mu}(t)$ are the mean and variance predicted by a Gauss closure of the nonlinear system (17.4). This is a very remarkable difference to other approaches, for example Pugachev and Sinitsyn (1987), where $\bar{m}(t)$ and $\bar{\mu}(t)$ are higher order approximations for the mean and variance of the nonlinear system. While the first approach leads, due to the linearity of the Fokker–Planck equation, to a system of linear equations for the determination of higher order approximations, the latter approach will yield a system of nonlinear equations. Moreover, numerical tests have revealed that the first approach seems to converge faster.

The generalized Hermitian polynomials are orthogonal to the functions

$$G_r(z) = (-1)^r \exp(\frac{1}{2}z^T \bar{\mu}(t)^{-1}z)\frac{\partial^r}{\partial y_1^{r_1}\dots\partial y_n^{r_n}}\exp(-\frac{1}{2}z^T\bar{\mu}(t)^{-1}z),\ y = \bar{\mu}(t)^{-1}z \tag{17.12}$$

with respect to the weight $\bar{p}(x,t)$, that is

$$\int_{\mathbb{R}^n} G_r(x - \bar{m}(t))H_s(x - \bar{m}(t))\bar{p}(x,t)dx = r_1!\dots r_n!\ \delta_{r_1 s_1}\dots\delta_{r_n s_n}. \tag{17.13}$$

Multiplying (17.9) with $G_r(x - \bar{m}(t))$ and integrating, we see that $b_r(t)$ is the expectation of $G_r(x - \bar{m}(t))$, i.e.

$$b_r(t) = E[G_r(x - \bar{m}(t))]. \tag{17.14}$$

Differentiating this equation with respect to time and making use of the Fokker–Planck equation leads to

$$\begin{aligned}
\dot{b}_r &= \bar{E}[\dot{G}_r + \frac{\partial G_r}{\partial x_i}f_i + \frac{1}{2}\frac{\partial^2 G_r}{\partial x_i \partial x_j}g_{ij}] \\
&\quad + \sum_{\|s\|\geq 1}\frac{1}{s!}\bar{E}[(\dot{G}_r + \frac{\partial G_r}{\partial x_i}f_i + \frac{1}{2}\frac{\partial^2 G_r}{\partial x_i \partial x_j}g_{ij})H_s]b_s, \tag{17.15}
\end{aligned}$$

where $\bar{E}[\cdot]$ denotes the expectation with respect to the Gaussian p.d.f. $\bar{p}(x,t)$. The solution procedure consists of two steps. At first, equation (17.15) is truncated at second order and the first and second moments of the Gauss closure are calculated iteratively until convergence. In a second step, the converged values are assigned to $\bar{m}(t)$ and $\bar{\mu}(t)$ and corrections are calculated by truncating (17.15) at higher order.

A special difficulty arises with the expectations that have to be carried out in equation (17.15). Evaluating the expectations with numerical or symbolic integration techniques would be very time–consuming, especially for higher dimensional systems. Therefore, integral formulas for at least piecewise polynomial system functions were developed and incorporated in a general automata program.

The adapted formulation of the Gram–Charlier approximation is especially well suited to obtain covariance and power spectral density functions. For more details, the reader is referred to Sperling (1986).

17.4　Generalized Gram–Charlier–Series C Approximation

The GC–A approximation has the disadvantage that the approximation of the p.d.f. is not necessarily positive. Moreover, for polynomial system functions, the maximum entropy principle yields a stationary p.d.f. that is an exponential function of a polynomial. This motivates a series expansion of $\ln p(x, t)$ instead of $p(x, t)$. In this way, a generalization of the GC–C is obtained. Let us consider the expansion

$$\ln p(x, t) = \sum_{\|r\| \geq 0} c_r(t) G_r(x - \bar{m}(t)). \tag{17.16}$$

As above, the parameters $\bar{m}(t)$ and $\bar{\mu}(t)$ that are needed to construct the Hermitian polynomials are obtained by a Gauss closure for the nonlinear system (17.4).

We insert equation (17.16) into the logarithmic Fokker–Planck equation and obtain

$$\sum_{\|r\| \geq 0} \left(\dot{c}_r G_r - c_r (r_i \dot{\bar{m}}_i G_{r, r_i-1} + \frac{1}{2} r_i (r_j - \delta_{ij}) \dot{\bar{\mu}}_{ij} G_{r, r_i-1, r_j-1}) \right)$$

$$= -\frac{\partial f_i}{\partial x_i} + \frac{1}{2} \frac{\partial^2 g_{ij}}{\partial x_i \partial x_j}$$

$$+ \sum_{\|r\| \geq 1} r_i c_r G_{r, r_i-1} \left(-f_i + \frac{1}{2} (\frac{\partial g_{ij}}{\partial x_i} + \frac{\partial g_{ji}}{\partial x_i}) \right)$$

$$+ r_i (r_j - \delta_{ij}) c_r G_{r, r_i-1, r_j-1} \frac{1}{2} g_{ij}$$

$$+ \frac{1}{2} \sum_{\|r\| \geq 1} \sum_{\|\tilde{r}\| \geq 1} r_i \tilde{r}_j c_r c_{\tilde{r}} G_{r, r_i-1} G_{\tilde{r}, \tilde{r}_i-1} g_{ij}. \tag{17.17}$$

The abbreviation r, r_i-1 denotes the multiindex $(r_1, \ldots, r_{i-1}, r_i-1, r_{i+1}, \ldots, r_n)$. After multiplication with $H_s(x - \bar{m}(t))$ and integration with respect to the weight $\bar{p}(x, t)$, we obtain the following system of differential equations

$$s! \dot{c}_s = -\bar{E}[(\frac{\partial f_i}{\partial x_i} + \frac{\partial^2 g_{ij}}{\partial x_i \partial x_j}) H_s] + s!(s_i + 1) \dot{\bar{m}}_i c_{s, s_i+1}$$

$$- s!(s_i + 1)(s_j + 1 + \delta_{ij}) \dot{\bar{\mu}}_{ij} c_{s, s_i+1, s_j+1}$$

$$+ \sum_{\|r\| \geq 1} \left(r_i \bar{E}[G_{r,r_i-1}(-f_i + \frac{1}{2}(\frac{\partial g_{ij}}{\partial x_j} + \frac{\partial g_{ji}}{\partial x_j}))H_s] \right.$$

$$\left. + \frac{1}{2} r_i(r_j - \delta_{ij}) \bar{E}[G_{r,r_i-1,r_j-1}g_{ij}H_s] \right) c_r$$

$$+ \frac{1}{2} \sum_{\|r\| \geq 1} \sum_{\|\tilde{r}\| \geq 1} r_i \tilde{r}_j \bar{E}[G_{r,r_i-1}G_{\tilde{r},\tilde{r}_j-1}g_{ij}H_s]c_r c_{\tilde{r}}. \qquad (17.18)$$

For purely external excitation, this system reduces to

$$s!\dot{c}_s = -\bar{E}[\frac{\partial f_i}{\partial x_i}H_s] + s!(s_i + 1)\dot{\bar{m}}_i c_{s,s_i+1}$$

$$- s!(s_i + 1)(s_j + 1 + \delta_{ij})\dot{\bar{\mu}}_{ij}c_{s,s_i+1,s_j+1}$$

$$+ \sum_{\|r\| \geq 1} \left(-r_i \bar{E}[G_{r,r_i-1}f_i H_s] + \frac{1}{2} r! g_{ij}\delta_{r,r_i-1,r_j-1;s} \right) c_r$$

$$+ \frac{1}{2} \sum_{\|r\| \geq 1} \sum_{\|\tilde{r}\| \geq 1} r_i \tilde{r}_j g_{ij} \bar{E}[G_{r,r_i-1}G_{\tilde{r},\tilde{r}_j-1}H_s]c_r c_{\tilde{r}}. \qquad (17.19)$$

Here, $\delta_{r;s}$ indicates the product of Kronecker deltas $\delta_{r_1 s_1} \delta_{r_2 s_2} \dots \delta_{r_n,s_n}$. Equations (17.18) and (17.19) constitute infinite dimensional systems of differential equations for the expansion coefficients. They have to be truncated at a suitable order.

A possible initial condition for equation (17.18) is the Gaussian p.d.f.. From the expansion (17.16), we have

$$c_s(t) = \frac{1}{s!} \bar{E}[\ln p(x,t)H_s(x - \bar{m}(t))]. \qquad (17.20)$$

Letting

$$\ln p(x, t = 0) = -\frac{1}{2}(\ln((2\pi)^n \|\mu\|) + (x - m)^T \mu^{-1}(x - m)), \qquad (17.21)$$

we obtain

$$c_0(t = 0) = -\frac{1}{2}(n(1 + \ln(2\pi)) + \ln(\|\mu\|)), \qquad (17.22)$$

$$c_s(t = 0) = -\frac{1}{2\, s!} \bar{E}[(x - m)^T \mu^{-1}(x - m)H_s(x - \bar{m})], \|s\| \geq 1. \qquad (17.23)$$

For the stationary solution, the truncation of (17.18) or (17.19) yields a system of algebraic equations that has a nonlinearity of quadratic type. Note that the coefficient $c_0(t)$ does not appear in the second member of (17.18) and (17.19). This is due to the fact that the corresponding second member of the logarithmic Fokker–Planck equation (17.8) contains only derivatives of

$\phi(x,t) = \ln p(x,t)$ and not the function $\phi(x,t)$ itself. Thus, in order to obtain the stationary solution, c_0 has to be calculated from the normalization condition

$$\int_{\mathbb{R}^n} p(x)dx = \int_{\mathbb{R}^n} \exp\left(c_0 + \sum_{\|r\| \geq 1} c_r G_r(x - \bar{m})\right)dx = 1, \qquad (17.24)$$

which yields

$$c_0 = -\ln\left(\int_{\mathbb{R}^n} \exp\left(\sum_{\|r\| \geq 1} c_r G_r(x - \bar{m})\right)dx\right). \qquad (17.25)$$

Another approach to the stationary solution of (17.18) or (17.19) would be the numerical integration of the equations, starting from the p.d.f. obtained by performing a Gauss closure for the stationary system. In this case, the numerical effort can be considerably reduced, if the system functions are time–independent. Then, systems (17.18) and (17.19) have constant coefficients.

Relations between Gram–Charlier series A and C were considered by Muscolino, Ricciardi and Vasta (1997). Starting from the identity

$$\int_{\mathbb{R}^n} G_r(x - \bar{m}(t))\left(\sum_{i=1}^{n} \frac{\partial(\ln p(x,t))}{\partial x_i} p(x,t)\right)dx$$

$$= -\int_{\mathbb{R}^n} r_i G_{r,r_i-1}(x - \bar{m}(t))p(x,t)dx \qquad (17.26)$$

and inserting for the second member the GC–A expansion (17.9) we get from the orthogonality condition (17.13)

$$\int_{\mathbb{R}^n} G_r(x - \bar{m}(t))\left(\sum_{i=1}^{n} \frac{\partial(\ln p(x,t))}{\partial x_i} p(x,t)\right)dx = -r_i b_{r,r_i-1}(t). \qquad (17.27)$$

Using (17.16), this leads to

$$\int_{\mathbb{R}^n} G_r(x - \bar{m}(t))\left(\sum_{\|\tilde{r}\| \geq 1} c_{\tilde{r}}(t)\tilde{r}_i G_{\tilde{r},\tilde{r}_i-1}(x - \bar{m}(t))p(x,t)\right)dx = -r_i b_{r,r_i-1}(t)$$

$$\qquad (17.28)$$

and after inserting the expansion (17.9) once again, we arrive at

$$\sum_{\|\tilde{r}\| \geq 1} \tilde{r}_i\Big(\bar{E}[G_r(x - \bar{m}(t))G_{\tilde{r},\tilde{r}_i-1}(x - \bar{m}(t))]c_{\tilde{r}}(t)$$

$$+ \sum_{\|s\| \geq 1} \frac{b_s(t)}{s!} \bar{E}[G_r(x - \bar{m}(t))G_{\tilde{r},\tilde{r}_i-1}(x - \bar{m}(t))H_s(x - \bar{m}(t))]\Big)c_{\tilde{r}}(t)$$

$$= -r_i b_{r,r_i-1}(t). \qquad (17.29)$$

For $\|r\| \geq 1$, this is a linear system of equations for the coefficients $c_{\tilde{r}}(t)$. After a truncation of the GC–C at a suitable order, say ν, it involves the coefficients $b_s(t)$ up to order $2\nu - 1$, since one can show that $\bar{E}[G_r(x - \bar{m}(t))H_s(x - \bar{m}(t))\prod_{i=1}^{n} x^{q_i}] = 0$ if there exists an i, $1 \leq i \leq n$ such that $s_i > r_i + q_i$. Again, for the stationary solution, the coefficient c_0 has to be calculated from the normalization condition.

17.5   Examples

This section compares the p.d.f. obtained by Gram–Charlier series A and C expansions. From the relation between the two series, it follows that one should compare the result of a GC–C of order ν to that of a GC–A of order $2\nu - 1$.

First order statistics such as mean and variance have also been compared and there seems to be no remarkable difference. However, if first order statistics are calculated from a GC–C expansion, a numerical integration is necessary. Apart from the higher computational efforts, this may introduce additional numerical errors. Thus, the use of a GC–C expansion for the calculation of first order statistics can not be recommended.

For the first examples, the mechanical system under consideration is a white noise excited single degree of freedom oscillator having velocity proportional damping and a spring with a piecewise linear characteristic. The system is described by

$$
\begin{aligned}
dX_1(t) &= X_2(t)dt, \\
dX_2(t) &= -cX_2(t)dt - \phi(X_1(t))dt + dW(t),
\end{aligned}
\tag{17.30}
$$

where $dW(t)$ has the properties stated in (17.2) and (17.3) and $\phi(x)$ is an odd piecewise linear function. It can easily be seen that in this case the stationary p.d.f. is the product of a zero–mean Gaussian distribution with variance $\mu = \frac{I}{2c}$ and the stationary p.d.f. of the equation

$$
d\tilde{X}_1(t) = -\phi(\tilde{X}_1(t))dt + d\tilde{W}(t),
\tag{17.31}
$$

where $d\tilde{W}(t)$ are the increments of a Wiener process with intensity $\frac{I}{c}$. This system possesses the stationary p.d.f.

$$
p(x) = C\exp(-\frac{2c}{I}\int_{\mathbb{R}}\phi(x)dx),
\tag{17.32}
$$

where C is a normalization constant.

Figure 17.1 shows a typical result for a p.d.f. of a hardening spring. The exact solution is compared to an order 10 GC-A approximation and an order 6 GC-C approximation and to the statistical linearization (SL), which for the case under consideration is equivalent to a second order truncation of the Gram Charlier series A or C. Equation (17.18) produces a slightly better approximation than (17.29). The GC–A expansion leads to negative values and an oscillating behaviour in the region $\frac{\|x\|}{\sigma} \geq 4$. For GC-C approximations of order greater than 6, the nonlinear algebraic equations could not be solved by an iterative technique and the calculation of the GC–C expansion from equation (17.29) yielded a positive leading coefficient and thus a p.d.f. that tends to infinity if $x \to \infty$.

For softening spring forces, the situation looks very different. A typical result is presented in Figure 17.2. The GC–A approximates the exact p.d.f. very well, while the GC–C expansion has a second maximum in the tail region. This effect was observed for different intensities, different slopes of the spring characteristic, and different locations of the change of slope. It was noticed that the SL leads to good approximations for the variances of the system.

In general, sudden changes in the p.d.f. can not be captured by a Gram–Charlier expansion. The approximations in Figure 17.3 indicate an oscillating behaviour.

For systems with two or more dimensions, equation (17.29) lead nearly in all cases to a singular matrix, so that in practice, the coefficients of the GC–C could not be obtained from (17.29). Equation (17.18) may work in many cases, but may — due to the nonlinearity — also yield meaningless results.

Figure 17.4 compares the Gram–Charlier series A and C approximation of the stationary p.d.f. of the oscillator

$$
\begin{aligned}
dX_1(t) &= X_2(t)dt, \\
dX_2(t) &= -X_1(t)dt - (0.5 + 1.75\|X_2(t)\|)X_2(t)dt + dW(t), \quad (17.33)
\end{aligned}
$$

where the intensity I of the increments of the Wiener process is equal to 1. The type of nonlinearity occurs for example in the modelling of the rolling motion of ships. For this case, a simulation study revealed that the stationary p.d.f. was best approximated by a GC–A.

The last example is concerned with large stationary crossvariances. The equation

$$
\begin{aligned}
dX_1(t) &= -200X_1(t)dt - 10X_2(t)dt - 1000X_1^3(t)dt - X_1(t)X_2(t)dt \\
dX_2(t) &= -4X_2(t)dt + dW(t), \quad (17.34)
\end{aligned}
$$

where $I = 8$, represents a system with a cubic nonlinearity that is parametrically excited by a filtered noise. The accuracy of the approximations of the joint stationary p.d.f. $p(x_1, x_2)$ depends on the value of x_2. For $x_2 = 0$, the GC–C yields the best result, when compared to simulations (Figure 17.5). But for $x_2 = \frac{4}{3}\sigma_{22}$, Figure 17.6, the GC–C overestimates the p.d.f..

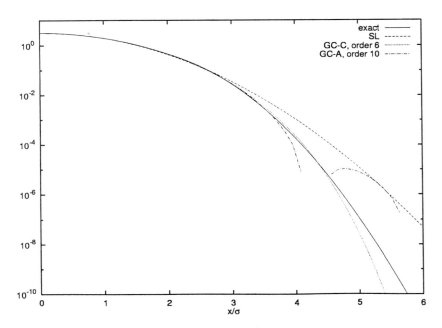

Figure 17.1: Stationary p.d.f. for $dX(t) = -k(X(t))dt + dW(t)$. $k(X) = 192 \cdot X + 31.16$, $X < -0.28$; $k(X) = 80 \cdot X$, $-0.28 \leq X \leq 0.28$; $k(X) = 192 \cdot X - 31.36$, $X > 0.28$; $I = 2.5$

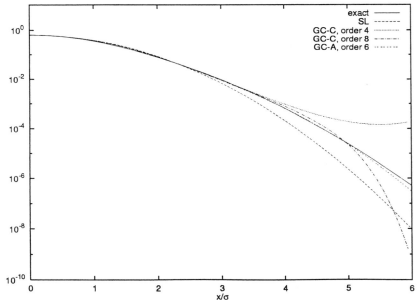

Figure 17.2: Stationary p.d.f. for $dX(t) = -k(X(t))dt + dW(t)$. $k(X) = -0.5 \cdot X - 0.5$, $X < -1$; $k(X) = X$, $-1 \leq X \leq 1$; $k(X) = 0.5 \cdot X + 0.5$, $X > 1$; $I = 1$

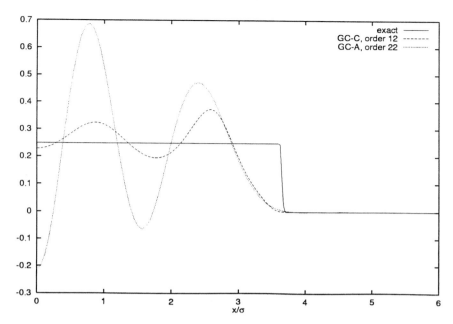

Figure 17.3: Stationary p.d.f. for $dX(t) = -k(X(t))dt + dW(t)$. $k(X) = 10 \cdot X + 20$, $X < -2$; $k(X) = 0$, $-2 \le X \le 2$; $k(X) = 10 \cdot X - 20$, $X > 2$; $I = 0.06$

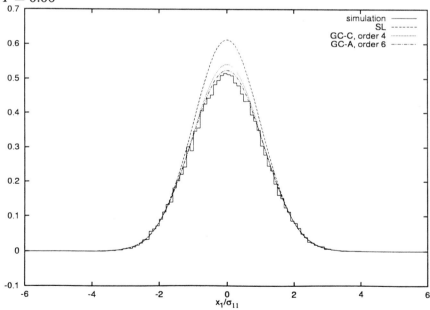

Figure 17.4: Joint stationary p.d.f. $p(x_1, x_2 = 0)$ for system (33)

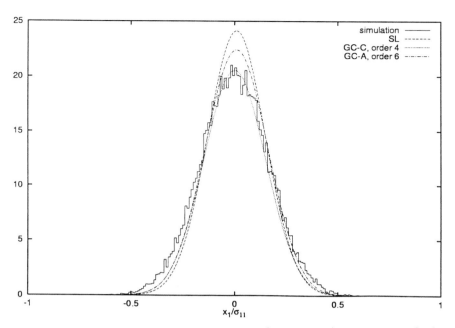

Figure 17.5: Joint stationary p.d.f. $p(x_1, x_2 = 0)$ for system (34)

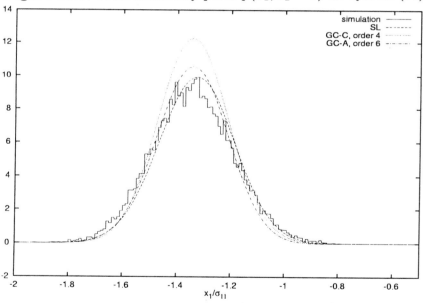

Figure 17.6: Joint stationary p.d.f. $p\left(x_1, x_2 = \frac{4}{3}\,\sigma_{22}\right)$ for system (34)

17.6 Conclusions

The present paper develops a generalized GC–C approximation technique for the p.d.f. and compares the result with the generalized GC–A expansion of Sperling (1979, 1986). Both approaches work efficiently only for few degrees of freedom. The presented approach yields a p.d.f. that is always positive. It suffers however from the typical shortcomings of polynomial expansions, i.e. that sudden changes of the exact p.d.f. will lead to an oscillating behaviour of the approximations.

In certain cases, especially for hardening springs, it was found that the GC–C may lead to considerable improvements in the approximation of the p.d.f. when compared with the GC–A approach. The latter should be preferred if only information about first and second order statistics and not about the p.d.f. itself is needed. Also, for softening spring forces, the approximations obtained by the generalized GC–A yielded better results than the GC–C.

The nonlinearity of the system of equations that determines the coefficients of the GC–C expansion may lead to ambiguous solutions and a careful examination of the obtained solution is necessary. The determination of the solution from the linear system of equations (17.29) seems in practice only be possible for unidimensional problems.

Acknowledgements. I would like to acknowledge stimulating discussions with Prof. L. Sperling, Otto–von–Guericke University of Magdeburg, and the authors, Muscolino, Ricciardi and Vasta (1997). I also wish to express my gratitude to Prof. M. Di Paola at Palermo for his hospitality.

References

1. Bernard, P. (1995). Some remarks concerning convergence of orthogonal polynomial expansions, In *Probabilistic Methods in Applied Physics* (Eds., P. Krée and W. Wedig), pp. 327–334, Berlin: Springer-Verlag.

2. Hampl, N. C. and Schuëller, G. I. (1987). Probability densities of the response of non-linear structures under stochastic dynamic excitation, In *U.S.-Austrian Joint Seminar on Stochastic Structural Mechanics* (Eds., Y. K. Lin and G. I. Schuëller), pp. 165–184, Boca Raton.

3. Muscolino, G., Ricciardi, G. and Vasta, M. (1997). Stationary and non-stationary probability density function for nonlinear oscillators, *International Journal of Non-Linear Mechanics*, **32**, 1051–1064.

4. Pugachev, V. S. and Sinitsyn, I. N. (1987). *Stochastic Differential Equations. Analysis and Filtering*, Chichester: John Wiley & Sons.

5. Sperling, L. (1979). Analyse stochastisch erregter nichtlinearer systeme mittels linearer differentialgleichungen für verallgemeinerte quasimomentenfunktionen, *Zeitschrift für angewandte Mathematik un Mechanik*, **59**, 169–176.

6. Sperling, L. (1986). Approximate analysis of nonlinear stochastic differential equations using certain generalized quasi-moment functions, *Acta Mechanica*, **59**, 183–200.

18

A Unified Approach to the Reliability of Recurrent Structures

---

Valeri Gorlov and Peter Tittmann

HTW Mittweida, Mittweida, Germany

Abstract: The paper deals with new methods for the calculation of the all-terminal reliability in recurrent graph structures. A recurrent graph structure is defined by an operation of joining smaller subgraphs. It is shown that many known regular graphs can be presented in this way and that the all-terminal reliability of recurrent graphs can be computed in polynomially bounded time. This approach can be extended to network reliability problems in the richer class of so-called tree-like graphs and to the K-terminal reliability problem.

Keywords and phrases: Network reliability, graph theory, connectivity, set partitions

---

18.1 Introduction

Let $G = (V, E)$ be an undirected loopless graph with vertex set V and edge set E. The edges of G are assumed to fail independently with known probabilities. The K-terminal reliability $R(G, K)$ is the probability that a specified subset $K \subseteq V$ belongs to one connected component of G. In the case $K = V$ this probability is called the all-terminal reliability or connectedness probability of G.

We investigate in this paper the computation of $R(G, K)$ for recurrent structures. A graph is called recurrent if it is composed of smaller subgraphs in a regular manner. More precisely, let $G = (V' \cup U \cup V, E)$ and $I = (W \cup V, F)$ be two undirected graphs with $|V| = |V'|$. We call I the initiator graph and G the generator graph. Now a recurrent structure G_n of length n is defined recursively as follows. G_0 corresponds to the initiator graph I. For $n \geq 1$

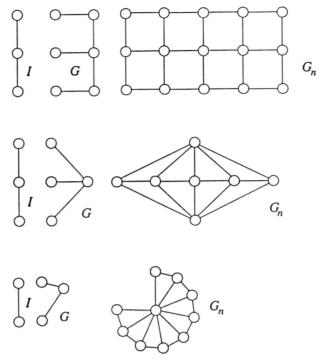

Figure 18.1: Recurrent structures

the graph G_n is obtained from G_{n-1} by identifying the vertex sets V of G_{n-1} and V' of a disjoint copy of G. In this process, a given one-to-one correspondence of the vertices of these two sets is assumed. The resulting set is renamed into V. We obtain from $I = (\{s,t\}, \{\{s,t\}\})$ and $G = (\{u,v,w,x\}, \{\{u,w\}, \{v,x\}, \{w,x\}\})$ by identifying the vertices s,u and t,v, respectively, the graph $G_1 = (\{u,v,w,x\}, \{\{u,v\}, \{u,w\}, \{v,x\}, \{w,x\}\})$. Thus in this case, G_n becomes the $2 \times n$ grid. We denote this kind of joining graphs by the $*$-operator. A multiple join of a graph with itself is written as a power. Thus we write $G_n = I * G^n$. Figure 18.1 shows some examples of recurrent structures.

We show that the all-terminal reliability of a recurrent structure G_n can be computed within a time polynomially dependent on the length n. The basic tool for that purpose is the decomposition of a graph by separating vertex sets. The idea of decomposing a network was (perhaps first) presented by Rosenthal (1977). Arnborg and Proskurowski (1989) showed the application of decomposition methods for the reliability analysis of k-trees in polynomially bounded time. Their approach is based on partitions of separating vertex sets. The nice paper by Bienstock (1988) gives a survey of the combinatorial tools for decomposition methods in network reliability. A collection of results for the reliability of recurrent structures can be found in the book by Koslov and Ushakov (1975).

18.2 The Decomposition of a Graph

Let $G = (V, E)$ be a connected graph with two subgraphs $G^1 = (V^1, E^1)$ and $G^2 = (V^2, E^2)$ such that $V = V^1 \cup V^2$, $V^1 \cap V^2 = U$, $E^1 \cup E^2 = E$, and $E^1 \cap E^2 = \emptyset$. According to section 1 we can write $G = G^1 * G^2$. In this case U is called a *separating vertex set* of G. Bienstock (1988) showed that the all-terminal reliability of G is determined by all-terminal reliabilities of G^1 and G^2:

$$R(G) = \sum_{\pi \in \mathcal{P}(U)} \sum_{\sigma \in \mathcal{P}(U)} R\left(G^1_\pi\right) a^{-1}(\pi, \sigma) R\left(G^2_\sigma\right). \qquad (18.1)$$

Here $\mathcal{P}(U)$ is the partition lattice of the separating vertex set and G^1_π denotes the graph obtained from G^1 by merging all vertices which are contained in one and the same block of π into one vertex. G^2_σ is defined analogously. The partial order in $\mathcal{P}(U)$ is defined as follows: Let $\pi \le \sigma$ iff σ refines π. Thus the least element $\mathbf{0}$ in $\mathcal{P}(U)$ is a partition with only one block - the set U itself. The greatest element $\mathbf{1}$ in $\mathcal{P}(U)$ is the partition consisting of $|U|$ one-element blocks. The number of terms of Equation (18.1) is the number of partitions of an u-set given by the Bell number:

$$B(u) = \sum_k \left\{ {u \atop k} \right\} \sim u! \frac{e^{e^r - 1}}{r^u \sqrt{2\pi r (r+1) e^r}}, \quad u = |U|, \ r \, e^r = u.$$

Here $\left\{ {u \atop k} \right\}$ denotes the Stirling number of the second kind. The derivation of the asymptotic behavior for $B(u)$ is presented in Odlyzko (1995).

Define the function:

$$a(\pi, \sigma) = \begin{cases} 1 \text{ if } \pi \wedge \sigma = \mathbf{0} \\ 0 \text{ otherwise.} \end{cases}$$

Wilf (1968) showed that this function is invertible and Doubilet (1972) found the explicit representation of the inverse:

$$a^{-1}(\pi, \sigma) = \sum_{\tau \in \mathcal{P}(U)} \frac{\mu(\pi, \tau) \mu(\sigma, \tau)}{\mu\left(\hat{0}, \tau\right)}. \qquad (18.2)$$

Let ζ be the Riemannian zeta function of $\mathcal{P}(U)$ and p_i the number of blocks of σ contained in the i-th block of π. The Möbius function μ has in the partition lattice the explicit representation:

$$\mu(\pi, \sigma) = \zeta(\pi, \sigma) \cdot (-1)^{|\sigma| - |\pi|} \prod_{i=1}^{|\pi|} (p_i - 1)!.$$

Bienstock (1988) observed that the complexity of equation (18.1) can be reduced if the graph is symmetric. The best reduction possible by symmetry yields a sum with

$$p(u) = \sum_k p(u,k) \sim \frac{1}{4u\sqrt{3}} \exp\left(\pi\sqrt{\frac{2u}{3}}\right)$$

terms. This is the number of partitions of an integer, which is relevant only in the case that all partitions of the separating vertex set are induced with the same probability.

Equation (18.1) can be generalized for the K-terminal reliability. We observe that Equation (18.1) remains true if all the separating vertices are terminal vertices. Otherwise, we introduce the set $\underline{\mathcal{P}}(U)$ of labeled partitions of U. A *labeled partition* is a set partition with an optional label assigned to each block. Thus, we obtain:

$$|\underline{\mathcal{P}}(U)| = \sum_k \left\{ {u \atop k} \right\} 2^k. \tag{18.3}$$

A subgraph G^i is said to introduce the labeled partition $\underline{\pi} \in \underline{\mathcal{P}}(U)$, iff the partition $\pi \in \mathcal{P}(U)$ is induced by the connected components of G^i and exactly those components contain vertices of K which are assigned to labeled blocks of $\underline{\pi}$. Now, let $P(G^i, \underline{\pi})$ be the probability that the subgraph G^i induces the labeled partition $\underline{\pi} \in \underline{\mathcal{P}}(U)$ and:

$$b(\underline{\pi}, \underline{\sigma}) = \begin{cases} 1, & \text{if } \underline{\pi} \wedge \underline{\sigma} \text{ has exactly one labelled block} \\ 0, & \text{otherwise.} \end{cases}$$

Here $\underline{\pi} \wedge \underline{\sigma}$ is defined as the minimum of π and σ where a block is labeled iff at least one of its elements is in a labeled block of π or of σ. We obtain for the K-terminal reliability:

$$R(G,K) = \sum_{\underline{\pi} \in \underline{\mathcal{P}}(U)} \sum_{\underline{\sigma} \in \underline{\mathcal{P}}(U)} P(G^1, \underline{\pi}) \, b(\underline{\pi}, \underline{\sigma}) \, P(G^2, \underline{\sigma}). \tag{18.4}$$

Unfortunately, the function b has no inverse for $u \geq 3$. Thus we obtain no equivalent of Equation (18.1) for $R(G,K)$. Tittmann (1990) presented decomposition formulae for other cases (*st*-reliability, directed graphs) and methods of approximate decomposition.

18.3   The All-Terminal Reliability of Recurrent Structures

Now we return to recurrent structures. By analyzing the states of the edges of the recursively repeating part of the graph, we obtain a system of recurrence equations. The only tool which is necessary is the well-known factoring formula:

$$R\left(G\right) = p_e R\left(G_e\right) + \left(1 - p_e\right) R\left(G_{-e}\right).$$

G_e and G_{-e} denote the graph obtained from G by contracting and removing the edge e, respectively. The Figure 18.2 shows an example graph $L_{2,n}$ and graphs induced by the states of the edges e, f, g.

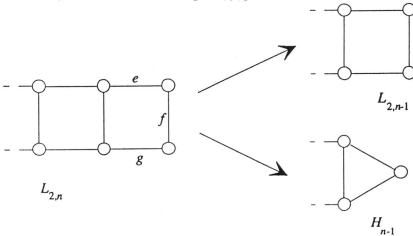

Figure 18.2: Edge factoring for the graph $L_{2,n}$

As a result of the first decomposition step a second graph H_n has to be analyzed. The second step is illustrated in Figure 18.3.

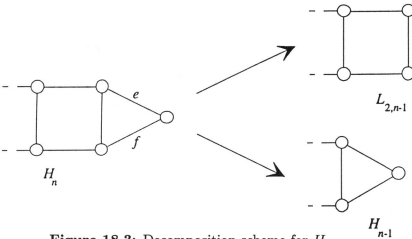

Figure 18.3: Decomposition scheme for H_n

We introduce the abbreviation:

$$\begin{aligned} l_n &= R\left(L_{2,n}, p\right) \\ h_n &= R\left(H_n, p\right). \end{aligned}$$

Considering the probabilities of each decomposition step we obtain the following system of equations:

$$
\begin{aligned}
l_n &= 3p^2 (1-p) l_{n-1} + p^3 h_{n-1} \\
h_n &= 2p (1-p) l_{n-1} + p^2 h_{n-1} \\
l_1 &= p \\
h_1 &= 1.
\end{aligned}
\tag{18.5}
$$

The equation system (18.5) can also be written as a recurrence equation for l_n:

$$
l_n = 3p^2 (1-p) l_{n-1} + p^5 + 2p^4 (1-p) \sum_{k=2}^{n-2} p^{2k-2} l_{n-k} \text{ for } n \geq 2
$$

$$
l_1 = p, \; l_2 = 4p^3 - 3p^4.
$$

The solution is:

$$
R(L_{2,n}, p) = \frac{p^{2n-1}}{2^n \alpha} [(4 - 3p + \alpha)^n - (4 - 3p - \alpha)^n]
\tag{18.6}
$$

with

$$
\alpha = \sqrt{12 - 20p + 9p^2}.
$$

The asymptotic development of $R(L_{2,n})$ is obtained by the dominant term of the solution:

$$
R(L_{2,n}, p) \sim \frac{1}{p\alpha} \left(\frac{p^2 (4 - 3p + \alpha)}{2} \right)^n.
$$

There are some questions arising from the presented example:

1. Is there a general procedure for finding the system of recurrent equations?

2. How many equations are necessary?

3. Is there a general way to determine the asymptotic behaviour of the reliability of a recurrent graph G_n?

The remaining part is devoted to the answer of these questions. As introduced in Section 18.1, we use a construction process for recurrent graphs starting from an initiator graph and recursively adding a generator graph. For the first step - joining the initiator and the generator - we could immediately employ the decomposition formula (18.1). However, for the next step we need not only the single value $R(G)$ but all the values $R(G_\pi^1)$, $\pi \in \mathcal{P}(V)$. We define a quadratic matrix in order to obtain these values. The number of rows of the matrix corresponds to the number of partitions of V and V'. In the following an additional linear order in $\mathcal{P}(V) = \{\pi_1, \ldots, \pi_r\}$, and $\mathcal{P}(V') = \{\sigma_1, \ldots, \sigma_r\}$, $r = |\mathcal{P}(V)|$, is assumed where π_r and σ_r are the greatest elements of $\mathcal{P}(V)$ and $\mathcal{P}(V')$, respectively. Let G_{ij} be the graph obtained from the generator graph

G by merging all vertices contained in one and the same block of $\sigma_i \in \mathcal{P}(V')$ and by merging all vertices contained in one and the same block of $\pi_j \in \mathcal{P}(V)$. Define Q to be the matrix of the all-terminal reliabilities of the graphs G_{ij}:

$$Q = (R(G_{ij}))_{i,j=1,...,r}.$$

The matrix A^{-1} of the same dimension is defined by the entries a^{-1} according to Equation (18.2):

$$A^{-1} = \left(a^{-1}(\pi_i, \pi_j)\right)_{i,j=1,...,r}.$$

The reliabilities of the initiator I are represented by the start vector

$$\mathbf{s} = (R(I_{\pi_1}), ..., R(I_{\pi_r})).$$

Let $\mathbf{e} = (0,...,0,1)^T$ be a r-dimensional unit vector and $P = A^{-1}Q$. We obtain from equation (18.1) by induction:

$$R(G_n) = R(I * G^n) = \mathbf{s} P^n \mathbf{e}. \tag{18.7}$$

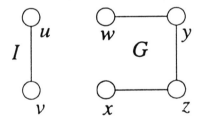

Figure 18.4: Initiator and generator for $L_{2,n}$

As an application let us consider again the ladder structure $L_{2,n}$ represented in Figure 18.2. The initiator and generator are shown in Figure 18.4. All edges are assumed to operate with probability p. We obtain:

$$
\begin{aligned}
\mathbf{s} &= (1, p) \\
A^{-1} &= \begin{pmatrix} 0 & 1 \\ 1 & -1 \end{pmatrix} \\
Q &= \begin{pmatrix} 2p - p^2 & 3p^2 - 2p^3 \\ p^2 & p^3 \end{pmatrix} \\
\mathbf{e} &= \begin{pmatrix} 0 \\ 1 \end{pmatrix}.
\end{aligned}
$$

The evaluation of Equation (18.7) for $n = 4$ gives:

$$R(L_{2,4}) = 153\,p^8 - 464\,p^9 + 534\,p^{10} - 276\,p^{11} + 54\,p^{12}.$$

We observe that the matrix $P = A^{-1}Q$ is the matrix of coefficients of the system of recurrence equations. Thus, the number of equations corresponds to the dimension of the matrix. In the general case this dimension is $B(u)$ where u denotes the cardinality of the separating set (or the set of couple vertices). For planar graphs this number reduces to the number t_u of noncrossing partitions of an u-set, which is given by the Catalan number:

$$C_n = \frac{1}{n+1}\binom{2n}{n} \sim \frac{4^{n-1}}{\sqrt{\pi n^3}}.$$

A noncrossing partition of a linear ordered set $\{1, 2, 3, 4, ...\}$ never contains the blocks $\{1, 3\}$ and $\{2, 4\}$ simultaneously. If the planar graph is in addition symmetric with respect to a symmetry axis in direction of increasing n then the number of equations reduces to:

$$\frac{1}{2}\left[\frac{1}{n+1}\binom{2n}{n} - \binom{n}{\lfloor \frac{n}{2}\rfloor}\right] \sim \frac{C_n}{2}.$$

To show this result, we use the bijection between noncrossing partitions of a set of cardinality u and plane trees with $u + 1$ vertices. A plane tree is a single vertex or a specified vertex (the root) together with an ordered sequence of plane trees. This recursive definition of plane trees is the key for counting them. Let $P(z)$ be the ordinary generating function for the number of plane trees. The recursive definition now translates into the equation:

$$P(z) = z + zP(z) + zP^2(z) + zP^3(z) + ... = \frac{z}{1 - P(z)}.$$

The solution is just the sequence of Catalan numbers. A symmetric partition of U is a partition that remains the same if we reverse the linear order of the elements of U. This property corresponds to a plane tree that remains unchanged under reflection of the embedding plane. Thus counting symmetric noncrossing partitions can be done by counting symmetric plane trees. A symmetric tree can be recursively defined as a single vertex or the root that is the father of a symmetric plane tree and a sequence of pairs of identical plane trees. Thus the ordinary generating function $S(z)$ for the number of symmetric plane trees satisfies:

$$S(z) = 1 + \frac{z\,S(z)}{1 - P(z^2)}.$$

Here a pair of identical plane trees is counted by doubling the weight of each vertex in a plane tree. By series expansion we obtain the number of symmetric plane trees:

$$s_n = \binom{n}{\lfloor \frac{n}{2}\rfloor}.$$

Consequently, the dimension of the matrices A and Q for symmetric planar graphs is:

$$s_n = \frac{1}{2}\left[\frac{1}{n+1}\binom{2n}{n} - \binom{n}{\lfloor\frac{n}{2}\rfloor}\right] \sim \frac{C_n}{2}.$$

In other cases of symmetry the dimension can be obtained by counting the number of orbits with respect to group action. The minimum dimension is achieved if the group acting on U is the full symmetric group S_u. In this case all partitions of the same type are induced with the same probability. Thus the dimension of P corresponds to the number of different (number) partitions of u:

$$p(u) \sim \frac{1}{4u\sqrt{3}}\exp\left(\pi\sqrt{\frac{2u}{3}}\right).$$

The following table gives the first values for the dimension of P for general graphs, planar graph, and graphs with symmetry.

n	1	2	3	4	5	6	7	8	9	10	11
$B(n)$	1	2	5	15	52	203	877	4140	21147	115975	678570
$p(n)$	1	2	3	5	7	11	15	22	30	42	55
C_n	1	2	5	14	42	132	429	1430	4862	16796	58786
s_n	1	2	4	10	26	76	232	750	2494	8524	29624

The asymptotic behavior of $R(G)$ can also be derived from Equation (18.7). The investigation of the matrix $P = A^{-1}Q$ shows that the entries are probabilities:

$$p_{ij} = P\left(G_{\pi_j}, \pi_i\right).$$

The value p_{ij} is the probability that the graph obtained from G by the fusion of all blocks of $\pi_j \in \mathcal{P}(V)$ induces the partition $\pi_i \in \mathcal{P}(V')$. The events corresponding to these probabilities are disjoint. Hence, we have:

$$\sum_i p_{ij} \leq 1.$$

This sum equals the probability that G_{π_j} is connected. If all column sums of P are properly lower than 1, then the entries of P^n tend to 0 if $n \to \infty$. Consequently, the all-terminal reliability of all recurrent structures tends to zero if the separating vertex sets V and V' are disjoint. The solution of the system of recurrent equations that determines the all-terminal reliability of a recurrent graph can be given in matrix form by $\mathbf{s}^T P^n \mathbf{e}$. Thus the eigenvalues of maximal absolute value of P permit an asymptotic estimation of $R(G)$.

To demonstrate the power of the presented approach, we computed the reliability polynomial of a 6×20 grid with the help of MAPLE V. However the complete representation of the resulting polynomial would require many pages. The coefficient of p^{214} in $R(\text{Grid}_{6,20}, p)$ is:

$$139\,058\,368\,483\,155\,218\,031\,377\,816\,656\,950\,558\,051\,299\,499.$$

The complete result and other reliability polynomials may be found in the Internet (http://www.mpi.htwm.de/Forschung/reliability.html).

18.4   Generalizations and Open Problems

The natural generalization of the all-terminal reliability is the K-terminal reliability. In this case we have to consider labelled partitions of U rather than partitions. However, since there is no simple order structure in the set of labelled partitions the lattice-theoretic tools employed for $R(G)$ fail now. Despite this fact, the direct application of Equation (18.4) is still possible. The number of terms is the number of labelled partitions of U given by (18.3). For the two-terminal reliability the set of labelled partitions can be reduced to the set $\mathcal{P}_1(U)$ of all labelled partitions with exact one labelled block. Numerical calculations indicate that the matrix B defined by $b(\underline{\pi}, \underline{\sigma})$ for all $\underline{\pi}, \underline{\sigma} \in \mathcal{P}_1(U)$ is regular up to $u = 7$. To prove that this is true for all $u \in \mathcal{N}$ is an open problem.

Another way to make the proposed approach wider applicable is a generalization of the structures under consideration. The simplest way is the usage of more than one generator. If G is replaced by a finite sequence $\{G_i\}$ of generators we obtain the following composition:

$$H = G_0 * G_1 * \ldots * G_n.$$

Let U_l be the vertex set separating G_{l-1} from G_l for $l = 1, \ldots, n$. Here it is generally not necessary to assume $|U_l| = |U_k|$ for different values k and l. The matrix $A_l = \left(a_{ij}^{(l)}\right)$ for $l = 1, \ldots, n$ is defined by:

$$a_{ij}^{(l)} = \begin{cases} 1, & \text{if } \pi_i \wedge \pi_j = \mathbf{0} \\ 0 & \text{else} \end{cases} \qquad \pi_i, \pi_j \in \mathcal{P}(U_l).$$

We assign to each graph $G_1, \ldots, G_{n-1}$ a matrix of all-terminal reliabilities:

$$Q_l = \left(q_{ij}^{(l)}\right), \ q_{ij}^{(l)} = R(G_{l,ij}), \ \pi_i \in \mathcal{P}(U_l), \ \pi_j \in \mathcal{P}(U_{l+1}).$$

The all-terminal reliabilities of the first graph G_0 and of the last graph G_n are represented by vectors:

$$\mathbf{q}_0 = (R(G_{0,\pi_1}), \ldots, R(G_{0,\pi_r}))^T, \ \pi_1, \ldots, \pi_r \in \mathcal{P}(U_1)$$
$$\mathbf{q}_n = (R(G_{n,\pi_1}), \ldots, R(G_{n,\pi_s}))^T, \ \pi_1, \ldots, \pi_s \in \mathcal{P}(U_n).$$

Thus we obtain the following result for the all-terminal reliability of H:

$$R(G_0 * \ldots * G_n) = \mathbf{q}_0^T A_1^{-1} Q_1 A_2^{-1} Q_2 \cdots A_{n-1}^{-1} Q_{n-1} A_n^{-1} \mathbf{q}_n.$$

This equation permits the computation of the all-terminal reliability of all graphs which exhibit a suitable sequence of separating vertex sets.

A graph that forms a partial k-tree is a more general structure than the above introduced linear recurrent graphs. A k-tree is recursively defined as follows. The complete graph on $k + 1$ vertices is a k-tree. A graph which is obtained from a k-tree G by inserting a new vertex that is connected by k edges to all vertices of a k-clique of G is again a k-tree. We obtain a partial k-tree by removing edges from a k-tree. Thus a partial k-tree can be reduced to a single vertex by subsequently plucking off vertices of degree at most k. It can be shown that many NP-hard problems in graphs become polynomially solvable if the problem is restricted to partial k-trees Arnborg and Proskurowski (1989).

For our purposes, we generalize the definition of k-trees. Let $\omega(G)$ be a measure of the complexity of G. This measure should be related to the computational effort necessary for the reliability analysis of G. In the case of all-terminal reliability we may choose the number of spanning trees or the domination of the graph. Johnson (1982) showed that these numbers correspond to the complexity of decomposition algorithms for $R(G)$ using bridge and parallel reductions, respectively. Let C and k be positive integer constants. We say that the graph G is tree-like of order (C, k) if there exists a finite sequence of separating vertex sets $\{U_i\}$ with $|U_i| \leq k$ in G such that all components G_j induced by the separation $\{U_i\}$ satisfy $\omega(G_j) \leq C$.

The all-terminal reliability of a tree-like graph can be obtained by analysing the subgraphs induced by the separating vertex sets. In this case we start the computation with the leaves of the tree. Let:

$$G = H \cup \bigcup_{i=0}^{n} G_i$$

and

$$H \cap G_i = (U_i, \emptyset) \text{ for } i = 0, ..., n.$$

The subgraphs G_i and G_j have only vertices of $U_i \cap U_j$ in common for $i \neq j$. Intuitively spoken, the branches $G_0, ..., G_n$ of the tree-like graph meet each other in H. We generalize the notion of the minimum of partitions to partitions defined on different sets. Assume $\pi_i \in \mathcal{P}(U_i)$, $i = 0, ..., n$ and let $\tilde{\pi}_i \in \mathcal{P}(\bigcup_{i=0}^{n} U_i)$ be the partition obtained by augmenting π with singletons according to all elements which are not contained in U_i. Then we define:

$$\bigwedge_{i=0}^{n} \pi_i = \bigwedge_{i=0}^{n} \tilde{\pi}_i = \max\left\{ \tau \in \mathcal{P}\left(\bigcup_{i=0}^{n} U_i\right) : \tau \leq \tilde{\pi}_i \text{ for } i = 0, ..., n \right\}.$$

We obtain as a consequence of the splitting formula:

$$R(G) = \sum_{\pi_0 \in \mathcal{P}(U_0)} \cdots \sum_{\pi_n \in \mathcal{P}(U_n)} P(G_0, \pi_0) P(G_1, \pi_1) \cdots P(G_n, \pi_n) R(H_{\pi_0 \wedge ... \wedge \pi_n}).$$

Now let $a_i(\pi, \sigma)$ be the minimum function assigned to the i-th separating vertex set and $\mathcal{P} = \mathcal{P}(U_0) \times \mathcal{P}(U_1) \times \ldots \times \mathcal{P}(U_n)$. Then the all-terminal reliability of G can be represented as a function of the all-terminal reliabilities of its subgraphs:

$$R(G) = \sum_{(\pi_0,\ldots,\pi_n)\in\mathcal{P}} \sum_{(\sigma_0,\ldots,\sigma_n)\in\mathcal{P}} \prod_{i=0}^{n} a_i^{-1}(\pi_i, \sigma_i) R(G_{i,\sigma_i}) R(H_{\pi_0 \wedge \ldots \wedge \pi_n}).$$

This formula can be rearranged in order to provide the possibility of recurrent computation. Let $\pi_0 \in \mathcal{P}(U_0)$. Then:

$$R(G'_{\pi_0}) = \sum_{(\pi_1,\ldots,\pi_n)\in\mathcal{P}'} \sum_{(\sigma_1,\ldots,\sigma_n)\in\mathcal{P}'} \prod_{i=1}^{n} a_i^{-1}(\pi_i, \sigma_i) R(G_{i,\sigma_i}) R(H_{\pi_0 \wedge \ldots \wedge \pi_n}).$$

Here we used $\mathcal{P}' = \mathcal{P}(U_1) \times \mathcal{P}(U_2) \times \ldots \times \mathcal{P}(U_n)$ and:

$$G' = H \cup \bigcup_{i=1}^{n} G_i.$$

The application of splitting techniques to tree-like graphs permits a much faster computation of complex structures provided that there are algorithms which find a suitable sequence of separating vertex sets in polynomially bounded time. Single separating vertex sets may be found easily using max-flow algorithms. However, the problem of minimizing the computational effort for the reliability analysis of graphs by multiple splitting is still open.

References

1. Arnborg, S. and Proskurowski, A. (1989). Linear time algorithms for NP-hard problems restricted to partial k-trees, *Discrete Applied Mathematics*, **23**, 11–24.

2. Bienstock, D. (1988). Some lattice-theoretic tools for network reliability analysis, *Mathematics of Operations Research*, **13**, 467–478.

3. Doubilet, P. (1972). On the foundations of combinatorial theory VII. Symmetric functions through the theory of distribution and occupancy, *Studies in Applied Mathematics*, **51**, 377–396.

4. Johnson, Rubin (1982). Some combinatorial aspects of network reliability, *Ph.D. Thesis*, University of California, Berkeley.

5. Koslov, B. A. and Ushakov, I. A. (1975). *Handbook of Reliability Computations*, Sovietski Radio Moscow, (in Russian), (German Translation, Akademie-Verlag, 1978).

6. Odlyzko (1995). Asymptotic enumeration methods, In *Handbook of Combinatorics. Vol. II* (Eds., R. L. Graham, M. Grötschel and L. Lovász), Amsterdam, The Netherlands: Elsevier.

7. Rosenthal, A. (1977). Computing the reliability of complex networks, *SIAM Journal of Applied Mathematics*, **32**, 384–393.

8. Tittmann, P. (1990). Decomposition and reduction methods for the reliability analysis of networks, *Ph.D. Thesis*, Mittweida, (in German).

9. Wilf, H. (1968). The Möbius function in combinatorial analysis and chromatic graph theory, In *Proof Techniques in Graph Theory* (Ed., F. Harary), pp. 179–188, New York: Academic Press.

PART IV
Process Control

19

Testing for the Existence of a Change-Point in a Specified Time Interval

Dietmar Ferger

Dresden University of Technology, Dresden, Germany

Abstract: Let $X_1, \ldots, X_n$ be a sequence of independent random variables with changing distribution after an unknown "time" $[n\theta]$, where $\theta \in (0,1)$ is the parameter of interest. We investigate the test-problem $H_0 : \theta \in \Theta_0$ against $H_1 : \theta \notin \Theta_0$, where Θ_0 is a closed subset of $(0,1)$ representing a known time span. Our procedures involve U-statistic type processes. They are shown to be consistent on the alternative H_1 and to be level-α tests for H_0. The determination of critical values requires a bootstrap approximation.

Keywords and phrases: Change of distributions, functional limit theorems for two-time parameter U-Statistic-type processes, bootstrap approximation

19.1 Introduction

Change-point problems are strongly related to classical questions in quality control. Suppose that the quality of an output may be assessed by some measurable characteristic X. Then one often wants to signal a possible disorder of the regular stochastic mechanism. Detecting a change may help to draw the necessary conclusions. Such a situation can be modelled by independent random elements $X_1, \ldots, X_n$ in a sample space $(\mathcal{X}, \mathcal{F})$ such that $X_1, \ldots, X_{[n\theta]}$ have a common distribution ν_1 and the remaining $X_{[n\theta]+1}, \ldots, X_n$ have a common distribution $\nu_2 \neq \nu_1$, for some $\theta \in [0,1)$. Here, as usual, $[x]$ denotes the integer part of the real number x. The so-called *change-point* θ as well as the distributions $\nu_1 \neq \nu_2$ are assumed to be unknown. Then one basic question is whether there is a change at all. In other words, we are concerned with the test problem $H_0^* : \theta = 0$ versus $H_1^* : \theta \in (0,1)$. Many statistical papers are dealing with this classical change-point problem. Csörgő and Horváth (1988) and Bhattacharya (1991) review the literature up to the end of the eighties.

However, besides the question whether a change has occurred or not, in many situations it is more natural to ask if a change has taken place within a certain period of time or not. As an example, consider an environmental system which was exposed to an external pollution during a given space of time. Then we like to know whether the pollution has effects on a certain population living in this system. Formally, we are faced with the test situation

$$H_0 : \theta \in \Theta_0 \quad \text{versus} \quad H_1 : \theta \notin \Theta_0$$

where Θ_0 is a given closed subset of the open unit interval $(0, 1)$ and only $\theta \in (0, 1)$ are considered. Our approach to this *nonstandard* test problem generalizes Ferger's (1994c) procedure in the classical set-up H_0^* vs. H_1^* which corresponds to the degenerate interval $\Theta_0 = \{0\}$.

19.2   A Family of Tests

In view of asymptotic investigations we need to formulate our model by means of triangular arrays. So let $X_{1n}, \ldots, X_{nn}$, $n \in I\!N$, be a double-indexed scheme of rowwise independent random elements defined on a common probability space $(\Omega, \mathcal{A}, P)$ with values in a measurable space $(\mathcal{X}, \mathcal{F})$. We suppose that for all $n \in I\!N$

$$\mathcal{L}(X_{in}) = P \circ X_{in}^{-1} = \begin{cases} \nu_1 & , \quad 1 \leq i \leq [n\theta] \\ \\ \nu_2 & , \quad [n\theta] < i \leq n \end{cases}.$$

As in Ferger (1994c) our basic test-statistic will be based on the empirical process

$$r_n(t) = n^{-2} \sum_{i=[nt]+1}^{n} \sum_{j=1}^{[nt]} K(X_{in}, X_{jn}), \quad 0 \leq t \leq 1.$$

Here

$$K : \mathcal{X} \, x \, \mathcal{X} \to I\!R$$

is a $\mathcal{F} \otimes \mathcal{F}$-measurable and antisymmetric mapping, which will be called *kernel* in the sequel. Antisymmetry means $K(x, y) = -K(y, x)$ for all $x, y \in \mathcal{X}$. Ferger (1995) proved under a certain second-moment condition on the kernel K that

$$\lim_{n \to \infty} \sup_{0 \leq t \leq 1} |r_n(t) - \bar{r}_n(t)| = 0 \quad P_\theta - \text{almost surely (a.s.)} \tag{19.1}$$

for all $\theta \in [0, 1)$. (The notation P_θ is used to stress the fact that θ is the true change-point.) The deterministic mean process $\bar{r}_n(t) := E_\theta \, r_n(t)$, $0 \leq t \leq 1$, is easily seen to be

$$\bar{r}_n(t) = \bar{r}_n(\lambda, \theta, t) := \begin{cases} n^{-2}[nt](n - [n\theta])\lambda &, \quad 0 \le t \le [n\theta]n^{-1} \\ n^{-2}(n - [nt])[n\theta]\lambda &, \quad [n\theta]n^{-1} < t \le 1 \end{cases}, \quad (19.2)$$

with

$$\lambda = \lambda(K) = \lambda(K, \nu_1, \nu_2) = \int K \, d\nu_2 \otimes \nu_1.$$

Since under the composite hypothesis H_0, the true parameter θ is not specified, θ needs to be replaced by

$$\theta_n = \arg\max_{t \in \Theta_0} |r_n(t)|, \quad (19.3)$$

which, under H_0, is the most plausible estimator of the true θ. Indeed, Ferger (1995) proved that

$$\theta_n - \theta = O(n^{-1}) \quad \text{as } n \to \infty \quad P_\theta - \text{a.s.}$$

for all $\theta \in \Theta_0$. Recall that by (19.2)

$$\bar{r}_n(\theta) = n^{-2}[n\theta](n - [n\theta])\lambda,$$

whence in view of (19.1) λ may be estimated by

$$\lambda_n = \frac{1}{\theta_n(1 - \theta_n)} r_n(\theta_n).$$

After all, we put

$$\Gamma_n(t) = n^{\frac{1}{2}}(r_n(t) - \bar{r}_n(\lambda_n, \theta_n, t)), \quad 0 \le t \le 1,$$

and by (19.1) come up with the following decision rule:

Reject the hypothesis H_0 if $\|\Gamma_n\|_p$ exceeds a critical value c_p. (19.4)

Here, $\|\cdot\|_p$, $1 \le p \le \infty$, denotes the L_p-norm on the Skorokhod-space $D[0,1]$ [confer, e.g., Billingsley (1968, Chapter 3)]: For $f \in D[0,1]$

$$\|f\|_p = \left(\int_{[0,1]} |f(t)|^p \, dt \right)^{\frac{1}{p}}, \quad 1 \le p < \infty,$$

and

$$\|f\|_\infty = \sup_{0 \le t \le 1} |f(t)|.$$

If $\alpha \in (0,1)$ denotes the level of significance, the critical value $c_p = c_{p,\alpha}$ in (19.4) is given by the $(1 - \alpha)$-quantile of $\|\Gamma\|_p$, where Γ is a Gaussian process specified in (19.14) below:

$$c_p = c_{p,\alpha} := \min \left\{ x \in I\!\!R : P(\|\Gamma\|_p \leq x) \geq 1 - \alpha \right\}. \tag{19.5}$$

Formally, our test (19.4) now may be described by the critical function

$$t_n = 1_{\left\{ \|\Gamma_n\|_p \leq c_{p,\alpha} \right\}}. \tag{19.6}$$

The following theorem states that t_n is a consistent asymptotic level-α test.

Theorem 19.2.1 *Let K be an antisymmetric kernel satisfying the moment-condition*

$$\int |K|^s \, d\nu_i \otimes \nu_j < \infty \quad \forall \, 1 \leq i, \, j \leq 2$$

for some $s > 2$. If $\lambda \neq 0$ and $\Theta_0 \subseteq (0,1)$ is closed, then for all $\alpha \in (0,1)$

$$\lim_{n \to \infty} P_\theta(t_n = 1) = \alpha \quad \forall \, \theta \in \Theta_0 \tag{19.7}$$

and

$$\lim_{n \to \infty} P_\theta(t_n = 1) = 1 \quad \forall \, \theta \notin \Theta_0. \tag{19.8}$$

By construction the test t_n in (19.6) depends on the kernel K. In view of Theorem 19.2.1 the quantity λ has the following statistical interpretation: A kernel K with $\lambda \neq 0$ is able to separate between the underlying (unknown) distributions ν_1, ν_2 provided they are not equal. Indeed, for K fixed, $\lambda = \lambda(K, \nu_1, \nu_2)$ measures the distance between ν_1 and ν_2. In Ferger (1997) it is shown that to each pair $\nu_1 \neq \nu_2$ one can construct explicitly kernels K with $\lambda \neq 0$. On the other hand, $\lambda = 0$ for all antisymmetric K, whenever $\nu_1 = \nu_2$. This easily follows from Fubini's Theorem. Consequently, before a set of data can be analysed, the statistician has to choose an antisymmetric kernel such that, hopefully, $\lambda = \lambda(K, \nu_1, \nu_2) \neq 0$. In many cases, if some prior information about the underlying distributions ν_1, ν_2 is available, this is possible. Typically, $\nu_2 = \nu_1 \circ T^{-1}$, where $T : \mathcal{X} \to \mathcal{X}$ is a certain measurable transformation. The following examples are taken from Ferger (1997). Here it is useful to mention that $\lambda(\alpha K) = \alpha \lambda(K)$ for all scalars $\alpha \in I\!\!R$. Especially for $\alpha \neq 0$ we have that

$$\lambda(\alpha K) \neq 0 \iff \lambda(K) \neq 0. \tag{19.9}$$

Example 19.2.1 (Multivariate location-model)
Let $\mathcal{X} = I\!\!R^k$, $T(x) = x + d$, $d \in I\!\!R^k$, $d \neq 0$. Then for $a \in I\!\!R^k$ fixed

$$K(x,y) = a'(x - y)$$

yields $\lambda(K) = a'd$, which, e.g., is strictly positive, if $a = d$.

Example 19.2.2 (Multivariate scale-model)

Let $\mathcal{X} = \mathbb{R}^k$, $T(x) = Ax$, A is a kxk-matrix, $A \neq I$. For a fixed kxk-matrix $B \neq 0$ put

$$K(x,y) = x'Bx - y'By.$$

Then

$$\lambda(K) = \text{trace}\left((A'BA - B)\sum\right) + \mu'(A'BA - B)\mu,$$

where μ denotes the expectation vector and $\sum$ the covariance matrix of ν_1.

A rich source of examples, if one relies on a parametric model is

Example 19.2.3 (Exponential family)

Suppose $\nu_i = f_i\mu$, $i = 1,2$, for some σ-finite measure μ on $(\mathcal{X}, \mathcal{F})$, where

$$f_i(x) = \exp\left\{\sum_{j=1}^k Q_j(\gamma_i)T_j(x) + Q_0(\gamma_i) + T_0(x)\right\}, \quad i = 1,2,$$

with $\gamma_1 \neq \gamma_2$ belonging to some parameter space $\Gamma \neq \emptyset$ and $Q_j : \Gamma \to \mathbb{R}$ are arbitrary and $T_j : \mathcal{X} \to \mathbb{R}$ are measurable mappings. Then

$$K(x,y) = \sum_{j=1}^k (Q_j(\gamma_2) - Q_j(\gamma_1))(T_j(x) - T_j(y))$$

induces

$$\lambda(K) = I(\nu_1, \nu_2) + I(\nu_2, \nu_1),$$

where I denotes the Kullback-Leibler Information. Thus $\lambda(K)$ is strictly positive, since $\nu_1 \neq \nu_2$.

The following Examples 19.2.4–19.2.7 are applications of Example 19.2.3.

Example 19.2.4 (Univariate normal distribution)

Let $\mathcal{X} = \mathbb{R}$, $\nu_i = N(\mu_i, \sigma_i^2)$, $i = 1,2$, with $(\mu_1, \sigma_1^2) \neq (\mu_2, \sigma_1^2) \in \mathbb{R}x(0,\infty)$. Then

$$K(x,y) = \left(\frac{\mu_2}{\sigma_2^2} - \frac{\mu_1}{\sigma_1^2}\right)(x - y) + \frac{1}{2}\left(\frac{1}{\sigma_1^2} - \frac{1}{\sigma_2^2}\right)(x^2 - y^2).$$

In the normal shift model $(\mu_1 \neq \mu_2, \sigma_1^2 = \sigma_2^2)$ we obtain by (19.9)

$$K(x,y) = x - y$$

and in the normal scale model $(\mu_1 = \mu_2, \sigma_1^2 \neq \sigma_2^2)$ by (19.9)

$$K(x,y) = x^2 - y^2.$$

Example 19.2.5 (Binomial distribution)

Let $\mathcal{X} = \{1, \ldots, N\}$ and $\nu_i = \text{Bin}(N, p_i)$, $i = 1, 2$, with $p_1 \neq p_2 \in (0, 1)$. Then by (19.3) and (19.9)

$$K(x, y) = x - y$$

is an appropriate kernel.

Example 19.2.6 (Gamma distribution)

Let $\mathcal{X} = (0, \infty)$ and ν_i be the Gamma distribution with parameters b_i and p_i, $i = 1, 2$, where $(b_1, p_1) \neq (b_2, p_2) \in (0, \infty)^2$. Then

$$K(x, y) = (b_1 - b_2)(x - y) + (p_2 - p_1) \log \frac{x}{y}.$$

If, e.g., $b_1 = b_2$ and $p_1 \neq p_2$, then by (19.9)

$$K(x, y) = \log x - \log y$$

yields a non vanishing λ.

Recall that the Gamma distribution with parameters b and p has the density

$$f(x) = \frac{b^p}{\Gamma(p)} x^{p-1} e^{-bx}, \quad x > 0.$$

Consequently, it includes the Exponential distribution ($p = 1$), the Erlang distribution ($p \in \mathbb{N}$) and the $\mathcal{X}^2$-distribution $\left(b = \frac{1}{2}, p \in \frac{\mathbb{N}}{2}\right)$.

Example 19.2.7 (Beta distribution)

Let $\mathcal{X} = (0, 1)$ and $\nu_i = \text{Beta}(p_i, q_i)$ with $(p_1, q_1) \neq (p_2, q_2) \in (0, \infty)^2$. Then

$$K(x, y) = (p_2 - p_1) \log \frac{x}{y} + (q_2 - q_1) \log \frac{1 - x}{1 - y}.$$

Example 19.2.8 (Multivariate normal distribution)

Let $\mathcal{X} = \mathbb{R}^d$, $d \in \mathbb{N}$, $\nu_i = N_d(\mu_i, S_i)$ with $\mu_i \in \mathbb{R}^d$ and S_i is a symmetric, positive definite $d{\times}d$-matrix. Then

$$K(x, y) = x' \left(S_1^{-1} - S_2^{-1}\right) x - y' \left(S_1^{-1} - S_2^{-1}\right) y - 2 \left(\mu_1' S_1^{-1} - \mu_2' S_2^{-1}\right) (x - y).$$

Especially, in the location-model ($\mu_1 \neq \mu_2, S_1 = S_2$)

$$K(x, y) = (\mu_2 - \mu_1)' S_1^{-1} (x - y)$$

and in the shift-model ($\mu_1 = \mu_2, S_1 \neq S_2$)

$$K(x, y) = x' \left(S_1^{-1} - S_2^{-1}\right) x - y' \left(S_1^{-1} - S_2^{-1}\right) y.$$

Further examples may be found in Ferger (1994c). In general, the critical value c_p in (19.5) involves unknown quantities. To make the test t_n in (19.6) practical, c_p is replaced by a bootstrap-approximation. In the next section we will see that the corresponding bootstrap version of t_n remains a consistent and asymptotic level-α test for H_0 vs. H_1.

19.3 The Bootstrap-Test Family

For all integers $1 \leq k \leq n - 1$ let

$$\nu_{1,k,n} = \frac{1}{k} \sum_{1 \leq i \leq k} \delta_{X_{in}}$$

and

$$\nu_{2,k,n} = \frac{1}{n - k} \sum_{k < i \leq n} \delta_{X_{in}}$$

be the empirical measures of the two subsamples $(X_i)_{1 \leq i \leq k}$ and $(X_i)_{k < i \leq n}$. (Here δ_x denotes the Dirac-measure at the point $x \in \mathcal{X}$.)

For short we write $\underline{X}_n = (X_1, \ldots, X_n)$. Introduce the *bootstrap variables* $X^*_{1n}, \ldots, X^*_{nn}$ on $(\Omega, \mathcal{A}, P)$ which are $P(\cdot|\underline{X}_n)$-independent with

$$\mathcal{L}(X^*_{in}|\underline{X}_n) = P(\cdot|\underline{X}_n) \circ X^{*-1}_{in} = \begin{cases} \nu_{1,k_n,n} & , & 1 \leq i \leq k_n \\ & & \\ \nu_{2,k_n,n} & , & k_n < i \leq n \end{cases}, \qquad (19.10)$$

where $k_n := n\theta_n$ and θ_n is the estimator (19.3). The existence of a probability space $(\Omega, \mathcal{A}, P)$ which is rich enough to carry all random variables, is ensured by a canonical construction, confer Ferger (1994b). Put

$$\Gamma^*_n(t) = n^{\frac{1}{2}} (r^*_n(t) - \bar{r}_n(\lambda^*_n, \theta^*_n, t)), \quad 0 \leq t \leq 1,$$

where r^*_n, λ^*_n and θ^*_n are the corresponding statistics pertaining to the bootstrap variables, that is

$$r^*_n(t) = n^{-2} \sum_{i=[nt]+1}^{n} \sum_{j=1}^{[nt]} K(X^*_{in}, X^*_{jn}), \quad 0 \leq t \leq 1,$$

$$\theta^*_n = \arg\max_{t \in \Theta_0} |r^*_n(t)|$$

and

$$\lambda^*_n = \frac{1}{\theta^*_n(1 - \theta^*_n)} r^*_n(\theta^*_n).$$

Finally, we define

$$c_p^* = c_{p,\alpha}^*(\underline{X}_n) := \min\left\{x \in \mathbb{R} : P(\|\Gamma_n^*\|_p \leq x | \underline{X}_n) \geq 1 - \alpha\right\}$$

and herewith

$$t_n^* = 1_{\left\{\|\Gamma_n\|_p > c_p^*\right\}}.$$

The test t_n^* is the bootstrap version of t_n. Let H denote the distribution function of $\|\Gamma\|_p$. We can prove

Theorem 19.3.1 *Let $\mathcal{X}$ be a separable metric space and let K be antisymmetric, bounded and continuous. If $\lambda \neq 0$, then for all $\alpha \in (0,1)$*

$$\lim_{n\to\infty} P_\theta(t_n^* = 1) = \alpha \quad \forall \, \theta \in \Theta_0 \tag{19.11}$$

provided $c_{p,\alpha}$ is the unique solution of $H(x-) \leq 1 - \alpha \leq H(x)$.
   Furthermore

$$\lim_{n\to\infty} P_\theta(t_n^* = 1) = 1 \quad \forall \, \theta \notin \Theta_0. \tag{19.12}$$

Dümbgen (1991) uses the same resampling method for constructing bootstrap confidence sets. In practice one can approximate the random critical value $c_{p,\alpha}^*(\underline{X}_n)$ via Monte-Carlo-Simulation as follows: We choose at random with equal probability $\frac{1}{k_n}$ one of the first $k_n = n\theta_n$ data and obtain a datum, say $X_{1n}^* \in \{X_{1n}, \ldots, X_{k_n n}\}$. After putting X_{1n}^* back into the subsample $\{X_{1n}, \ldots, X_{k_n n}\}$ we repeat this procedure independently k_n times and get data $X_{1n}^*, \ldots, X_{k_n n}^*$, which by construction are independent with distribution $\nu_{1,k_n,n}$. In the same way we generate further independent random variables X_{in}^*, $i = k_n + 1, \ldots, n$ with common distribution $\nu_{2,k_n,n}$. More precisely, let $(U_{in} : 1 \leq i \leq n, n \in \mathbb{N})$ be a triangular array of rowwise $P(\cdot | \underline{X}_n)$-independent integer-valued random variables with

$$P(U_{in} = l | \underline{X}_n) = \frac{1}{k_n} \quad \forall \, 1 \leq l, \, i \leq k_n$$

and

$$P(U_{in} = l | \underline{X}_n) = \frac{1}{n - k_n} \quad \forall \, k_n < l, \, i \leq n.$$

If we define

$$X_{in}^*(\omega) := X_{U_{in}(\omega),n} \,, \quad \omega \in \Omega,$$

then the X_{in}^* are $P(\cdot | \underline{X}_n)$-independent with conditional distribution (19.10). With this method we obtain a resampling variable

$$T_n^* := \|\Gamma_n^*\|_p.$$

Repeating the resampling part independently $m \in \mathbb{N}$ times yields m copies of T_n^*: $T_{n,1}^*, \ldots, T_{n,m}^*$ $P(\cdot | \underline{X}_n)$-independent with common conditional distribution function $P(T_{n,m}^* \leq x | \underline{X}_n) = P(\|\Gamma_n^*\|_p \leq x | \underline{X}_n)$.

Let $\widehat{T}^*_{n,m}(1-\alpha)$ denote the $(1-\alpha)$-sample quantile of $T^*_{n,1}, \ldots, T^*_{n,m}$:

$$\widehat{T}^*_{n,m}(1-\alpha) := H^{-1}_{n,m}(1-\alpha),$$

where

$$H_{n,m}(x) = \frac{1}{m}\sum_{i=1}^{m} 1_{\{T^*_{n,i} \le x\}}, \quad x \in \mathbb{R},$$

is the empirical distribution function of the $T^*_{n,i}$. By Theorem 2.3.1 in Serfling (1980) we obtain immediately

Proposition 19.3.1 *Let $H_n(x|\underline{X}_n) = P(\|\Gamma^*_n\|_p \le x|\underline{X}_n)$ and assume that*

$$H_n(x - |\underline{X}_n) \le 1 - \alpha \le H_n(x|\underline{X}_n)$$

has a unique solution with probability one, then

$$\lim_{m\to\infty} \widehat{T}^*_{n,m}(1-\alpha) = c^*_{p,\alpha}(\underline{X}_n) \quad P_\theta - \text{a.s.} \quad \text{for all } \theta \in (0,1).$$

Notice that we are free in the choice of m.

19.4　The Proofs

The proofs of Theorems 19.2.1 and 19.3.1 are based on a functional limit theorem for the bivariate process $b_n = \{b_n(s,t) : 0 \le s, t \le 1\}$ defined by

$$b_n(s,t) =$$

$$= 1_{\{t \le s\}} n^{-\frac{3}{2}} \left[\sum_{i=[nt]+1}^{[ns]} \sum_{j=1}^{[nt]} K(Z_{in}, Z_{jn}) + \sum_{i=[ns]+1}^{n} \sum_{j=1}^{[nt]} (K(Y_{in}, Z_{jn}) - l_n) \right] +$$

$$+ 1_{\{t > s\}} n^{-\frac{3}{2}} \left[\sum_{i=[nt]+1}^{n} \sum_{j=1}^{[ns]} (K(Y_{in}, Z_{jn}) - l_n) + \sum_{i=[nt]+1}^{n} \sum_{j=[ns]+1}^{[nt]} K(Y_{in}, Y_{jn}) \right].$$

Here, $Z_{1n}, \ldots, Z_{nn}$ and $Y_{1n}, \ldots, Y_{nn}$, $n \in \mathbb{N}$, are two independent sequences of i.i.d. random elements defined on $(\Omega, \mathcal{A}, P)$ with values $(\mathcal{X}, \mathcal{F})$ having distributions $\mathcal{L}(Z_{1n}) = \nu_{1n}$ and $\mathcal{L}(Y_{1n}) = \nu_{2n}$. Moreover

$$l_n = \int K \, d\nu_{2n} \otimes \nu_{1n}.$$

Ferger (1994a) proved, that under a certain moment - and stability assumption

$$b_n \xrightarrow{\mathcal{L}} b. \tag{19.13}$$

Here, (19.13) means that the sequence (b_n) of random elements in the space $D([0,1]^2)$ converges in distribution in the sense of Neuhaus (1971) to a centered Gaussian process b, which has continuous sample paths with probability one. The covariance function of b is given in (2.10) of Ferger (1994a). The stability assumption especially is fullfilled if $\mathcal{X}$ is a topological space, K is bounded and continuous and $(\nu_{in})_{n \in \mathbb{N}}$ converges in the weak topology to a measure ν_i, $i = 1, 2,$.

PROOF OF THEOREM 19.2.1. For the proof of (19.7) it suffices to show that

$$\Gamma_n \xrightarrow{\mathcal{L}} \Gamma := b(\theta, \cdot) - g(\theta, \cdot)b(\theta, \theta) \quad \forall\, \theta \in \Theta_0, \tag{19.14}$$

where

$$g(\theta, t) = 1_{[0,\theta]}(t)\theta^{-1}t + 1_{(\theta,1]}(t)(1-\theta)^{-1}(1-t).$$

Indeed, by (19.14) and the Continuous Mapping Theorem (CMT) $\|\Gamma_n\|_p \xrightarrow{\mathcal{L}} \|\Gamma\|_p$. But $\mathcal{L}(\|\Gamma\|_p)$ is continuous on $\mathbb{R}$ by Theorem 1 and the Corollary in Lifshits (1982). This immediately yields (19.7). To prove (19.14) we decompose the process Γ_n properly:

$$\Gamma_n(t) = \Gamma_{n1}(t) + \Gamma_{n2}(t) + \Gamma_{n3}(t)$$

with

$$\Gamma_{n1}(t) = n^{\frac{1}{2}}(r_n(t) - \bar{r}_n(\lambda, \theta, t)) + n^{\frac{1}{2}}(\bar{r}_n(\lambda, \theta, t) - \bar{r}_n(\tilde{\lambda}_n, \theta, t))$$

$$\Gamma_{n2}(t) = n^{\frac{1}{2}}(\bar{r}_n(\tilde{\lambda}_n, \theta, t) - \bar{r}_n(\lambda_n, \theta, t))$$

$$\Gamma_{n3}(T) = n^{\frac{1}{2}}(\bar{r}_n(\lambda_n, \theta, t) - \bar{r}_n(\lambda_n, \theta_n, t))$$

and

$$\tilde{\lambda}_n = \frac{1}{[n\theta](n - [n\theta])} \sum_{i=[n\theta]+1}^{n} \sum_{j=1}^{[n\theta]} K(X_{in}, X_{jn}).$$

Since

$$\Gamma_{n1} \stackrel{\mathcal{L}}{=} b_n(\theta, \cdot) - g_n(\theta, \cdot)b_n(\theta, \theta) \quad \forall\, n \in \mathbb{N} \quad \forall\, \theta \in \Theta_0,$$

where

$$g_n(\theta, t) = \begin{cases} [n\theta]^{-1}[nt] & , \ 0 \le t \le [n\theta]n^{-1} \\[2mm] (n - [n\theta])^{-1}(n - [nt]) & , \ [n\theta]n^{-1} < t \le 1 \end{cases},$$

it follows by the extended CMT [cf., e.g., Billingsley (1968, Theorem 5.5)] that

$$\Gamma_{n1} \xrightarrow{\mathcal{L}} \Gamma. \tag{19.15}$$

Furthermore, after some algebra

$$\sup_{0 \le t \le 1} |\Gamma_{n2}(t)| \le C_0 n^{-\frac{1}{2}} \left| n\theta_n - \frac{[n\theta]}{n} \right| |r_n(\theta)| + C_1 n^{\frac{1}{2}} |r_n(\theta) - r_n(\theta_n)|, \quad (19.16)$$

where C_0 and C_1 are positive constants depending on Θ_0. By Lemma 2.2 of Ferger (1994b) which also holds for θ_n as long as $\theta \in \Theta_0$, and by (19.1) the first summand is $o_P(1)$. As to the second summand, we have for all ε, $\delta > 0$:

$$P\left(n^{\frac{1}{2}} |r_n(\theta) - r_n(\theta_n)| > \varepsilon \right) \le$$

$$\le P\left(|b_n(\theta, \theta) - b_n(\theta, \theta_n)| > \tfrac{\varepsilon}{2} \right) + P\left(n^{\frac{1}{2}} |\bar{r}_n(\lambda, \theta, \theta) - \bar{r}_n(\lambda, \theta, \theta_n)| > \tfrac{\varepsilon}{2} \right) \le$$

$$\le P(\omega_n(\delta) > \tfrac{\varepsilon}{2}) + P(|\theta - \theta_n| > \delta) + P\left[C_3 n^{-\frac{1}{2}} \left| \theta_n - \frac{[n\theta]}{n} \right| > \tfrac{\varepsilon}{2} \right],$$

where $\omega_n(\delta)$ denotes the oscillation modulus of the process $b_n(\theta, \cdot)$ and C_3 is a positive constant depending on $|\lambda|$. By Lemma 2.1 of Ferger (1994b)

$$\lim_{\delta \to 0} \limsup_{n \to \infty} P\left(\omega_n(\delta) > \frac{\varepsilon}{2} \right) = 0,$$

whence the second summand in (19.16) is also stochastically negligible. In particular we obtain that

$$\sup_{0 \le t \le 1} |\Gamma_{n2}(t)| = o_P(1). \quad (19.17)$$

Since

$$\sup_{0 \le t \le 1} |\Gamma_{n3}(t)| \le C_4 |\lambda_n| \left| \theta_n - \frac{[n\theta]}{n} \right| = o_P(1), \quad (19.18)$$

(19.14) follows from (19.15), (19.17) and (19.18).

For the proof of (19.8) let be $\theta \notin \Theta_0$. By (19.1)

$$\lim_{n \to \infty} \sup_{0 \le t \le 1} |r_n(t) - r(\lambda, \theta, t)| = 0 \quad P_0 - \text{a.s.} \quad \forall \, \theta \in [0, 1],$$

where

$$r(\lambda, \theta, t) = \lim_{n \to \infty} \bar{r}_n(\lambda, \theta, t) = \begin{cases} \lambda(1 - \theta)t & , \quad 0 \le t \le \theta \\ \\ \lambda\theta(1 - t) & , \quad \theta < t \le 1 \end{cases}$$

By compactness of Θ_0, θ_n converges to $\theta_1 := \arg\max_{t \in \Theta_0} |r(t)|$ a.s., where $\theta_1 \in \{\inf \Theta_0, \sup \Theta_0\} \subseteq \Theta_0$, because $\theta \notin \Theta_0$.

Especially $\forall \, \theta \notin \Theta_0$

$$\theta_n \to \theta_1 \ne \theta \quad \text{a.s.}$$

and thus

$$\lambda_n \to \lambda_1 := \frac{1}{\theta_1(1-\theta_1)} r(\lambda,\theta,\theta_1) \quad \text{a.s.} \; .$$

But for $1 \le p < \infty$ and, e.g., $\theta_1 > \theta$, we have

$$\|\Gamma_n\|_p \ge n^{\frac{1}{2}} \left\{ \int_\theta^{\theta_n} |r_n(t) - \bar{r}_n(\lambda_n,\theta_n,t)|^p \, dt \right\}^{\frac{1}{p}}$$

for all sufficiently large $n \in I\!N$.

Note that the second factor converges almost surely to

$$\left\{ \int_\theta^{\theta_1} |\lambda\theta(1-t) - \lambda_1(1-\theta_1)t|^p \, dt \right\}^{\frac{1}{p}} > 0.$$

Therefore

$$\|\Gamma_n\|_p \to +\infty \quad \text{a.s.} \quad \forall\, \theta \notin \Theta_0, \tag{19.19}$$

which proves (19.8). The case $p = \infty$ is treated similarly. ∎

PROOF OF THEOREM 19.3.1. Observe that by Lemma 3.2 of Ferger (1994b) $\nu_{i,n} \to_w \nu_i$, $i=1,2$, with probability one. Following the proof of (19.14) we obtain that

$$\Gamma_n^* \xrightarrow{\mathcal{L}} \Gamma \text{ under } P(\cdot|\underline{X}_n) \quad \text{a.s.} \quad \forall\, \theta \in \Theta_0,$$

from which we can infer as in the proof of Theorem 2.3.1 in Serfling (1980), that for all $1 \le p \le \infty$

$$c_{p,\alpha}^*(\underline{X}_n) \to c_{p,\alpha} \quad \text{a.s.} \quad \forall\, \theta \in \Theta_0. \tag{19.20}$$

Thus by (19.14)

$$\|\Gamma_n\|_p - c_{p,\alpha}^* \xrightarrow{\mathcal{L}} \|\Gamma\|_p - c_p \;\; \forall\, \theta \in \Theta_0,$$

which proves (19.11). Finally (19.19) and (19.20) yield (19.12). ∎

Acknowledgement. I am deeply grateful to David Groneberg and Thomas Linn (University of Giessen) for many helpful discussions.

References

1. Bhattacharya, P. K. (1991). Estimation in a change-point model, Manuscript, University of California at Davis.

2. Billingsley, P. (1968). Convergence of probability measures, New York: John Wiley & Sons.

3. Csörgő, M. and Horváth, L. (1988). Nonparametric methods for change-point problems, In *Handbook of Statistics* (Eds., P. R. Krishnaiah and C. R. Rao), Vol. 7, pp. 403–425, Amsterdam, The Netherlands: Elsevier.

4. Dümbgen, L. (1991). The asymptotic behavior of some nonparametric change-point estimators, *Annals of Statistics*, **19**, 1471–1495.

5. Ferger, D. (1994a). An extension of the Csörgő-Horváth functional limit theorem and its applications to change-point problems, *Journal of Multivariate Analysis*, **51**, 338–351.

6. Ferger, D. (1994b). Asymptotic distribution theory of change-point estimators and confidence intervals based on bootstrap approximation, *Mathematical Methods of Statistics*, **3**, 362–378.

7. Ferger, D. (1994c). On the power of nonparametric change-point tests, *Metrika*, **41**, 277–297.

8. Ferger, D. (1995). Change-point estimators based on weighted empirical processes with applications to the two-sample problem in general measurable spaces, *Habilitationsschrift*, University of Giessen (in German).

9. Ferger, D. (1997). A new approach to the two sample problem in general measurable spaces, *Preprint*, University of Giessen.

10. Ferger, D. and Stute, W. (1992). Convergence of change-point estimators, *Stochastic Processes and their Applications*, **42**, 345–351.

11. Lifshits, M. A. (1982). On the absolute continuity of distributions of functionals of random processes, *Theory of Probability and its Applications*, **27**, 600–607.

12. Neuhaus, G. (1971). On weak convergence of stochastic processes with multidimensional time parameter, *Annals of Mathematical Statistics*, **42**, 1285–1295.

13. Serfling, R. J. (1980). *Approximation Theorems of Mathematical Statistics*, New York: John Wiley & Sons.

20

On the Integration of Statistical Process Control
and Engineering Process Control in Discrete
Manufacturing Processes

Rainer Göb

Universität Würzburg, Würzburg, Germany

Abstract: *Statistical process control* (SPC) and *engineering process control* (EPC) or *automatic process control* (APC) are two approaches to the control of industrial manufacturing processes. After a discussion of the considerable differences of these approaches in history, theory and industrial implementation, the paper presents a general model of production processes which require the application both of SPC control and EPC control. Some special cases of this model are considered explicitly. For these cases the formulae for the process output as a function of the process parameters and of random deviations are developed.

Keywords and phrases: Statistical process control, engineering process control, automatic process control, deterministic trend, random walk with drift, shift of model parameters

---

20.1   Introduction – Two Simple Examples

Statistical process control (SPC) and *engineering process control* (EPC) or *automatic process control* (APC) are two approaches to the control of industrial manufacturing processes. These approaches exhibit considerable differences in history, theory and industrial implementation. To get an idea about the basic principles let us study two simple examples of SPC and EPC.

20.1.1 An example from statistical process control

Consider a manufacturing process which at equidistant time points 1,2,3,...
produces discrete items 1,2,3,... with random quality characteristics $\xi_1, \xi_2, \xi_3, ...$
The ξ_i are assumed as i.i.d. with normal distribution $\mathcal{N}(\mu; \sigma^2)$. Let T be a
target value of the quality characteristic. We distinguish two states of process
operation. In the satisfactory state the quality characteristic is on target up to
a zero mean normally distributed random deviation, i.e. $\mu = E[\xi_i] = T$. In the
unsatisfactory state the expected value of the quality characteristic differs from
the target, i.e. $\mu = E[\xi_i] \neq T$. Process operation is monitored by a simple
Shewhart $\bar{X}$ chart. At equidistant time points $h, 2h, 3h, ...$ ($h \geq 1$) samples of
size n are taken from the production. The absolute deviation $|\bar{\xi}_k - T|$ of the
sample mean

$$\bar{\xi}_k \quad = \quad \frac{1}{n} \sum_{i=0}^{n-1} \xi_{kh+i}$$

from the process target T is compared with a control limit c. If $|\bar{\xi}_k - T| \leq c$
no action is taken on the process. If $|\bar{\xi}_k - T| > c$ the process is stopped and
checked for irregularities. From the point of view of mathematical statistics the
application of such an $\bar{X}$ chart amounts to successive applications of the Gauß
test for the null hypothesis $\mu = T$ against the alternative $\mu \neq T$.

20.1.2 An example from engineering process control

Consider a manufacturing process which at time t has a certain characteristic
ξ'_t. ξ'_t can be the characteristic of some output of the process but it may also be
a side condition of production, e.g. temperature, pressure etc. The time t may
be discrete or continuous. Let T be a target value of the characteristic. As a
function of t the characteristic is assumed to exhibit a linear trend according
to the model

$$\xi'_t \quad = \quad T + dt + \varepsilon_t$$

where the ε_t are i.i.d. with normal distribution $\mathcal{N}(0; \sigma^2)$. To keep the characteristic on target in expectation, i.e. to obtain controlled values ξ_t of the
characteristic with $E[\xi_t] = T$, the process is permanently readjusted. This
permanent action on the process is represented by a control variable κ_t and the
control equation

$$\xi_t \quad = \quad \xi'_t + \kappa_t .$$

The intended property of ξ_t is obtained for $\kappa_t = -dt$, then

$$\xi_t \quad = \quad T + \varepsilon_t.$$

20.2 Comparison of SPC and EPC

Examples 20.1.1 and 20.1.2 give an idea about the different attitudes of SPC and EPC towards the control of industrial manufacturing processes. Let us contrast the differences in more detail.

20.2.1 History and range of application of SPC and EPC

Statistical process control dates back to the work of Walter Shewhart in the 1920s at Bell Telephone. From the beginning, SPC was developed and applied in the *parts industries* (discrete manufacturing industries), i.e. in industries concerned with the production of discrete items (e.g. ball bearings, electric bulbs etc.). EPC originated in a different industrial segment, which may be called *process industry* or *substance industry*. This type of industry is concerned with the production of continuous substances (liquids, powders etc.) in continuous processes. Such production environments used various adjustment strategies, often administered by automatic controllers. Hence the term *automatic process control* (APC), which is sometimes used as a synonym of EPC. Nowadays there is a growing tendency to introduce EPC methods also in discrete manufacturing systems.

20.2.2 Quality criteria in the parts and process industries

In the parts industry, interest is in reproducing discrete units (ball bearings, screws, transistors) which mostly are rated in a conforming/nonconforming scheme. Continuous scale quality characteristics (length, weight etc.) are only used to establish a conforming/nonconforming scheme by means of specification ranges. By contrast, the continuous process industries are most often concerned with continuous measures for product quality, for instance with measures of quantity or purity of substance.

20.2.3 Technical properties of production processes in parts and process industries

In the parts industry, production processes are normally stable. Side conditions and raw materials are under control. Disturbances only arise from shocks acting on the production facilities, but such shocks occur only rarely. In case of occurrence, however, they have massive consequences. By contrast, process industries are often unable to set up stable production processes. Here, the processes exhibit inherent drifts or trends (in temperature, pressure etc.) which have to be counteracted by permanent adjustment.

20.2.4 Statistical tools of SPC and EPC

In terms of statistical theory, the essential tool of SPC is a *test of significance*. Corresponding to the assumed stability of production processes, the null hypothesis is chosen as *"The process is in-control"* and the alternative as *"The process is out-of-control"*. The essential statistical tool of EPC is a *prediction*. Given a stochastic model and given a sequence of data from the process, the amount of a later deviation from target is predicted so as to infer the necessary adjustment to counteract the deviation.

20.2.5 Process changes in SPC and EPC

Any approach to process control needs a model of *process changes*, i.e. a model for the changes of the process parameters (output mean, output variance, output proportion nonconforming) through production time. On this topic, the traditional approaches to SPC and EPC differ significantly, corresponding to their origins in different types of industries.

SPC models use to identify process changes as abrupt *shifts* of the process parameters due to assignable causes, *disturbances* or *shocks* which affect the manufacturing facilities. These shift models generally share four basic assumptions:

- The magnitude of shifts is large relative to the process noise variance.

- Shifts are rare events, the period till the first shift occurrence or between two successive shifts is large.

- Shifts can result from a variety of assignable causes. Detection of a specific assignable cause requires expert engineering knowledge on the production process, it is time consuming, and expensive.

Under these assumptions, a constant automatic adjustment strategy (i.e., an EPC) obviously is not the appropriate remedy against process variation.

EPC models which originate in continuous process industries consider process changes in the form of *continuous drifts*. In contrast to the shift models of SPC, the basic assumptions of the EPC approach are:

- The drift is slow. Measured in a short time interval the drift effect is small relative to the noise variance.

- The drift permanently continues through production time.

- The drift is an inherent property of the production process. Expert engineering knowledge about the production process provides knowledge about the structure of the drift process. To a certain degree, the drift effect can be estimated and predicted.

Under these assumptions constant automatic adjustments are a reasonable control strategy.

20.2.6 Process monitoring in SPC and EPC

In SPC, processes are monitored by sampling. Corresponding to the assumed stability of production processes, the intervals between successive samples are relatively large. EPC often requires permanent monitoring of production. Often data from production are collected and processed automatically.

20.2.7 Actions on the production process in SPC and EPC

SPC needs *inspective* and *corrective* actions on the production process. If the null hypothesis "*The process is in-control*" is rejected the process is inspected (and often even stopped) to search for actual shifts of the process parameters. If changes in the process parameters are verified the technical reasons for these shifts have to be investigated and removed by a *renewal* or a *repair* of the production facilities. Inspections, repairs and renewals are time consuming and expensive, they require skilled staff, machinery and material. However, these actions on the process are relatively rare, i.e. the intervals of ongoing operation between successive actions are long.

EPC needs *compensating* actions (adjustments) to counteract the drift effect. These adjustments occur permanently but they require only minor expenses in time and money. They follow a repetitive pattern or a simple algorithm that can often be left to an automatic controller.

20.2.8 The structure of process models in SPC and EPC

An essential difference of the process models in SPC and EPC is the distinction between the *open loop* behaviour (behaviour without control actions) and *closed loop* behaviour (behaviour in the presence of control actions), which is crucial for EPC and unnecessary for SPC. Since we are interested in discrete manufacturing processes, here and in the sequel we restrict attention to discrete time, i.e. the output quality characteristic is observed at discrete time points 1,2,... only.

In SPC, detection of an assignable cause and subsequent corrective action occurs only rarely. If it occurs, it amounts to a complete *renewal* of the manufacturing process. Hence it is useful to split the entire production run into the periods (*renewal cycles*) between two successive corrective actions (renewals) and to consider each renewal cycle along a separate time axis 1,2... with corresponding output quality characteristics ξ_1, ξ_2,.... . The effect of control actions is not reflected in the output model.

In standard EPC, control actions are taken regularly at each time point 1,2,... . Without this permanent compensatory actions the process would exhibit a completely different behaviour. A model of the process behaviour without control is indispensable for the design and evaluation of control rules. Thus we have to distinguish the *open loop* output quality characteristics ξ_1', ξ_2', ,... of the process without control (left alone) from the *closed loop* output quality characteristics ξ_0, ξ_1, ξ_2,... of the process subject to adjustments. In a simple model, the adjustment which transforms the open loop quality characteristic into the closed loop quality characteristic is, as in example 20.1.2, expressed by a control variable κ_t which acts according to the model

$$\xi_t = \xi_t' + \kappa_t.$$

20.3 Models of Process Changes in SPC and EPC

In this section we consider the mathematical models of manufacturing processes in SPC and EPC which reflect their different ideas about the structure of process changes.

20.3.1 Process changes in SPC models

SPC is designed for manufacturing processes which exhibit discrete parameter shifts, occurring at random time points. Thus the most general model SPC model considers the sequence $(\xi_t')_{I\!N_0}$ of output quality characteristics as a function

$$\xi_t' = \mu_t + \varepsilon_t. \tag{20.1}$$

In this formula $(\mu_t)_{I\!N_0}$ is the sum

$$\mu_t = \mu^\star + \sum_{i=1}^{N_t} \delta_i \tag{20.2}$$

of a target value $\mu^\star$ and a *mark accumulator process* $\left(\sum_{i=1}^{N_t} \delta_i\right)_{I\!N_0}$ based on an underlying marked point process, see Snyder (1975), Chapters 3,...,7. The occurrences 1,2,... of the point process represent the successive shifts, the associated real-valued marks $\delta_1, \delta_2, ...$ represent the sizes of the shifts. The counting process $(N_t)_{I\!N_0}$ gives the number of shifts (occurrences) in the time interval $[0; t)$. $(\varepsilon_t)_{I\!N_0}$ is a white noise process (a sequence of i.i.d. random variables with mean 0) independent of $(\mu_t)_{I\!N_0}$.

A simple and popular instance of a marked point process is one with shifts occurring according to a *Poisson process* $(N_t)_{I\!N_0}$. Most investigations on control charts use further simplifications. For instance, deterministic absolute values $|\delta_i| = \Delta$ ($\Delta > 0$) of the shifts are frequently assumed. Many approaches assume a *single* shift of a given absolute value Δ which occurs after a random (often assumed exponentially distributed) time ν. In this case we have

$$\xi_t' = \begin{cases} \mu^\star + \varepsilon_t, & \text{if } t \leq \nu, \\ \mu^\star + \gamma\Delta + \varepsilon_t, & \text{if } t > \nu, \end{cases} \tag{20.3}$$

where the random variable γ is the sign of the deviation from target with

$$P(\gamma = 1) = p, \quad P(\gamma = -1) = 1 - p \quad \text{with } p \in [0; 1] .$$

In case of one-sided shifts we have $p = 0$ or $p = 1$, in case of two-sided shifts it usually assumed that $p = 0.5$.

20.3.2 Process changes in EPC models

EPC is designed for manufacturing processes which exhibit continuous parameter drifts. Below we give two simple but typical models for open loop output variables ξ'_t which correspond to this idea. Throughout T is a target value for the output quality characteristic and (ε_t) is a white noise process in the sense introduced in paragraph 20.3.1, i.e. a sequence of i.i.d. random variables with mean 0.

Deterministic trend

In many cases, for instance if the drifting behaviour is caused by aging of a tool, see Quesenberry (1988), a simple regression model of the form

$$\xi'_t = T + \vartheta_t + \varepsilon_t \quad \text{with } \vartheta_t = dt \tag{20.4}$$

can be used to model the drift.

Random walk with drift

In the deterministic trend model the mean of the output varies but the variance remains constant. A simple model for a situation where the variance increases linearly with time is the random walk with drift model determined by the recursion

$$\xi'_t - \xi'_{t-1} = d + \varepsilon_t, \quad \xi_0 = T \tag{20.5}$$

or by the explicit formula

$$\xi'_t = T + \vartheta_t + \sum_{i=1}^{t} \varepsilon_i \quad \text{with } \vartheta_t = dt. \tag{20.6}$$

20.4 Process Control in SPC and EPC

In this section we consider some details of the specific control procedures of SPC and EPC which correspond to the process models described in the previous Section 20.3. As explained in Section 20.2, SPC considers relatively stable processes which require *preventive monitoring* but few actions. In contrast, EPC considers drifting processes which require *permanent adjustment*.

20.4.1   Process control as process monitoring in SPC

In SPC process changes are identified as relatively rare, discrete and abrupt shifts of the process parameters from a satisfactory state S_I to an unsatisfactory state S_{II}. Corresponding to this idea of process changes the primary objective of SPC is not to act on the process but to monitor the process in order to detect signals of unsatisfactory values of the process parameters. Actions on the process are implemented only after such signals have been identified. The monitoring tool is a *control chart* which consists in successive applications of a test of significance of the null hypothesis "*Process parameter at present time t is in S_I*" ("$\mu_t \in S_I$" in terms of the model of paragraph 20.3.1) against the alternative "*Process parameter at present time t is in S_{II}*" ("$\mu_t \in S_{II}$" in terms of the model of paragraph 20.3.1). The tests are based on random samples from the process. Generally the intervals between successive samples are relatively large. For instance, in the model underlying the standard two-sided Shewhart $\bar{X}$ chart, compare example 20.1.1, μ_t is the mean of a normal distribution $\mathcal{N}(\mu_t; \sigma^2)$, $S_I = \{\mu_0\}$, $S_{II} = I\!\!R \backslash \{\mu_0\}$, and the control chart consists in successive applications of the Gauß test of the null hypothesis "$\mu_t = \mu_0$" against the alternative "$\mu_t \neq \mu_0$".

20.4.2   Process control as process adjustment in EPC

EPC considers instable processes where the parameters are subject to a drift or trend. Correspondingly, EPC requires permanent adjustment of the process. In the mathematical model the adjustment is expressed by a *control variable* κ_t which transforms the open loop output characteristic ξ_t' into the closed loop output characteristic ξ_t. For simplicity's sake we confine ourselves to the simple additive model

$$\xi_t \quad = \quad \xi_t' + \kappa_t \ . \tag{20.7}$$

In a further simplification we restrict attention to a discrete control scheme where the value κ_t of the control variable at time point t is determined as a function

$$\kappa_t \quad = \quad f_t(\xi_{t-1}, \xi_{t-2}, ...; \kappa_{t-1}, \kappa_{t-2}, ...) \tag{20.8}$$

of past closed loop output quality characteristics and all past values of the control variable. As the criterion to evaluate the control variable we use the square deviation from the target T, i.e. we use the sequence (κ_t) which minimizes

$$E\left[(\xi_t - T)^2\right] \quad = \quad E\left[(\xi_t' + \kappa_t - T)^2\right]$$

among all sequences satisfying (20.8).

For the two models presented in paragraph 20.3.2, above, this *minimum mean square error* (MMSE) control variable can be inferred from Theorem 4.1

in Åström (1970). In the deterministic trend model 20.3.2 we obtain the MMSE control variable

$$\kappa_t \quad = \quad dt. \tag{20.9}$$

In the random walk with drift model 20.3.2 the MMSE control variable is determined by the recursion

$$\kappa_{t+1} - \kappa_t \quad = \quad T - \xi_t - d. \tag{20.10}$$

Note that the controller (20.10) is a feedback controller which uses information from the process, whereas the controller (20.9) is deterministic.

Application of these control variables according to (20.7) leads in both cases to closed loop output characteristics

$$\xi_t \quad = \quad \xi'_t + \kappa_t \quad = \quad T + \varepsilon_t \tag{20.11}$$

which are on target up to the white noise component ε_t.

20.5 Problems of the Integration of SPC and EPC

In this section we discuss the history and the problems of approaches to an integration of SPC and EPC.

20.5.1 History of SPC/EPC integration

Interest in SPC and EPC integration originated in the 1950's in chemical industries. Chemical manufacturing processes are often subject to inherent drifts. Thus they are the typical area of application of EPC schemes which are put into place to compensate for such drifting behavior. However, abrupt, large shifts in the quality characteristic indicate major failures or errors in the process that can not be, in general, compensated for by the EPC controller. For this reason, following a trend in industry, many authors have suggested to "add" an SPC chart at the output of an EPC-controlled production process to detect large shifts. There is no clear methodology, however, that models such integration efforts in a formal and general way.

In contrast, interest in SPC/EPC integration in discrete parts manufacturing is more recent. In these type of production processes, drifting behaviour, due to aging of the process for instance, is also observed, but it has generally been ignored by SPC approaches. A typical example of this is a metal machining process, where the performance of the cutting tool deteriorates (in many cases, almost linearly) with time. Some years ago, when market competition was not so intense, specifications were wide enough for a production process to drift without producing a large proportion of nonconforming product. With

increasing competition, quality specifications have become more rigorous, and
drifting behaviour instead of being tolerated is actively compensated for by
simple EPC schemes.

20.5.2   Models proposed in the literature for SPC/EPC integration

Academic interest on the area of SPC/EPC integration has occurred as a natural reaction to the requirements of industrial practice. Diverse authors have discussed the different aims and strategies of SPC and EPC [e.g. Barnard (1963),
MacGregor (1988, 1990), Box and Kramer (1992), Montgomery et al. (1994)].
However, only heuristic approaches were suggested in this discussion. Propo-
nents of either side admit that many control problems in modern manufacturing
processes cannot be solved by either SPC or EPC alone. Concluding from this,
methods from each field are recommended as auxiliary tools in a scheme originally developed either for SPC or for EPC applications alone. Consequently
none of these approaches have been really successful from a methodological
point of view. The models used were originally designed either for proper SPC
or EPC applications but not for an integrating view.

Few specific models have been proposed for the integration of SPC and EPC.
Among these models we find ASPC, see the subsequent paragraph 20.5.2, and
run-to-run control procedures, see the subsequent paragraph 20.5.2.

Algorithmic statistical process control (ASPC)

Vander Weil et al. (1992) [see also Tucker et al. (1993)] model the observed
quality characteristic ξ_t of a batch polimerization process at time t as

$$\xi_t \quad = \quad \mu \mathrm{II}_{(t_0;+\infty)}(t) + b\kappa_t + \rho_t \tag{20.12}$$

where

$$\mathrm{II}_B(t) \quad = \quad \begin{cases} 1, & \text{if } t \in B, \\ 0, & \text{if } t \in \mathbb{R} \setminus B, \end{cases} \tag{20.13}$$

is the indicator function of a set $B \subset \mathbb{R}$. Thus the first term on the right
represents a shift of magnitude μ that occurs at time t_0, κ_t is the compensatory
variable, and the noise term ρ_t is a stationary ARMA(1,1) stochastic process.
In what the authors refer to as *Algorithmic Statistical Process Control* (ASPC),
process shifts are monitored by a CUSUM chart whereas the ARMA noise
is actively compensated for by an EPC scheme. Using a similar approach,
Montgomery et al. (1994) presented some simulation results.

Basic weakness of the ASPC approach: there is no explicit stochastic model
for the time t_0 of shift occurrence.

Run to run process control

Sachs et al. (1995), [see also Ingolfsson et al., (1993)] assume instead a simple linear regression model with no dynamic effects for controlling certain semiconductor manufacturing processes. The model is

$$\xi_t = \mu + b\kappa_t + \varepsilon_t . \tag{20.14}$$

By using a control chart on the residuals of model (20.14) – called "Generalized SPC" by the authors–, their method applies two different types of EPC schemes: an EWMA-based controller if the observed deviation from target is "small" (called "gradual control" by the authors) and a Bayesian controller that determines the moment and magnitude of larger deviations in case they occur. Other authors [Butler and Stefani (1994), Del Castillo and Hurwitz (1997)] have extended model (20.14) to the case were deterministic trends and ARMA(1,1) noise exist.

Basic weakness of the run-to-run models: the classical rationale of SPC applications is a shift, stochastic in time of occurrence and/or in magnitude. Again, this is not reflected by the run-to-run models.

20.6 A General Model for the Integration of SPC and EPC

From a methodological point of view, the approaches for SPC/EPC integration presented in paragraph 20.5.2 are not satisfactory. In ASPC the stochastic model is incomplete. In run-to-run process control the position of SPC is not represented adequately.

To prepare a satisfactory model, we recall that simultaneous application of SPC and EPC procedures to one and the same manufacturing process only makes sense if the process exhibits both kinds of changes considered in Section 20.3, above: discrete and abrupt variation by shift, which represents the position of SPC, as well as continuous variation by drift, see paragraph 20.3.2, which represents the position of EPC. Consequently, an integrative model for SPC and EPC should contain components corresponding to the two types of process change models given in Section 20.3: a mark accumulator process component to justify the use of SPC, see paragraph 20.3.1, and a drift component to justify the use of EPC, see paragraph 20.3.2. Thus a unifying model for SPC and EPC can be expressed by the following model for the uncontrolled (open loop) process output ξ'_t:

$$\xi'_t = F_t\Big((\mu_s^{(1)})_{s\leq t}, ..., (\mu_s^{(K)})_{s\leq t}, (\vartheta_s^{(1)})_{s\leq t}, ..., (\vartheta_s^{(M)})_{s\leq t}, (\varepsilon_s)_{s\leq t}\Big) \tag{20.15}$$

where $(\mu_s^{(1)}), ..., (\mu_s^{(K)})$ are K different processes of type (20.2) (mark accumulator process plus target value) representing the effect of shifts to be treated by SPC, where $(\vartheta_s^{(1)}), ..., (\vartheta_s^{(M)})$ are M different drift processes which represent the effect of continuous drifts to be treated by EPC, see formulae (20.4) and (20.5), and where (ε_s) is a white noise sequence, i.e. a sequence of i.i.d. random variables with mean 0. For some applications it is necessary to choose all past values $(\mu_s^{(i)})_{s \leq t}$, $(\vartheta_s^{(j)})_{s \leq t}$, $(\varepsilon_s)_{s \leq t}$ as the arguments of the functions F_t to allow for possible cumulative effects of $\mu_t^{(i)}, \mu_{t-1}^{(i)}, ..., \vartheta_t^{(j)}, \vartheta_{t-1}^{(j)}, ..., \varepsilon_t, \varepsilon_{t-1}, ...$ on ξ_t'. In the subsequent examples we shall restrict attention to *linear* functions F_t. However, in view of the complexity of EPC models it seems reasonable to admit arbitrary functions F_t in the basic model.

20.7 Special Models for the Integration of SPC and EPC

Equation (20.15) gives a generic framework for a process model which integrates the positions of SPC and EPC. Let us now consider an important example with one drift component (ϑ_s), i.e. $M = 1$, and with two shift components $(\mu_s^{(1)})$, $(\mu_s^{(2)})$, i.e. $K = 2$.

In many cases an abrupt shift can be modeled as a translation of the output value ξ_t'. To provide for this possibility we assume a shift process $(\mu_s^{(1)})_{I\!N_0}$ with a target $\mu_1^\star = \mu_0^{(1)} = 0$, and we replace T by $T + \mu_t^{(1)}$ in the drift models 20.3.2 and 20.3.2.

Usually the models for drift processes $(\vartheta_t)_{I\!N_0}$ which are used in EPC depend on parameters. In examples 20.3.2 and 20.3.2 this is the drift parameter d which expresses the amount of deviation from target per time unit. These parameters can be subject to shifts during production. To provide for this possibility in our examples we assume a shift process $(\mu_s^{(2)})_{I\!N_0}$ with a target $\mu_2^\star = \mu_0^{(2)} = d$, and we replace d by $\mu_t^{(2)}$ in the drift models 20.3.2 and 20.3.2.

Thus we obtain the following models.

20.7.1 Additive disturbance and shift in drift parameter in the deterministic trend model

If we consider the deterministic trend model under the assumption of possible shifts in the mean and of possible shifts of the drift parameter we obtain the output equation

$$\xi_t' = \mu_t^{(1)} + T + \sum_{i=1}^{t} \mu_i^{(2)} + \varepsilon_t. \tag{20.16}$$

Equation (20.16) constitutes a special case of equation (20.15) with $K = 2$, $M = 1$,

$$\vartheta_t = dt, \qquad F_t\left((\mu_s^{(1)})_{s\leq t}, (\mu_s^{(2)})_{s\leq t}, (\vartheta_s)_{s\leq t}\right) = \mu_t^{(1)} + \sum_{i=1}^{t}(\mu_i^{(2)} - d) + \vartheta_t + \varepsilon_t .$$

Application of the control variable $\kappa_t = -dt$ (which is MMSE in the deterministic trend model *without* shifts) to the open loop output (20.16) leads to the closed loop equation

$$\xi_t = \xi_t' + \kappa_t = \mu_t^{(1)} + T + \sum_{i=1}^{t}(\mu_i^{(2)} - d) + \varepsilon_t . \qquad (20.17)$$

20.7.2 Additive disturbance and shift in drift parameter in the random walk with drift model

If we consider the random walk with drift model under the assumption of possible shifts in the mean and of possible shifts of the drift parameter we obtain the output equation

$$\xi_t' = \mu_t^{(1)} + T + \sum_{i=1}^{t}\mu_i^{(2)} + \sum_{i=1}^{t}\varepsilon_i = \mu_t^{(1)} + \sum_{i=1}^{t}\mu_i^{(2)} - dt + \vartheta_t + \sum_{i=1}^{t}\varepsilon_i .$$
$$(20.18)$$

In the scheme of Equation (20.15) we have $K = 2$, $M = 1$,

$$\vartheta_t = dt, \ F_t\left((\mu_s^{(1)})_{s\leq t}, (\mu_s^{(2)})_{s\leq t}, (\vartheta_s)_{s\leq t}\right) = \mu_t^{(1)} + \sum_{i=1}^{t}(\mu_i^{(2)} - d) + \vartheta_t + \sum_{i=1}^{t}\varepsilon_i .$$

Application of the MMSE control variable defined by the recursion (20.10) (which is MMSE in the random walk model *without* shifts) to the open loop output (20.18) leads to the closed loop equation

$$\xi_t = \xi_t' + \kappa_t = T + \mu_t^{(1)} - \mu_{t-1}^{(1)} + \mu_t^{(2)} - d + \varepsilon_t \qquad \text{for } t \geq 2. \qquad (20.19)$$

From (20.18) and (20.19) we obtain the following explicit expression for the control variable:

$$\kappa_t = -d - \mu_{t-1}^{(1)} - \sum_{i=1}^{t-1}\mu_i^{(2)} - \sum_{i=1}^{t-1}\varepsilon_i . \qquad (20.20)$$

20.8   Discussion of SPC in the Presence of EPC

EPC control schemes are designed for fixed and known parameter values of the underlying models. It is to be expected that they are not able to handle sudden parameter shifts in a satisfactory way, so that supplementary SPC schemes are required to detect these abrupt changes. To investigate this conjecture in the models exposed in the previous Section 20.7 we consider simply structured parameter shifts of type (20.3), which are appropriate for many applications.

20.8.1   Effect of simple shifts on EPC controlled processes

With respect to the shift sequences $\mu_t^{(1)}$ and $\mu_t^{(2)}$ we assume

$$\mu_t^{(l)} = \begin{cases} \mu_l^\star, & \text{if } t \le \nu_l, \\ \mu_l^\star + \gamma_l \Delta_l, & \text{if } t > \nu_l, \end{cases} \qquad (20.21)$$

where $\mu_1^\star = 0$, $\mu_2^\star = d$ are the target values, $\Delta_l > 0$ is the absolute shift size, ν_l is the random time till occurrence of the shift, and γ_l is the random sign of the shift.

Under these assumptions the closed loop output equation (20.17) of the deterministic trend model becomes

$$\xi_t = T + \gamma_1 \Delta_1 \mathrm{I\!I}_{(\nu_1; +\infty)}(t) + \left(t - \lfloor \nu_2 \rfloor \right) \gamma_2 \Delta_2 \mathrm{I\!I}_{(\nu_2; +\infty)}(t) + \varepsilon_t. \qquad (20.22)$$

Applying the same assumptions to the closed loop output (20.19) of the random walk with drift model we obtain for $t \ge 2$

$$\xi_t = T + \gamma_1 \Delta_1 \mathrm{I\!I}_{(\nu_1; \nu_1+1]}(t) + \gamma_2 \Delta_2 \mathrm{I\!I}_{(\nu_2; +\infty)}(t) + \varepsilon_t. \qquad (20.23)$$

For the control variable of the random walk with drift model we obtain by inserting (20.21) into (20.20)

$$\kappa_t = -td - \gamma_1 \Delta_1 \mathrm{I\!I}_{(\nu_1+1; +\infty)}(t) - \gamma_2 \Delta_2 \left(t - 1 - \lfloor \nu_2 \rfloor \right) \mathrm{I\!I}_{(\nu_2+1; +\infty)}(t) - \sum_{i=1}^{t-1} \varepsilon_i. \qquad (20.24)$$

20.8.2   Shifts occurring during production time

The controller (20.9) of the deterministic trend model has no feedback from the output and is thus principally not able to compensate for random shifts. As it is obvious from (20.22), an additive shift takes the process mean away from its target T to $T + \gamma_1 \Delta_1$, but the output at least remains stable in its mean. A shift in the drift parameter is more harmful. After such a shift, the output mean has a trend component $\left(t - \lfloor \nu_2 \rfloor\right)\gamma_2\Delta_2$ for $t > \nu_2$. It is obvious that in the presence of possible shifts such a process should be monitored by a supplementary SPC scheme.

The feedback controller of (20.10) for the random walk with drift is able to react both on additive shifts and on shifts in the drift parameter. As obvious from (20.23), due to the delay of one time period in the controller's action, an additive shift only leads to a single outlier of the output ξ_t at $t = \lfloor \nu_1 + 1 \rfloor$ but remains without effect at further time points. A shift in the drift parameter can be more harmful. After such a shift, the output mean is constantly off target T at $T + \gamma_2\Delta_2$. However, this shift in the mean has serious consequences only if $|\gamma_2\Delta_2|$ is large or if the cost of being off target is large. In such cases it is reasonable to monitor the process by supplementary SPC procedures.

20.8.3   Effect of a biased drift parameter estimate

In the approach of the model presented in Section 20.7, a biased estimate of the drift parameter d can be interpreted as a shift in the drift parameter which coincides with the setup of the process. This is quite useful as it allows to study the effect of mistakenly use a biased trend parameter estimate in an EPC scheme. Let $\widehat{d}$ be the biased estimate which is used instead of d in the control equations (20.9) and (20.10). Then we can describe the situation by (20.22) and (20.23) by letting

$$\Delta_2 = |d - \widehat{d}|, \quad P(\gamma_2 = 1) = 1, \text{ if } d \geq \widehat{d}, \quad \text{and} \quad P(\gamma = -1) = 1, \text{ if } d < \widehat{d}.$$

The effect of this type of parameter shifts in the trend and random walk models is exactly the same as in paragraph 20.8.2, above.

20.8.4   Effect of constraints in the compensatory variable

An important aspect in practice, usually not addressed in the literature on SPC-EPC integration, is that the compensatory variable must usually be constrained to lie within certain region of operation, i.e.

$$A \leq \kappa_t \quad \text{or} \quad \kappa_t \leq B \quad \text{or} \quad A \leq \kappa_t \leq B$$

for all instants t. In particular, integral controllers such as equation (20.10) can compensate for shifts of any size *provided* that the controllable factor is unconstrained.

It is useful to consider what would happen if the EPC schemes given by equations (20.9) and (20.10) are applied to a constrained input process. Since the drift is linear, the control variable κ_t moves in the opposite direction than the drift to keep ξ_t on target. However, at some point, the controller hits a boundary (either A or B), and remains there afterwards. In the control engineering literature this is referred to as "saturation" of the EPC scheme.

Effect of constraints under the deterministic trend model

Let us discuss the case of a constrained control variable in the deterministic trend model. For simplicity's sake we only discuss the case $d > 0$ with a lower bound $A < 0$. The case $d < 0$ with a corresponding upper bound $B > 0$ is completely analogous.

The relation among the control variable κ_t of equation (20.9) and the constrained control variable $\tilde{\kappa}_t$ is

$$\tilde{\kappa}_t \quad = \quad \begin{cases} \kappa_t = -dt, & \text{if } t \leq \frac{-A}{d}, \\ A, & \text{if } t > \frac{-A}{d}. \end{cases} \tag{20.25}$$

Hence the output $\tilde{\xi}_t$ of the process under constrained control is

$$\tilde{\xi}_t \quad = \quad \mu_t^{(1)} + T + \sum_{i=1}^t \mu_i^{(2)} + \tilde{\kappa}_t + \varepsilon_t \quad = \tag{20.26}$$

$$\begin{cases} \mu_t^{(1)} + T + \sum_{i=1}^t \mu_i^{(2)} - dt + \varepsilon_t, & \text{if } t \leq \frac{-A}{d}, \\ \mu_t^{(1)} + T + \sum_{i=1}^t \mu_i^{(2)} + A + \varepsilon_t, & \text{if } t > \frac{-A}{d}. \end{cases}$$

Under the simple shift components of type (20.21) the output $\tilde{\xi}_t$ satisfies the right-hand side of (20.22) for $t \leq \frac{-A}{d}$. For $t > \frac{-A}{d}$ we obtain

$$\tilde{\xi}_t \quad = \quad T + \gamma_1 \Delta_1 \mathrm{I\!I}_{(\nu_1;+\infty)}(t) + \left(t - \lfloor \nu_2 \rfloor\right)\gamma_2 \Delta_2 \mathrm{I\!I}_{(\nu_2;+\infty)}(t) + dt + A + \varepsilon_t .$$
$$\tag{20.27}$$

Obviously, the arguments in favor of supplementary application of SPC schemes in the trend model which are put forward in Section 20.8.2 also hold in the case of constrained controllers.

Effect of constraints under the random walk model

In the random walk model we also restrict attention to the case $d > 0$ with a lower bound $A < 0$.

Different from the situation for the deterministic trend model the time ρ till hitting or falling below the lower bound A is stochastic and defined by

$$\rho = \min\{t \mid \kappa_t \leq A\} =_{(20.24)}$$

$$\min \left\{ t \mid \sum_{i=1}^{t-1} (\varepsilon_i + d) \geq -A - d - \gamma_1 \Delta_1 \mathrm{I\!I}_{(\nu_1+1;+\infty)}(t) \right.$$

$$\left. -\gamma_2 \Delta_2 \left(t - 1 - \lfloor \nu_2 \rfloor\right) \mathrm{I\!I}_{(\nu_2+1;+\infty)}(t) \right\} . \tag{20.28}$$

From (20.28) the distribution of ρ can be found by first determining the conditional distribution under ν_i and γ_i and then integrating with respect to the corresponding densities. Considering ρ in view of (20.28) as a first transition time of the sequence $\left(\sum_{i=1}^{t-1}(\varepsilon_i + d)\right)_{I\!N}$ of sums of i.i.d. random variables, methods of sequential analysis, see Siegmund (1985), can be used to approximate the conditional distribution and the conditional moments of ρ.

The relation among the control variable κ_t of (20.24) and the constrained control variable $\tilde{\kappa}_t$ is

$$\tilde{\kappa}_t \;\; = \;\; \begin{cases} \kappa_t, & \text{if } t < \rho, \\ A, & \text{if } t \geq \rho. \end{cases} \tag{20.29}$$

Hence the output $\tilde{\xi}_t$ of the process under constrained control is

$$\tilde{\xi}_t \;\; = \;\; \mu_t^{(1)} + T + \sum_{i=1}^{t}\mu_i^{(2)} + \sum_{i=1}^{t}\varepsilon_i + \tilde{\kappa}_t \;\; = \tag{20.30}$$

$$\begin{cases} \mu_t^{(1)} - \mu_{t-1}^{(1)} + T + \mu_t^{(2)} - d + \varepsilon_t, & \text{if } t < \rho, \\ \mu_t^{(1)} + T + \sum_{i=1}^{t}\mu_i^{(2)} + \sum_{i=1}^{t}\varepsilon_i + A, & \text{if } t \geq \rho. \end{cases}$$

Under the simple shift components of type (20.21) the output $\tilde{\xi}_t$ satisfies the right-hand side of (20.23) for $2 \leq t < \rho+1$. For $t \geq 2$, $t \geq \rho+1$ we obtain

$$\tilde{\xi}_t \;\; = \;\; T + \gamma_1\Delta_1 I\!I_{(\nu_1;+\infty)}(t) + td + \gamma_2\Delta_2\Big(t - \lfloor \nu_2 \rfloor\Big) I\!I_{(\nu_2;+\infty)}(t) + \sum_{i=1}^{t}\varepsilon_i + A \;.$$

$$\tag{20.31}$$

In the unconstrained case supplementary application of SPC in the random walk model only makes sense in special cases, see Section 20.8.2. From (20.31) it is evident that in the constrained case supplementary application of SPC is much more interesting and perhaps indispensable.

20.8.5 Effect of using a wrong model

We now study what would happen if a wrong drift model is used.

If the feedback controller defined by (20.10) is used in the deterministic trend model the explicit expression for ξ_t is

$$\xi_t \;\; = \;\; T + \mu_t^{(1)} - \mu_{t-1}^{(1)} + \mu_t^{(2)} - d + \varepsilon_t - \varepsilon_{t-1}.$$

Under the simple shifts of type (20.21) we obtain

$$\xi_t \;\; = \;\; T + \gamma_1\Delta_1 I\!I_{(\nu_1;\nu_1+1]}(t) + \gamma_2\Delta_2 I\!I_{(\nu_2;+\infty)}(t) + \varepsilon_t - \varepsilon_{t-1} \;. \tag{20.32}$$

Whether there are parameter shifts or not, the output exhibits twice the variance compared with the case of using the correct model. This case occurs in Quesenberry's (1988) d_2 and d_3 rules.

If parameter shifts occur we have the following result: except the single outlier for $\nu_1 < t \leq \nu_1 + 1$, ξ_t is permanently off target for $t > \nu_2$ with absolute deviation Δ_2. This, and the uncertainty about the correctness of the assumptions of the model make advisable to use SPC methods in addition to the simple EPC schemes.

If on the contrary, the deterministic trend controller (20.9) is used in a random walk with drift process, the closed loop equation is

$$
\xi_t \quad = \quad T + \mu_t^{(1)} + \sum_{i=1}^{t} \mu_i^{(2)} + \sum_{i=1}^{t} \varepsilon_i - dt.
$$

Under the simple shifts of type (20.21) we obtain

$$
\xi_t \quad = \quad T + \gamma_1 \Delta_1 \, \mathrm{I\!I}_{(\nu_1;+\infty)}(t) + \left(t - \lfloor \nu_2 \rfloor \right) \gamma_2 \Delta_2 \, \mathrm{I\!I}_{(\nu_2;+\infty)}(t) + \sum_{i=1}^{t} \varepsilon_i \ . \quad (20.33)
$$

In this case, whether there are parameter shifts or not, the output exhibits variance that increases linearly with time compared with the case of using the correct model. Thus it is evident that using an EPC controller designed for a random walk with drift model is "safer" than using an EPC controller designed for a deterministic trend process in case we selected (by mistake) the wrong drift model.

Taking the shifts into account we have the following result: there is a shift in the mean for $\nu_1 < t$ and a shift which results in trend for $\nu_2 < t$. Again, given the uncertainty about the correctness of model assumptions it is obviously advisable to use additional SPC methods.

20.9   Conclusion

The models of Section 20.6 and 20.7 describe production processes which are subject to changes by abrupt shifts as well as to changes by continuous drifts. The discussion of Section 20.8 shows that in many cases the usual EPC controllers are not able to counteract the effect of disturbances resulting from shifts. Thus SPC procedures are required for the control of this type of production processes. However, the classical theory does not consider SPC in the presence of drifts. Thus further investigation is required on the evaluation of classical SPC charts in the presence of drifts and to design new appropriate SPC methods.

References

1. Åström, K. J. (1970). *Introduction to Stochastic Control Theory*, San Diego, CA: Academic Press.

2. Barnard, G. A. (1959). Control charts and stochastic processes, *Journal of the Royal Statistical Society, Series B*, **21**, 239–271.

3. Box, G. E. P. and Jenkins, G. (1976). *Time Series Analysis, Forecasting, and Control*, Revised edition, Oakland, CA: Holden Day.

4. Box, G. E. P. and Kramer, T. (1992). Statistical process monitoring and feedback adjustment–a discussion, *Technometrics*, **34**, 251–267.

5. Butler, S. W. and Stefani, J. A. (1994). A supervisory run-to-run control of a polysilicon gate etch using *in situ* ellipsometry, *IEEE Transactions on Semiconductor Manufacturing*, **7**, 193–201.

6. Del Castillo, E. and Hurwitz, A. (1997). Run to run process control: a review and some extensions, *Journal of Quality Technology*, **29**, 184–196.

7. Del Castillo, E. (1996). Some aspects of process control in semiconductor manufacturing, *Proceedings of the 4th Würzburg-Umea Conference in Statistics*, pp. 37–52.

8. Ingolfsson A. and Sachs, E. (1993). Stability and sensitivity of an EWMA controller, *Journal of Quality Technology*, **25**, 271–287.

9. MacGregor, J. F. (1988). On-line statistical process control, *Chemical Engineering Progress*, October, pp. 21–31.

10. MacGregor, J. F. (1990). A different view of the funnel experiment, *Journal of Quality Technology*, **22**, 255–259.

11. Montgomery, D. C. et al. (1994). Integrating statistical process control and engineering process control, *Journal of Quality Technology*, **26**, 79–87.

12. Quesenberry, C. P. (1988). An SPC approach to compensating a tool-wear process, *Journal of Quality Technology*, **20**.

13. Sachs, E., Hu, A. and Ingolfsson, A. (1995). Run by run process control: Combining SPC and feedback control, *IEEE Transactions on Semiconductor Manufacturing*, **8**, 26–43.

14. Siegmund, D. (1975). *Sequential Analysis*, New York: Springer-Verlag.

15. Snyder, D. L. (1995). *Random Point Processes*, New York: John Wiley & Sons.

16. Tucker, W. T., Faltin, F. W. and Vander Wiel, S. A. (1993). ASPC: An elaboration, *Technometrics*, **35**, 363–375.

17. Vander Wiel S. A., Tucker, W. T., Faltin, F. W. and Doganaksoy, N. (1992). Algorithmic statistical process control: Concepts and an application, *Technometrics*, **34**, 286–297.

21

Controlling a Process with Three Different States

Gudrun Kiesmüller

Universität Würzburg, Würzburg, Germany

Abstract: This paper is concerned with a process model which combines two fields: reliability theory and statistical process control. We consider three possible states of a production process: working on high quality, working on low quality and not working. Three different control strategies are presented and investigated. An economic approach is used in order to compare them and to determine the optimal control policy.

Keywords and phrases: Process-control, control-strategies, economic model, long run profit per item, optimization

21.1 Introduction

Since Duncan's pioneering paper on the economic design of control charts [Duncan (1956)], many mathematical models for controlling a process and proper scheduling of maintenance actions have been developed and investigated from an economic point of view. Different economic approaches are reviewed and developed by Lorenzen and Vance (1986) or von Collani (1981).

In quality control usually an *in-control* state and an *out-of-control* state for the production process are defined, where the out-of-control state is characterized by a larger probability for producing nonconforming items. In order to detect the out-of-control state control charts are used. See, for instance, von Collani (1989) for the design and evaluation of control charts.

In reliability theory the two different process-states have an other interpretation: the operating-state and the non-operating-state. The aim is to avoid an unexpected failure and therefore different strategies are investigated, see for instance Beichelt (1993).

In this paper we combine the two models and define an operating-state with two distinct substates, one with a low and the other with a larger nonconforming probability.

21.2   Process Model

The time for production of one item is assumed to be constant and used as time unit in the seque.

It is supposed that the quality of an item produced is determined by exactly one observable univariate characteristic denoted by ξ. Let ξ_i describe the *quality characteristic* of the i-th item produced, then the production process can be modelled with respect to the quality of the items by the sequence

$$\{\xi_i\}_{i\in\mathbb{N}}. \tag{21.1}$$

We assume that there are two quality classes defined by two-sided specifications with respect to ξ:

$$\xi_i \in [L, U] \quad \Leftrightarrow \quad \text{item } i \text{ conforming} \tag{21.2}$$

$$\xi_i \notin [L, U] \quad \Leftrightarrow \quad \text{item } i \text{ nonconforming.} \tag{21.3}$$

Thus we can introduce the following dichotomic function as *item quality indicator:*

$$\zeta = \begin{cases} 1 & \text{if} \quad \xi \in [L, U] \\ 0 & \text{if} \quad \xi \notin [L, U] \end{cases} \tag{21.4}$$

In order to derive a suitable process quality indicator we need some assumptions on the distribution of ξ.

(A1) The expectation of ξ is a random variable called ϑ.

(A2) The random variable ξ under the condition $\vartheta = \mu$ is normally distributed, i.e.

$$P(\xi_i \leq x | \vartheta = \mu) = \Phi(\frac{x - \mu}{\sigma}) \tag{21.5}$$

where $\Phi(x)$ denotes the distribution function of the normal distribution $\mathcal{N}(0, 1)$.

(A3) The variance σ^2 is known and constant. Therefore, without loss of generality, we set $\sigma^2 = 1$.

(A4) For $0 < i_1 < \ldots < i_n$ and under the condition $\vartheta_{i_1} = \mu_1, \ldots, \vartheta_{i_n} = \mu_n$ the random variables $\xi_{i_1}, \ldots, \xi_{i_n}$ are independent:

$$P(\xi_{i_1} \leq x_1, \ldots, \xi_{i_n} \leq x_n \mid \vartheta_{i_1} = \mu_1, \ldots, \vartheta_{i_n} = \mu_n) =$$
$$P(\xi_{i_1} \leq x_1 \mid \vartheta_{i_1} = \mu_1) \cdot \ldots \cdot P(\xi_{i_n} \leq x_n \mid \vartheta_{i_n} = \mu_n). \tag{21.6}$$

Because of the dichotomic item quality indicator the probability of producing a conforming item determines the quality of the process. This probability depends on the value μ of the random variable ϑ.

$$P(\zeta = 1 \mid \vartheta = \mu) = \Phi(U - \mu) - \Phi(L - \mu). \tag{21.7}$$

The random variable ϑ determines the quality of the output and therefore we call ϑ the *process quality indicator*. The value of ϑ which maximizes the probability of producing a conforming item is used as *target value* μ_0 and is given by

$$\mu_0 = \arg \max_{\mu} P(L \leq \xi \leq U \mid \vartheta = \mu) = \frac{L + U}{2}. \tag{21.8}$$

Without loss of generality we set $\mu_0 = 0$ being equivalent with $L = -U$.

We assume that it is possible to adjust the process quality indicator on an arbitrary value $\mu \in [-\mu_A; \mu_A]$. The value μ_A is given by technical reasons and is assumed to be known.

At time $t = 0$ the process is on target. Shocks occur at random timepoints t_i and shift the process quality indicator by η_i. If N_t denotes the random number of shocks in the interval $[0, t]$ then the process quality indicator at time t is given by:

$$\vartheta_t = \eta_1 + \ldots + \eta_{N_t}. \tag{21.9}$$

For the shift size η_i we make the following assumption:

(A5) The absolute value of η_i is constant and equal to η. So we get the following presentation:

$$\vartheta_t = \eta N_t. \tag{21.10}$$

The sign of η at the first shock determines the sign of the following shocks.

Therefore any shock increases the distance between the actual process quality indicator ϑ and the target value μ_0, and therefore, any shock leads to an increased probability of producing nonconforming items.

For the distribution of the random variable N_t we make the following assumption:

(A6) Shocks arrive according to a homogeneous Poisson process with rate $\lambda, \lambda > 0$. So the counting process N_t has the following distribution:

$$P(N_t = i) = \frac{(\lambda t)^i}{i!} \, e^{-\lambda t}. \tag{21.11}$$

It follows immediately, that the interarrival times $\{T_k\}$ are independent and exponentially distributed with parameter λ.

Generally small deviations of the process quality indicator from target do not justify an expensive corrective action and therefore are tolerable. If there are larger deviations production quality is unsatisfactory and a corrective action becomes profitable. If the deviation from target is too large then it is assumed that the process stops working. Thus we can define three different states of the process depending on the actual value of ϑ.

- State I if $0 \leq |\vartheta| \leq \mu_Z$: bringing the process quality indicator back on target is not profitable from an economic point of view.

- State II if $\mu_Z < |\vartheta| \leq \mu_A$: bringing the process quality indicator back on target is profitable from an economic point of view.

- State III if $\mu_A < |\vartheta|$: process failure, which can only be removed by a corrective action.

The parameter μ_Z is called the *state-boundary* and the parameter μ_A is called the *stop-boundary*. μ_A is assumed to be known because of technical reasons while the state-boundary μ_Z is optimized later.

For the given shock model with fixed shift size η it is possible to express the state-boundary and the stop-boundary by the minimal numbers of shocks which lead to State II or State III, respectively. We define

$$i_Z := \min\{i \mid i \in \mathbb{N}, i > |\frac{\mu_z}{\eta}|\} \tag{21.12}$$

and

$$i_A := \min\{i \mid i \in \mathbb{N}, i > |\frac{\mu_A}{\eta}|\}. \tag{21.13}$$

Thus we obtain the following representation of the process states:

$$
\begin{array}{lll}
0 \leq N_t \leq i_Z - 1 & \text{State I (S I)} & \tag{21.14} \\
i_Z \leq N_t \leq i_A - 1 & \text{State II (S II)} & \tag{21.15} \\
i_A \leq N_t & \text{State III (Failure).} & \tag{21.16}
\end{array}
$$

An illustration of the process states and a possible behaviour of the process quality indicator is given in Figure 21.1.

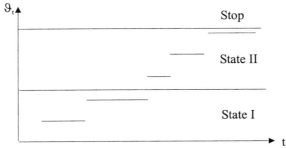

Figure 21.1: Process quality indicator

Usually it's not possible to have exact knowledge about the quality indicator ϑ_t at any time t, because continuous observation leads to a large expenditure of work and is too expensive. So only from time to time a determination of the actual value of the process quality indicator, which is called inspection, is done.

21.3 Control Model

State III is immediately recognized as the process stops working. After such a failure a process *renewal* is undertaken, after which the process starts operating in State I with the same stochastical properties as a new one.

In order to detect State II, actions have to be performed which are called *inspections*. It is assumed that an inspection reveals the actual state of the process with probability 1. Moreover, the inspections are performed periodically at time points $l \cdot h, l \in \mathbb{N}$, called *decision time points*. If State II is detected an *adjustment* is undertaken bringing the process quality indicator back on target. After an adjustment the process has the same properties as a new one, i.e. an adjustment renews the process, too. If an inspection reveals State I, no further action is undertaken. The process continues operating.

Clearly, any *control policy* considered here is completely determined by two quantitives, namely the inspection interval $h \in \mathbb{N} \cup \{+\infty\}$ and the state-boundary $i_Z \in \mathbb{N}, 0 < i_Z \le i_A$. The set of periodic control policies, denoted by $\mathcal{S}$, is given by:

$$\mathcal{S} := \{(h, i_Z) | h \in \mathbb{N} \cup \{+\infty\}, i_Z \in \mathbb{N}_0, 0 < i_Z \le i_A\}. \tag{21.17}$$

$\mathcal{S}$ includes three different types of control strategies:

- **Renewal-Strategy:**

$$\mathcal{S}_1 = \{(h, i_Z) | 0 < h < \infty; i_Z = i_A\} \cup \{(h, i_Z) | h = +\infty; 0 \le i_Z \le i_A\}. \tag{21.18}$$

  The directive for any $(h, i_Z) \in \mathcal{S}_1$ is essentially the same: After a process failure ($|\vartheta| > \mu_A$) perform immediately a process renewal.

- **Inspection-Strategy:**

$$\mathcal{S}_2 = \{(h, i_Z) | 0 < h < \infty; 1 < i_Z < i_A\}. \tag{21.19}$$

For any $(h, i_Z) \in \mathcal{S}_2$ a renewal is undertaken after a process failure. Additionally the process is inspected at the decision points $l \cdot h, l \in \mathbb{N}$. If the inspection reveals State II i. e. $\mu_Z < |\vartheta| \le \mu_A$ then an adjustment is performed.

- **Adjustment-Strategy:**

$$\mathcal{S}_3 = \{(h, i_Z) | 0 < h < \infty; i_Z = 1\}. \tag{21.20}$$

The Adjustment-Strategy is a special case of the Inspection-Strategy. If this strategy is used any detected deviation from target is corrected.

21.4   Long Run Profit Per Item

For comparing different control policies $(h, i_Z) \in \mathcal{S}$ we need to measure process profitability. To this end the process as well as the different interventions performed according to a given control policy are described economically. Let

g^+ :   average profit for a conforming item
g^- : $(< g^+)$ average profit for a nonconforming item
e :   average cost for an inspection
r_I: average cost for an adjustment
r_{II}:   average cost for a renewal

It is assumed that for technical reasons inspections are not more expensive than adjustments and renewals and that adjustments are not more expensive than renewals.

$$0 < e \leq r_I \leq r_{II}. \tag{21.21}$$

The consecutive corrective actions divide the process into independent and stochastically equivalent parts, which are called *renewal cycles*. The random duration of a renewal cycle is denoted by τ. The following random variables refer to one renewal cycle:

P: Profit taking into account the cost for the control policy
G: Profit without taking into account the cost for the control policy
A: Number of inspections
R_I: Number of adjustments
R_{II}:   Number of renewals
M :   Number of items produced

Next we select the long run profit per item produced denoted by Π as an appropriate measure for process profitability. Π as a function of $(h, i_Z) \in \mathcal{S}$ is used for comparing different control policies and to define and determine an optimal control policy. By the strong law of large numbers

$$\Pi(h, i_Z) = \frac{E[P]}{E[M]} \ . \tag{21.22}$$

holds with probability 1. The average profit is given by

$$E[P] = E[G] - eE[A] - r_I E[R_I] - r_{II} E[R_{II}]. \tag{21.23}$$

Having selected the production time of one item as time unit the number of produced items is equal to the integral part of the length of a renewal cycle. By neglecting the integralness of the number of produced items we obtain:

$$\Pi(h, i_Z) = \frac{E[G] - eE[A] - r_I E[R_I] - r_{II} E[R_{II}]}{E[\tau]} . \qquad (21.24)$$

We call a control plan $(h^*, i_Z^*) \in \mathcal{S}$ *optimal* with respect to Π if

$$\Pi(h^*, i_Z^*) \geq \Pi(h, i_Z) \quad \forall (h, i_Z) \in \mathcal{S} \qquad (21.25)$$

holds, and we call a strategy $\mathcal{S}_i, (i = 1, 2, 3)$ *optimal* if the optimal control plan (h^*, i_Z^*) is an element of $\mathcal{S}_i$.

A control policy $(h, i_Z) \in \mathcal{S}$ is only profitable if

$$\Pi(h, i_Z) > 0 \qquad (21.26)$$

holds.

21.5 Renewal-Strategy

There is essentially only one policy in $\mathcal{S}_1$ and therefore the optimal renewal policy is obviously $(h^*, i_Z^*) = (\infty, i_A)$. For comparing the optimal renewal-policy with the other strategies we have to determine $\Pi(\infty, i_A)$.

The random length τ of one renewal cycle is the sum of i_A independent and exponentially distributed random variables with parameter λ, i.e., τ has an Erlang distribution of order i_A with parameter λ implying

$$E[\tau] = \frac{i_A}{\lambda} . \qquad (21.27)$$

Next the expected profit without taking into account the cost for the control policy is determined. Let $g^{(j)}$ be the average profit for an item produced when j shocks have occurred:

$$g^{(j)} = g^+ - (g^+ - g^-)\Big(\Phi(-U + j\eta) + \Phi(-U - j\eta)\Big). \qquad (21.28)$$

Then the average profit $E[G]$ during one renewal cycle is given by:

$$E[G] = \sum_{j=0}^{i_A - 1} g^{(j)} E[\tau] = \frac{1}{\lambda} \sum_{j=0}^{i_A - 1} g^{(j)}. \qquad (21.29)$$

Thus we get for the long run profit per item for the optimal renewal-policy

$$\Pi(h^*, i_Z^*) = \pi(\infty, i_A) = \frac{1}{i_A} \Big(\sum_{j=0}^{i_A - 1} g^{(j)} - \lambda r_{II} \Big). \qquad (21.30)$$

21.6   Inspection-Strategy

For the control policies $(h, i_Z) \in \mathcal{S}_2$ we first determine the expectations for the numbers of adjustments $E[R_I]$ and the numbers of renewals $E[R_{II}]$. In one renewal cycle there is either one adjustment or one renewal undertaken, therefore we get:

$$R_I + R_{II} = 1 \Rightarrow E[R_I] + E[R_{II}] = 1. \tag{21.31}$$

which means that it is sufficient to determine one of these two expectations. The expectation of R_I is equal to the probability that an adjustment is undertaken. Hence:

$$
\begin{aligned}
E[R_I] &= P(R_I = 1) \\
&= \sum_{l=0}^{\infty} P(\text{Adjustment at time (l+1)h}) \\
&= \sum_{l=0}^{\infty} P(i_Z \leq N_{(l+1)h} < i_A \mid N_{lh} < i_Z) \cdot P(N_{lh} < i_Z) \\
&= \sum_{l=0}^{\infty} \sum_{i=0}^{i_Z-1} \sum_{j=i_Z}^{i_A-1} P(N_{(l+1)h} = j \mid N_{lh} = i) \cdot P(N_{lh} = i) \\
&= \sum_{l=0}^{\infty} \sum_{i=0}^{i_Z-1} \sum_{j=i_Z}^{i_A-1} P(N_h = j - i) P(N_{lh} = i) \\
&= \sum_{l=0}^{\infty} e^{-\lambda(l+1)h} \sum_{i=0}^{i_Z-1} \sum_{j=i_Z}^{i_A-1} \frac{(\lambda h)^{j-i}}{(j-i)!} \frac{(\lambda l h)^i}{i!} .
\end{aligned}
\tag{21.32}
$$

In the next step the average number of inspections is determined. There are i inspections in one renewal cycle if the transition from State I to State II takes place in the interval $(ih, (i+1)h]$. Let $H_j(x)$ denote the distribution function of an Erlang distribution of order j and parameter λ, and $\overline{H}_j(x)$ its survival function, then we get the following distribution for the number of inspections A

$$P(A = i) = H_{i_Z}((i+1)h) - H_{i_Z}(ih) \tag{21.33}$$

and hence

$$E[A] = \sum_{i=0}^{\infty} i \cdot W(A = i) = \sum_{i=1}^{\infty} W(A \geq i) = \sum_{i=1}^{\infty} \overline{H}_{i_Z}(ih). \tag{21.34}$$

In order to determine the expectation $E[\tau]$ we introduce a further random variable T, which describes the time from the state transition until the next control action after which the process quality indicator is brought back on target. $E[T]$ gives the expected time operating in State II during one renewal cycle, i.e.

$$E[\tau] = E[T_1 + T_2 + \ldots + T_{i_Z}] + E[T] = \frac{i_Z}{\lambda} + E[T]. \qquad (21.35)$$

Obviously we have for the random variable T: $0 \leq T \leq h$. Let $F_T(x)$ denote the distribution function of T, then for $x \in [0, h]$:

$$
\begin{aligned}
F_T(x) &= \sum_{l=0}^{\infty} \Big\{ H_{i_Z}((l+1)h) - H_{i_Z}((l+1)h - x) \\
&\quad + H_{i_A - i_Z}(x)\Big(H_{i_Z}((l+1)h - x) - H_{i_Z}(lh) \Big) \Big\}.
\end{aligned}
\qquad (21.36)
$$

The proof is very technical. In a first step the distribution function for the conditional random variable T under condition $R_I = 1$ is determined and then the distribution function for the conditional random variable T under condition $R_{II} = 1$. With (21.36) the expectation $E[T]$ is obtained:

$$
\begin{aligned}
E[T] &= \frac{1}{\lambda} \sum_{l=0}^{\infty} e^{-\lambda(l+1)h} \Big\{ \sum_{i=i_Z}^{i_A - 1} (i - i_Z + 1) \sum_{j=0}^{i_Z - 1} \frac{(\lambda l h)^j}{j!} \frac{(\lambda h)^{i-j+1}}{(i-j+1)!} \\
&\quad + (i_A - i_Z) \sum_{j=0}^{i_Z - 1} \frac{(\lambda l h)^j}{j!} \Big(e^{\lambda h} - \sum_{i=0}^{i_A - i_Z} \frac{(\lambda h)^i}{i!} \Big) \\
&\quad - (i_A - i_Z) \sum_{i=i_A - i_Z}^{i_A - 1} \Big(\frac{(\lambda(l+1)h)^{i+1}}{(i+1)!} - \sum_{j=0}^{i_A - i_Z} \frac{(\lambda h)^j}{j!} \frac{(\lambda l h)^{i-j+1}}{(i-j+1)!} \Big) \Big\}.
\end{aligned}
\qquad (21.37)
$$

It remains to calculate the average profit without taking into account the cost for control actions. The profit G is the sum of the profits made in State I and in State II. If G_T denotes the profit during State II we get

$$E[G] = \frac{1}{\lambda} \sum_{j=0}^{i_Z - 1} g^{(i)} + E[G_T]. \qquad (21.38)$$

Let g_m denote the profit derived from m items produced in State II, then

$$g_m = m g^+ - (g^+ - g^-) \sum_{i=1}^{m} p_{II,i} \qquad (21.39)$$

where $p_{II,i}$ is the probability that item i produced in State II is nonconforming:

$$p_{II,i} = e^{-\lambda i} \sum_{j=0}^{\infty} \frac{(\lambda i)^j}{j!} \Big(\Phi(-U + (i_Z + j)\eta) + \Phi(-U - (i_Z + j)\eta) \Big). \qquad (21.40)$$

Thus we get for $E[G_T]$:

$$E[G_T] = \sum_{m=1}^{h} g_m \Big(F_T(m+1) - F_T(m) \Big) \tag{21.41}$$

and with (21.38)

$$E[G] = \frac{1}{\lambda} \sum_{j=0}^{i_Z-1} g^{(j)} + \sum_{m=1}^{h} g_m \Big(\overline{F}_T(m) - \overline{F}_T(m+1) \Big). \tag{21.42}$$

Inserting (21.34), (21.32), (21.37) and (21.42) in (21.24) yields an explicit expression for the long run profit per item as function of $(h, i_Z) \in \mathcal{S}_2$.

21.7   Adjustment-Strategy

The Adjustment-Strategy $\mathcal{S}_3$ can be looked upon as a special case of the Inspection-Strategy, obtained by setting $i_Z = 1$. For $i_Z = 1$ the above expectations simplify to:

$$E[A] = \frac{e^{-\lambda h}}{1 - e^{-\lambda h}} \tag{21.43}$$

$$E[R_I] = \frac{1}{e^{\lambda h} - 1} \sum_{j=1}^{i_A-1} \frac{(\lambda h)^j}{j!} \tag{21.44}$$

$$E[R_{II}] = \frac{e^{\lambda h} - \sum\limits_{j=0}^{i_A-1} \frac{(\lambda h)^j}{j!}}{e^{\lambda h} - 1} \tag{21.45}$$

$$E[\tau] = \frac{1}{\lambda} \frac{1}{e^{\lambda h} - 1} \Big\{ i_A (e^{\lambda h} - \sum_{i=0}^{i_A} \frac{(\lambda h)^i}{i!}) + \sum_{i=1}^{i_A} \frac{(\lambda h)^i}{(i-1)!} \Big\} \tag{21.46}$$

$$\Pi(h,1) = \frac{(\frac{1}{\lambda} g^{(0)} - r_I + E[G_T])(e^{\lambda h} - 1) - (r_{II} - r_I)(e^{\lambda h} - \sum\limits_{j=0}^{i_A-1} \frac{(\lambda h)^j}{j!}) - e}{\frac{1}{\lambda} \Big\{ i_A (e^{\lambda h} - \sum\limits_{i=0}^{i_A} \frac{(\lambda h)^i}{i!}) + \sum\limits_{i=1}^{i_A} \frac{(\lambda h)^i}{(i-1)!} \Big\}}. \tag{21.47}$$

21.8   Numerical Examples

In the last three sections we have derived explicit expressions of $\Pi(h, i_Z)$ for the three different strategies. Because of the complicated structure of the formulae optimization is only possible by numerical methods.

In a first step the optimal control policy of each strategy is determined numerically, and then the overall optimal control policy is obtained by comparison.

Example 21.8.1 $\lambda = 0.01, \eta = 0.1, U = 1.96, i_A = 10,$
$g^+ = 10, g^- = 3, r_I = 110, r_{II} = 10000, e = 100$

Strategy	control policy		Objective Function
Renewal-Strategy	$h = +\infty$	$i_Z = 10$	$\Pi(+\infty, 10) = -0.58534007$
Adjustment-Strategy	$h = 311$	$i_Z = 1$	$\Pi(311, 1) = 9.37461375$
Inspection-Strategy	$h = 308$	$i_Z = 2$	$\Pi(308, 2) = 9.29169482$

In this example only the Adjustment- and the Inspection-Strategy are profitable. It is optimal to use an Adjustment-Strategy for controlling the process. This is not very surprising, because the cost for an adjustment is not much higher than for an inspection. The optimal control policy is $(311, 1)$.

Example 21.8.2 $\lambda = 0.01, \eta = 1, U = 1.96, i_A = 10,$
$g^+ = 1, g^- = 0.5, r_I = 100, r_{II} = 1000, e = 0.01$

Strategy	control policy		Objective Function
Renewal-Strategy	$h = +\infty$	$i_Z = 10$	$\Pi(+\infty, 10) = -0.3782$
Adjustment-Strategy	$h = 367$	$i_Z = 1$	$\Pi(367, 1) = 0.50833313$
Inspection-Strategy	$h = 14$	$i_Z = 4$	$\Pi(14, 4) = 0.55085898$

In this example again the Adjustment- and the Inspection-Strategy are profitable. But in this case the Inspection-Strategy is the better one. The optimal control policy is $(14, 4)$.

Example 21.8.3 $\lambda = 0.01, \eta = 1, U = 1.96, i_A = 10,$
$g^+ = 10, g^- = 9, r_I = 900, r_{II} = 1000, e = 800$

Strategy	control policy		Objective Function
Renewal-Strategy	$h = +\infty$	$i_Z = 10$	$\Pi(+\infty, 10) = 8.24350024$
Adjustment-Strategy	no optimal control policy		
Inspection-Strategy	no optimal control policy		

In this example there exist no optimal Adjustment- or Inspection-Strategy. The Renewal-Strategy is the best one and the optimal strategy is $(+\infty, 10)$.

References

1. Beichelt, F. (1993). *Zuverlässigkeits- and Instandhaltungstheorie*, Stuttgart: B. G. Teubner.

2. Duncan, A. J. (1956). The economic design of $\overline{X}$ control charts used to maintain current control of process, *Journal of the American Statistical Association*, **51**, 228–242.

3. Lorenzen, T. J. and Vance, L. C. (1986). The economic design of control charts: A unified approach, *Technometrics*, **28**, 3–10.

4. von Collani, E. (1981). Kostenoptimale prüfpläne für die laufende kontrolle eines normalverteilten merkmals, **Metrika**, **28**, 211–236.

5. von Collani, E. (1989). *The Economic Design of Control Charts*, Stuttgart: B. G. Teubner.

22

CUSUM Schemes and Erlang Distributions

Sven Knoth

Europe–University Viadrina, Frankfurt (Oder), Germany

Abstract: Exact expressions are derived for the Average Run Length (ARL) of CUSUM schemes when observations follow an Erlang distribution. Additionally, exact expressions for the quasi–stationary distribution are found. Based on the results for the ARL as function of the headstart and the quasi–stationary density function, the so–called average delay can be determined. Both functions are obtained by solving integral equations. Eventually, by using the right eigenfunction of the CUSUM transition kernel – the left one yields the quasi–stationary density function – a geometric approximation of the run length distribution is constructed.

Keywords and phrases: CUSUM schemes, average run length, average delay, quasi–stationary distribution, Erlang distribution, run length approximation

22.1 Introduction

CUSUM schemes are widely applied in quality control. They were introduced by Page (1954) and are used as a so–called change point detection method. Originally, methods like CUSUM were developed for the normal distribution. But, there are many papers on CUSUM schemes for other distributions or for the nonparametric case. Here, we determine the characteristics of CUSUM charts for detecting a change point in a series of Erlang distributed random variables.

First, recall the change point model in terms of the Erlang distribution. We observe a sequence $X_1, X_2, \ldots$ of independent distributed random variables with the common density function

$$f_\theta(x) \;=\; \theta^n \, \frac{x^{n-1}}{(n-1)!} \, e^{-\theta x} \, \mathbf{1}_{[0,\infty)}(x) \,, \tag{22.1}$$

where $n \in I\!N^+$, $\theta \in (0, \infty)$, and $\mathbf{1}_{[0,\infty)}(x)$ is the indicator function of $[0, \infty)$. In the sequel, the second Erlang parameter n in (22.1) is fixed. Otherwise, the parameter θ can change. In order to specify this situation we use the well–known change point model. We assume, that for some $m = 1, 2, \ldots, \infty$ the $X_1, X_2, \ldots X_{m-1}$ and the $X_m, X_{m+1}, \ldots$ are identically distributed with $\theta = \theta_0$ and $\theta = \theta_1$ respectively. The index m is called change point. In Figure 22.1 we see simulated data with $m = 50$, $n = 2$, $\theta_0 = 1$ and $\theta_1 = 2$. In practice, the change point is unknown and we wish to apply a detection rule, which will detect the change point soon after its occurrence without too many "false alarms". In this context the CUSUM scheme is in some sense optimal [cf. Lorden (1971), Moustakides (1986), and Ritov (1990)].

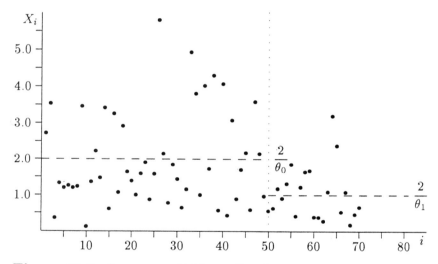

Figure 22.1: Sequence of Erlang distributed random variables with change point $m = 50$

In order to introduce the CUSUM scheme we use the "likelihood ratio" notation, i.e. we consider the sequence of the log–likelihood ratios for Erlang distributed variables X_i

$$Y_i = \ln \frac{f_{\theta_1}(X)}{f_{\theta_0}(X)} = n \ln \frac{\theta_1}{\theta_0} - (\theta_1 - \theta_0) X_i. \qquad (22.2)$$

Then, the CUSUM scheme is determined by two design parameters z and h and is given by the sequence

$$Z_0 = z \in [0, h),$$
$$Z_i = \max\{0, Z_{i-1} + Y_i\}, \ i = 1, 2, \ldots,$$

and the stopping rule

$$L = \min\{i : Z_i \geq h\}.$$

The characteristics of the CUSUM scheme are related to the properties of the stopping time L. In the beginning of the evaluation of such change point detection schemes the Average Run Length (ARL) was the usually used criterion. For its definition we need some new notations. We denote with $E^{(m)}$ and $P^{(m)}$ the expectation and the distribution function respectively, where m stands for the position of the change point m. Then the ARLs are defined by $E^{(\infty)}(L)$ and $E^{(1)}(L)$ for the in–control and out–of–control case respectively. The first case corresponds to the situation of no change point, while the second one stands for a change at the first position of the series. Hence, for the ARL we consider identically distributed series. Therefore, we write the ARL as $\mathcal{L}(z)$, where θ as the true (possibly different from θ_0 and θ_1) Erlang parameter is suppressed. The argument z is the initial CUSUM value known as *headstart*.

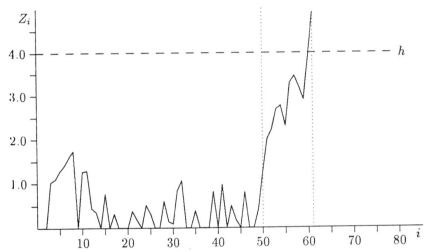

Figure 22.2: CUSUM scheme for the data of Figure 22.1

By looking at Figure 22.2, where the CUSUM scheme for the artificial data of Figure 22.1 is presented, we see that for $m \notin \{1, \infty\}$ the ARL seems to be not a suitable criterion for the detection speed. At the change point m the CUSUM series Z_i differs from the headstart $z_0 = 0$. Therefore, the classical ARL $\mathcal{L}(0)$ with $\theta = \theta_1$ forms an upper bound of the average delay of detection the change point m, i.e. we obtained a worst case criterion. Otherwise, for CUSUM schemes with headstarts $z_0 \neq 0$ the ARL $\mathcal{L}(z_0)$ provides only a roughly measure of the delay behaviour. Again, we have to use the worst case criterion $\mathcal{L}(0)$, or we investigate the average delay after a occurred change point itself. In order to model the last one Roberts (1966), Pollak and Siegmund (1985), Pollak (1985), and Mevorach and Pollak (1991) proposed the following criterion

$$\mathcal{D} = \lim_{m \to \infty} \mathcal{D}^{(m)} \quad \text{with} \tag{22.3}$$

$$\mathcal{D}^{(m)} = E^{(m)}(L - m + 1 \mid L \geq m)$$

as the measure for the speed of detection of the change point m. Here, it is called (conditional) average delay. The above formularization implicitly uses the quasi–ergodicity of the CUSUM sequence Z_i, i.e. the convergence of the conditional distribution $P^{(m)}(Z_m \leq z \,|\, L \geq m)$ for $m \to \infty$ to the quasi-stationary (i. e. conditional stationary) distribution.

Eventually, we are interested in more than the first moment of the stopping time L. In order to get a better insight in the CUSUM performance it is useful to consider the whole distribution function of L.

In the next sections we determine the above characteristics, i. e. the ARL, the average delay, and an approximation of the probability function of L for Erlang distributed random variables. In Section 22.2 we introduce and in Section 22.3 we solve the integral equations defining the above characteristics. Some numerical and conclusions complete this paper.

Before we go into details, results for the related exponential distribution ($n = 1$ in (22.1)) are reviewed. We start with the classical characteristic, the ARL. The first results for the ARL in the case of exponential distribution can be found in a paper on Poisson processes of Dvoretzky, Kiefer, and Wolfowitz (1953). A related work on the sequential probability test (SPRT) was written by Raghavachari (1965). In 1973, Lorden and Eisenberger provided good approximations for the ARL in the above two parameter cases. Five years later Khan (1978) presented improved Wald approximations for the SPRT and used them for computing the above CUSUM ARL. In both cases an approximation of the excess over the boundary was constructed. Lucas (1985) and Wang (1990) used the Markov chain approach of Brook and Evans (1972). By considering ladder variables Pollak and Siegmund (1986) obtained approximations of the ARL. Furthermore, we find another approximation in the paper of Tartakovskij and Ivanova (1992) which compared CUSUM with the Shiryaev–Roberts[1] rule. Eventually, Vardeman and Ray published exact formulas already in 1985. They solved the integral equation introduced here in Section 22.2 for the exponential distribution in a direct way. A special case (h small enough, i. e. $h < n \ln \frac{\theta_1}{\theta_0}$ with $n \in I\!\!R^+$) of CUSUM schemes for the gamma distribution is considered in Regula (1975).

Early results on quasi–stationary distributions are presented in Madsen and Conn (1973) for the SPRT for exponentially distributed random variables. Mevorach and Pollak (1991) consider quasi–stationary started CUSUM schemes. In the case of exponentially distributed random variables, they compute the average delay (22.3) by simulation. Besides, Pollak (1987) prove the asymptotic optimality of quasi–stationary started Shiryaev–Roberts rules for the average delay criterion (22.3). Furthermore, Pollak and Siegmund (1985) and Mevorach and Pollak (1991) demonstrate that in the finite case CUSUM and the Shiryaev–Roberts rule perform similarly.

Different methods are used to approximate the run length distribution. The

---

[1]The direct counterpart of CUSUM; it seems to be less popular than CUSUM

Markov chain approach of Brook and Evans (1972) is the most popular one. Next, we find some numerical procedures for fitting a geometric distribution in Woodall (1983), Waldmann (1986), and Gold (1989). This access is based on the quasi–stationary structure as well.

22.2 Transition Kernel and Integral Equations

In the sequel, the considered partial sequences $\{X_i\}$ follow the density $f_\theta(x)$ with fixed θ. We denote by $g(y)$ and $G(y)$ the density and the distribution function of the log–likelihood ratio Y_i given by (22.2) respectively. By using the modified Lebesgue measure μ due to Woodall (1983), i.e. $\mu(\{0\}) = 1$, and the kernel function [cf. Madsen and Conn (1973)]

$$M(x,y) = \begin{cases} G(-x) , & y = 0 \\ g(y-x), & y > 0 \\ 0 & , \text{ otherwise} \end{cases} , \ x \in [0,h) \qquad (22.4)$$

we obtain for the transition within the CUSUM sequence Z_i

$$P\left(Z_{i+1} \leq z \mid Z_i = x\right) = \int_0^z M(x,y)\,d\mu(y) ,$$

where $x, z \in [0,h)$. Then we get the ARL integral equation [see Page (1954) or Vardeman and Ray (1985)]

$$\mathcal{L}(z) = 1 + G(-z)\,\mathcal{L}(0) + \int_0^h g(x-z)\mathcal{L}(x)\,dx .$$

In order to derive similar equations for the quasi–stationary distribution we need some additional results. Thereby, we interpret the stopped sequence as absorbed in a state greater than h. Further, we define the sequence of iterated kernels

$$M_1(x,y) = M(x,y) ,$$
$$M_i(x,y) \underset{i>1}{=} \int_0^h M_{i-1}(x,z)M(z,y)\,d\mu(z) , \ x,y \in [0,h) .$$

Hence, for a headstart z_0 with $z \in [0,h)$

$$P_{z_0}(Z_i \leq z) = \int_0^z M_i(z_0,y)\,d\mu(y) . \qquad (22.5)$$

Now we introduce an important property for kernel functions $M(x,y)$. If $a, b \in (0,\infty)$ and an integer i_0 exist such that for the iterated kernel

$$a \leq M_{i_0}(x,y) \leq b , \quad \forall x,y \in [0,h) ,$$

then the kernel will be named "primitive". For primitive kernels the following proposition holds.

Proposition 22.2.1 [Harris (1963)] *Primitive kernels M possess a simply positive eigenvalue* λ*, which exceeds the other eigenvalues in magnitude. We denote the corresponding right and left eigenfunctions with* ϕ *and* ψ *respectively, i.e.*

$$\lambda\phi(x) = \int_0^h M(x,y)\,\phi(y)\,d\mu(y) \quad and$$

$$\lambda\psi(y) = \int_0^h M(x,y)\,\psi(x)\,d\mu(x).$$

Then ϕ *and* ψ *are uniformly positive, bounded, and apart from a factor unique in* x *and* y *on* $[0,h)$*. Standardizing* ϕ *and* ψ *by*

$$\int_0^h \phi(x)\,\psi(x)\,d\mu(x) = 1 \tag{22.6}$$

allows to choose a constant $\delta \in (0,1)$*, which does not depend on* x *and* y*, such that*

$$M_i(x,y) = \lambda^i\phi(x)\psi(y)\cdot[1+O(\delta^i)],\ i \to \infty.$$

PROOF. See Harris (1963, pp. 67 and 78–80). ∎

The standardizing of ψ into a density function, i.e. $\int_0^h \psi(x)\,d\mu(x) = 1$, transforms (22.6) into a condition of ϕ. For these ϕ and ψ the following corollary holds.

Corollary 22.2.1 *For* $i \gg 1$

$$P_{z_0}(L > i) \approx \lambda^i\phi(z_0),$$
$$P_{z_0}(L = i) \approx \lambda^{i-1}(1-\lambda)\phi(z_0).$$

PROOF.

$$P_{z_0}(L > i) \quad = \quad P_{z_0}(Z_i < h) = \lim_{\varepsilon\downarrow 0} P_{z_0}(Z_i \le h - \varepsilon)$$

$$= \quad \lim_{\varepsilon\downarrow 0}\int_0^{h-\varepsilon} M_i(z_0,y)\,d\mu(y) = \int_0^h M_i(z_0,y)\,d\mu(y),$$

$$\underset{\text{Prop. } 22.2.1}{\approx} \quad \lambda^i\phi(z_0).$$

The second approximation follows directly from the first one. ∎

Hence, we obtained an approximation for the distribution function of L. In addition to the sequence of the iterated kernels define a sequence of transition densities by

$$f_{z_0}^{(i)}(y) = \frac{M_i(z_0,y)}{\int_0^h M_i(z_0,x)\,d\mu(x)}.$$

Remember that $\mathcal{L}(z)$ is the ARL of a CUSUM scheme with headstart z. Hence the average delay in (22.3) can be written as

$$\mathcal{D} = \lim_{m \to \infty} \int_0^h f_{z_0}^{(m-1)}(y)\mathcal{L}(y)\,d\mu(y)\,.$$

Note that θ_0 is true for the sequence $\{f_{z_0}^{(m-1)}(y)\}_{m=1,2,\ldots}$, while the parameter θ for the ARL function $\mathcal{L}(y)$ is equal to the post change point parameter θ_1. The next proposition allows to determine the limit of $f_{z_0}^{(m-1)}(y)$.

Proposition 22.2.2 *For primitive kernels the sequence of transition densities converges uniformly to the (standardized) left eigenfunction ψ.*

PROOF. See Madsen and Conn (1973). ∎

Corollary 22.2.2

$$\mathcal{D} = \int_0^h \psi(y)\mathcal{L}(y)\,d\mu(y)\,.$$

PROOF. Apply Prop. 22.2.2 and Lebesgue's Theorem on Dominated Convergence. ∎

Remark 22.2.1 The CUSUM–Modification by $Z_0 \sim \psi$ [cf. Pollak (1985), Mevorach and Pollak (1991)], i.e., random headstart due the quasi–stationary distribution, yields

$$P_\psi(L = i) = (1 - \lambda)\,\lambda^{i-1}\,,$$
$$\mathcal{L}_{\theta_0}(\psi) = \frac{1}{1 - \lambda}\,,$$
$$\mathcal{L}_{\theta_1}(\psi) = \mathcal{D} = \mathcal{D}^{(m)}\,, \forall m \geq 1.$$

Thus, in the i.i.d. case L exactly follows the geometric distribution with parameter λ. Moreover, the sequence of delays $\mathcal{D}^{(m)}$ is constant over all change point positions.

Eventually, we obtained Fredholm integral equations of the second kind as definition equations for the ARL function $\mathcal{L}$ and the eigenfunctions ψ and ϕ. In the following section we solve these integral equations and combine the results by Corollary 22.2.2 to get the average delay $\mathcal{D}$.

22.3   Solution of the Integral Equations

First, we need the special shape of the functions G and g. Writing

$$\alpha = \theta/|\theta_1 - \theta_0| \quad \text{and}$$
$$\gamma = n\,|\ln(\theta_1/\theta_0)|\,,$$

we obtain for $\theta_0 < \theta_1$

$$g(y) = \alpha^n \frac{(\gamma - y)^{n-1}}{(n-1)!}\, e^{-\alpha(\gamma-y)}\, \mathbf{1}_{(-\infty,\gamma]}(y)\,,$$

$$G(y) = \begin{cases} \sum\limits_{i=0}^{n-1} \frac{[\alpha\,(\gamma-y)]^i}{i!}\, e^{-\alpha\,(\gamma-y)} & ,\, y < \gamma \\ 1 & ,\, y \geq \gamma \end{cases},$$

and for $\theta_0 > \theta_1$

$$g(y) = \alpha^n \frac{(\gamma + y)^{n-1}}{(n-1)!}\, e^{-\alpha(\gamma+y)}\, \mathbf{1}_{[-\gamma,\infty)}(y)\,,$$

$$G(y) = \left[1 - \sum_{i=0}^{n-1} \frac{[\alpha\,(\gamma+y)]^i}{i!}\, e^{-\alpha\,(\gamma+y)} \right] \mathbf{1}_{[-\gamma,\infty)}(y)\,.$$

First, we dissolve the modified Lebesgue measure to obtain the following integral equations for $\mathcal{L}$, ψ, and ϕ.

$$\mathcal{L}(z) = 1 + G(-z)\,\mathcal{L}(0) + \int_0^h g(x - z)\mathcal{L}(x)\,dx\,,$$

$$\lambda\,\psi(0) = G(0)\psi(0) + \int_0^h G(-x)\psi(x)\,dx\,,$$

$$\lambda\,\psi(y) \underset{y>0}{=} g(y)\,\psi(0) + \int_0^h g(y - x)\psi(x)\,dx\,,$$

$$\lambda\,\phi(x) = G(-x)\,\phi(0) + \int_0^h g(y - x)\phi(y)\,dy\,.$$

We have to distinguish between the two parameter constellations. In order to sketch the solution strategy we start with the left eigenfunction ψ for $\theta_0 < \theta_1$. Then it follows for $y > 0$

$$\lambda\,\psi(y) = \alpha^n \frac{(\gamma - y)^{n-1}}{(n-1)!} e^{-\alpha(\gamma-y)} \mathbf{1}_{(-\infty,\gamma]}(y)\psi(0) +$$

$$+ \int_0^h \frac{(\gamma - y + x)^{n-1}}{(n-1)!} e^{-\alpha(\gamma-y+x)} \mathbf{1}_{(-\infty,\gamma]}(y - x)\psi(x)\,dx\,.$$

Using the binomial formula (similarly to Kohlruss (1994) and Knoth (1997)) and introducing the constant $k = \alpha^n e^{-\alpha\gamma}$ we obtain

$$\psi(y) = \frac{k}{\lambda}\left[\frac{(\gamma-y)^{n-1}}{(n-1)!}\mathbf{1}_{(-\infty,\gamma]}(y)\psi(0)+ \right. \tag{22.7}$$

$$\left. +\sum_{s=0}^{n-1}\frac{[(i+1)\gamma-y]^s}{s!}\int_{\max\{0,y-\gamma\}}^{h}\frac{(x-i\gamma)^{n-1-s}}{(n-1-s)!}e^{-\alpha x}\psi(x)\,dx\right]e^{\alpha y}\,.$$

for $y \in (i\gamma, (i+1)\gamma]$ with $i = 0, 1, \ldots, N$. Now we define a series of constants for $i = 0, 1, \ldots, N$, $j = 0, 1, \ldots, n-1$ by

$$C_{00} = \psi(0) + \int_0^h e^{-\alpha x}\psi(x)\,dx\,,$$

$$C_{ij} \underset{i+j\neq0}{=} \int_{i\gamma}^{h}\frac{(x-i\gamma)^j}{j!}e^{-\alpha x}\psi(x)\,dx\,.$$

With these constants it follows from (22.7)

$$\psi(y) = \frac{k}{\lambda}\sum_{s=0}^{n-1}\frac{[(i+1)\gamma-y]^s}{s!}\times$$

$$\times\left[\int_{\max\{0,y-\gamma\}}^{i\gamma}\frac{(x-i\gamma)^{n-1-s}}{(n-1-s)!}e^{-\alpha x}\psi(x)\,dx + C_{i,n-1-s}\right]e^{\alpha y}\,.$$

In this way we obtain straightforward the first lemma.

Lemma 22.3.1 *($\theta_0 < \theta_1$)*

With $y \in (i\gamma, (i+1)\gamma]$, we obtain

$$\psi(y) = \sum_{r=0}^{i}\left(\frac{k}{\lambda}\right)^{r+1}\sum_{s=0}^{n-1}C_{i-r,n-1-s}\frac{[(i+1)\gamma-y]^{rn+s}}{(rn+s)!}e^{\alpha y} \tag{22.8}$$

PROOF. See Knoth (1997b). ∎

It remains to determine the constants $C_{00}, \ldots, C_{N,n-1}$, $\psi(0)$, and the dominating eigenvalue λ. For this we use the defining equations for C_{ij}, the equation $\lambda\psi(0) = \ldots$, and the density standardization by replacing ψ by its representation (22.8). For fixed λ the other constants are solutions of a system of linear equations. Finally, we solve using a secant method a nonlinear equation to obtain the dominating eigenvalue λ. Confer for detailed discussion and corresponding proofs to Knoth (1995). Additionally, in Knoth (1995) the equivalence between (22.8) in the case of $n = 1$, i.e. the exponential distribution, and the Mevorach and Pollak results is proved.

Turning to the case $\theta_0 > \theta_1$ we use the constants

$$\tilde{C}_{ij} \;=\; \frac{(h-i\gamma)^j}{j!}\,\psi(0) + \int_0^{h-i\gamma} \frac{(h-i\gamma-x)^j}{j!}\,e^{\alpha x}\psi(x)\,dx\,.$$

The next lemma shows the corresponding eigenfunction.

Lemma 22.3.2 *($\theta_0 > \theta_1$)*

For $y \in (h-(i+1)\gamma, h-i\gamma]$ we obtain

$$\psi(y) \;=\; \sum_{r=0}^{i} \left(\frac{k}{\lambda}\right)^{r+1} \sum_{s=0}^{n-1} \tilde{C}_{i-r,n-1-s} \frac{[y-h+(i+1)\gamma]^{rn+s}}{(rn+s)!}\,e^{-\alpha y} \quad (22.9)$$

PROOF. See Knoth (1997b) as well.   ∎

Example 22.3.1 We consider Erlang distributions with $n = 1, 2, 6$ and $\theta_0, \theta_1 \in \{1, 2\}$. The CUSUM decision interval h equals 4. Hence, we obtain for the above constants

$$\alpha \;=\; 1\,,\ \gamma \;=\; n\ln 2\,,\ N \;=\; \begin{cases} 5\,, & n=1 \\ 2\,, & n=2 \\ 0\,, & n=6 \end{cases}\,.$$

The dominating eigenvalues are presented in Table 22.1. Figure 22.3 shows the graphs of the left eigenfunction ψ for $\theta_0 < \theta_1$.

Table 22.1: Dominating eigenvalues λ for both parameter constellations

n	1	2	6
$\theta_0 < \theta_1$	0.997596	0.996490	0.995278
$\theta_0 > \theta_1$	0.998521	0.997831	0.997076

The bold dots in Figure 22.3 correspond to $\psi(0)$, i.e. they represent the conditional probabilities that $Z_i = 0$ for large i. The dotted lines mark the cut points of the intervals of (22.8) and (22.9).

Combining [cf. Knoth (1995, 1997) and Corollary 22.2.1] the results of ψ and the ARL function $\mathcal{L}$, provides a more complicated result for the average delay (22.3).

Continue with the right eigenfunction, i.e. with the geometric approximation of the run length distribution.

We introduce the constants B_{ij} in the following way.

$$B_{ij} \;=\; \sum_{t=0}^{j} \frac{[\alpha(h-i\gamma)]^t}{t!}\,\frac{\phi(0)}{\alpha^{j+1}} + \int_0^{h-i\gamma} \frac{(-x+h-i\gamma)^j}{j!}\,e^{\alpha x}\phi(x)\,dx\,.$$

Then we obtain for the right eigenfunction ϕ.

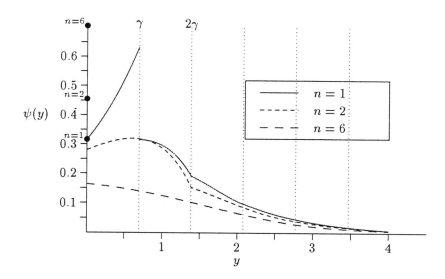

Figure 22.3: Quasi–stationary density function $\psi(y)$ for $\theta_0 < \theta_1$

Lemma 22.3.3 *($\theta_0 < \theta_1$)*

For $y \in (h - (i+1)\gamma, h - i\gamma]$

$$\phi(y) = \sum_{r=0}^{i} \left(\frac{k}{\lambda}\right)^{r+1} \sum_{s=0}^{n-1} B_{i-r,n-1-s} \frac{[y - h + (i+1)\gamma]^{rn+s}}{(rn+s)!} e^{-\alpha y} \quad (22.10)$$

PROOF. Analogously to Lemmas 22.3.1 and 22.3.2 see Knoth (1997b). ∎

Further, we consider the case $\theta_0 > \theta_1$ for ϕ. Let

$$\tilde{B}_{ij} = -\frac{\phi(0)e^{-i\alpha\gamma}}{\lambda^i \alpha^{j+1}} + \int_{i\gamma}^{h} \frac{x^j}{j!} e^{-\alpha x} \phi(x)\, dx.$$

The next lemma gives the representation of the right eigenfunction ϕ.

Lemma 22.3.4 *($\theta_0 > \theta_1$)*

If $y \in (i\gamma, (i+1)\gamma]$, then

$$\phi(y) = \frac{\phi(0)}{\lambda^{i+1}} + \sum_{r=0}^{i} \left(\frac{k}{\lambda}\right)^{r+1} \sum_{s=0}^{n-1} \tilde{B}_{i-r,n-1-s} \frac{[(i+1)\gamma - y]^{rn+s}}{(rn+s)!} e^{\alpha y} \quad (22.11)$$

PROOF. See Knoth (1997b). ∎

The constants in (22.10) and (22.11) are determined in the same way as for the left eigenfunction (we already calculated the dominating eigenvalue λ

for the left eigenfunction ψ). In order to standardize ϕ due to Proposition 22.2.1 we have to calculate integrals of the kind $\int_0^h \phi(x)\psi(x)\,dx$. It can be done analogously to the average delay integral [cf. Knoth (1995)]. Then, we obtain the geometric approximation of the run length distribution according to Corollary 22.2.1. The corresponding results are illustrated in Figure 22.4 for two different headstarts complemented with the quasi–stationary started CUSUM. Additionally, the CUSUM process is approximated by a Markov chain [Brook and Evans (1972)] with the goal to derive an approximation of the run length distribution by matrix operations. The nearly straight lines correspond to the geometric, the curved ones to the matrix approximation.

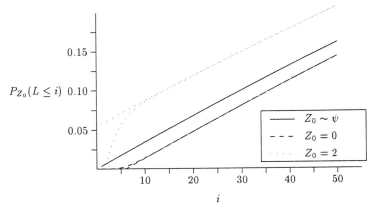

Figure 22.4: Approximations of the run length distribution function for
$$\theta_0 < \theta_1$$

Eventually, we consider the ARL function. Now we define a series of constants by

$$
D_{ij} = \sum_{t=0}^{j} \frac{[\alpha(h - i\gamma]^t}{t!} \frac{\mathcal{L}(0)}{\alpha^{j+1}} + \tag{22.12}
$$
$$
+ \int_0^{h-i\gamma} \frac{(-x + h - i\gamma)^j}{j!} e^{\alpha x} \mathcal{L}(x)\,dx - \frac{i}{\alpha^{j+1}} e^{\alpha(h-i\gamma)}.
$$

The solution of the integral equation of $\mathcal{L}$ in the case of $\theta_0 < \theta_1$ is shown in the following lemma.

Lemma 22.3.5 *For $z \in (h - (i+1)\gamma, h - i\gamma]$, $i = 0, 1, \ldots, N$, we obtain*

$$
\mathcal{L}(z) = i+1 + \sum_{r=0}^{i} k^{r+1} \sum_{s=0}^{n-1} D_{i-r,n-1-s} \frac{[z - h + (i+1)\gamma]^{rn+s}}{(rn + s)!} e^{-\alpha z} \tag{22.13}
$$

PROOF. See Knoth (1997a).   ∎

The straightforward way to determine the constants D_{ij} is inserting the solution with unknown constants into the equations (22.12). Analogously we can obtain $\mathcal{L}_\theta(0)$ using (22.8) for $z = 0$. For a more feasible manner we confer to the linear equation system given in Knoth (1995).

For the above example we present in Figure 22.5 the ARL function for the cases $\theta = \theta_0$ (so–called in–control state) and $\theta = \theta_1$ (out–of–control state).

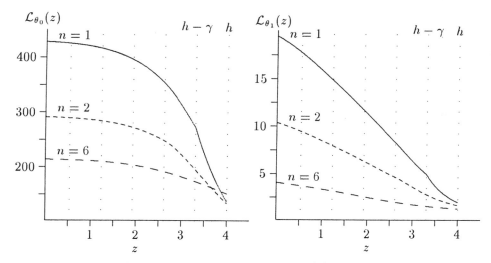

Figure 22.5: ARL function $\mathcal{L}(z)$ for $\theta_0 < \theta_1$

Looking at the case $\theta_0 > \theta_1$ we use the constants

$$E_{0j} = -\frac{\mathcal{L}(0)}{\alpha^{j+1}} + \int_0^h \frac{x^j}{j!} e^{-\alpha x} \mathcal{L}(x)\, dx\,,$$

$$E_{ij} \underset{i>0}{=} -\frac{i\, e^{-\alpha i\gamma}}{\alpha^{j+1}} + \int_{i\gamma}^h \frac{(x - i\gamma)^j}{j!} e^{-\alpha x} \mathcal{L}(x)\, dx\,.$$

Now we are able to show the solution of the integral equation for $\theta_0 > \theta_1$.

Lemma 22.3.6 *With $z \in (i\gamma, (i+1)\gamma]$, $i = 0, 1, \ldots, N$, we obtain*

$$\mathcal{L}(z) = \mathcal{L}(0) + i + 1 + \sum_{r=0}^{i} k^{r+1} \sum_{s=0}^{n-1} E_{i-r,n-1-s} \frac{[(i+1)\gamma - z]^{rn+s}}{(rn+s)!} e^\alpha \quad (22.14)$$

PROOF. See Knoth (1997a). ∎

22.4   Numerical Example

We consider the example of the previous section, i.e. $\theta \in \{1,2\}$, $h = 4$, and $n = 2$. The results obtained by solving the integral equations of Section 22.3 are compared with those computed by means of the Brook and Evans approach. In the last case we used four different matrix dimensions t. The corresponding results are summarized in Table 22.2.

Table 22.2: CUSUM characteristics for $\theta_0 < \theta_1$ and $\theta_0 > \theta_1$

method		$\mathcal{L}_{\theta_0}(0)$	$\mathcal{L}_{\theta_1}(0)$	λ_{θ_0}	$\mathcal{D}_{\theta_0}$	$\mathcal{D}_{\theta_1}$
Brook/Evans	$t = 10$	274.84	10.365	0.996281	268.92	9.196
	$t = 50$	290.54	10.373	0.996485	284.47	9.120
	$t = 100$	290.90	10.372	0.996490	284.82	9.120
section 22.3		291.00	10.371	0.996490	284.93	9.120

method		$\mathcal{L}_{\theta_0}(0)$	$\mathcal{L}_{\theta_1}(0)$	λ_{θ_0}	$\mathcal{D}_{\theta_0}$	$\mathcal{D}_{\theta_1}$
Brook/Evans	$t = 10$	454.20	7.551	0.997783	451.05	7.178
	$t = 50$	463.80	7.535	0.997829	460.67	7.163
	$t = 100$	464.09	7.535	0.997831	460.97	7.163
section 22.3		464.18	7.535	0.997831	461.07	7.163

22.5   Conclusions

The integral equation approach for Erlang distributed random variables yields explicit formulas for the CUSUM characteristics ARL, average delay, and the geometric approximation of the distribution function. Despite the complex nature of the above formulas we obtain a good insight in the CUSUM performance. The Brook and Evans (1972) approach as commonly used tool for analyzing change point methods provides simpler formulas, but requires more extensive calculations. Finally, the Erlang distribution may be used as a example with known exact CUSUM characteristics for examining approximation techniques of CUSUM schemes.

References

1. Brook, D. and Evans, D. A. (1972). An approach to the probability function of CUSUM–runlength, *Biometrika*, **59**, 539–549.

2. Dvoretzky, A., Kiefer, J. and Wolfowitz, J. (1953). Sequential decision problems for processes with continuous time parameter. Testing hypotheses, *Annals of Mathematical Statistics*, **242**, 254–264.

3. Gold, M. (1989). The geometric approximation to the CUSUM run length distribution, *Biometrika*, **76**, 725–733.

4. Harris, T. (1963). *The Theory of Branching Processes*, Berlin: Springer-Verlag.

5. Khan, R. A. (1978). Wald's approximations to the ARL in CUSUM–procedures, *Journal of Statistical Planning and Inference*, **2**, 63–77.

6. Knoth, S. (1995). Quasistationäre CUSUM-Verfahren bei Erlangverteilung. *Ph.D. thesis*, TU Chemnitz-Zwickau, Germany.

7. Knoth, S.(1997a). Exact average run lengths of CUSUM schemes for Erlang distributions *Technical Report*, **73**, Europe–University Viadrina, Frankfurt (Oder), Germany.

8. Knoth, S. (1997b). Quasi–stationarity of CUSUM schemes for Erlang distributions, *Technical Report*, **82**, Europe–University Viadrina, Frankfurt (Oder), Germany.

9. Kohlruss, D. (1994). Exact formulas for the OC and the ASN functions of the SPRT for Erlang distributions, *Sequential Analysis*, **13**, 53–62.

10. Lorden, G. (1971). Procedures for reacting to a change in distribution, *Annals of Mathematical Statistics*, **42**, 1897–1908.

11. Lorden, G. and Eisenberger, I. (1973). Detection of failure rate increases, *Technometrics*, **15**, 167–175.

12. Lucas, J. M. (1985). Counted data CUSUM's, *Technometrics*, **27**, 129–144.

13. Madsen, R. W. and Conn, P. S. (1973). Ergodic behaviour for nonnegative kernels, *Annals of Probability*, **1**, 995–1013.

14. Mevorach, Y. and Pollak, M. (1991). A small sample size comparison of the CUSUM and Shiryaev–Roberts approaches to change point detection, *The American Journal of Mathematics and Management Science*, **11**, 277–298.

15. Moustakides, G. V. (1986). Optimal stopping times for detecting changes in distributions, *The Annals of Statistics*, **14**, 1379–1387.

16. Page, E. S. (1954). Continuous inspection schemes, *Biometrika*, **41**, 100–115.

17. Pollak, M. (1985). Optimal detection of a change in distribution, *Technometrics*, **13**, 206–227.

18. Pollak, M. and Siegmund, D. (1985). A diffusion process and its applications to detection a change in the drift of Brownian motion, *Biometrika*, **72**, 267–280.

19. Pollak, M. and Siegmund, D. (1986). Approximations to the ARL of CUSUM–tests, *Technical Report*, Department of Statistics, Stanford University.

20. Raghavachari, M. (1965). Operating characteristic function and expected sample size of a Sequential Probability Ratio Test for the simple exponential distribution, *Bulletin of the Calcutta Statistics Association*, **14**, 65–73.

21. Regula, G. (1975). Optimal CUSUM procedure to detect a change in distribution for the gamma family, *Ph.D. thesis*, Case Western University, Cleveland, OH.

22. Riordan, G. (1982). *Kombinatornye Toždestva*, Moskva: Nauka.

23. Ritov, Y. (1990). Decision theoretic optimality of the CUSUM procedure, *The Annals of Statistics*, **18**, 1464–1469.

24. Roberts, S. W. (1966). A comparison of some control chart procedures, *Technometrics*, **8**, 411–430.

25. Tartakovskij, A. G. and Ivanova, I. A. (1992). Sravnenie nekotorych posledovatel'nych pravil obnaruženija razladki, *Probl. Peredači Inf.*, **28**, 21–29.

26. Vardeman, S. and Ray, D. (1985). Average run lengths for CUSUM schemes when observations are exponentially distributed, *Technometrics*, **27**, 145–150.

27. Waldmann, K.-H. (1986). Bounds for the distribution of the run length of one–sided and two–sided CUSUM quality control schemes, *Technometrics*, **28**, 61–67.

28. Wang, A.-L. (1990). Exponential CUSUM scheme for detecting a decrease in mean, *Sankhyā*, **52**, 105–114.

29. Woodall, W. H. (1983). The distribution of the run length of one–sided CUSUM procedures for continuous random variables, *Technometrics*, **24**, 295–301.

23

On the Average Delay of Control Schemes

H. G. Kramer and W. Schmid

Europe–University Viadrina, Frankfurt (Oder), Germany

Abstract: In this paper we consider control charts for time series. Several upper bounds are derived for the average delay of a generalized Shewhart chart. Furthermore, it is proved that for a Gaussian process the average delay is greater than or equal to the average delay for independent variables.

In an extensive computer study several EWMA and CUSUM charts are compared with each other. As measures for the performance the average run length and the maximal average delay are used. The target process is assumed to be an autoregressive process of order 1 and the out-of-control state is described by a change point model. It turns out that for both performance criteria the same ranking between the charts is obtained.

Keywords and phrases: Statistical process control, average delay, control charts, time series

23.1 Introduction

In the last years many authors have treated the problem of statistical process control for dependent variables. It was shown that the classical control schemes cannot be applied directly in the context of time series [e.g. Montgomery and Mastrangelo (1991), Maragah and Woodall (1992)]. Thus, it is necessary to model the underlying time series structure. Alwan and Roberts (1988) and Harris and Ross (1991) proposed an elegant procedure for such a situation. Since the residuals are independent variables for many processes (like ARMA models) all of the well-known control schemes like Shewhart, EWMA, and CUSUM procedures can be applied to the residuals. Another method was already discussed by Vasilopoulos and Stamboulis (1978). They showed that the classical Shewhart chart can easily be extended to time series. A modified EWMA chart was introduced in Schmid (1997a) while CUSUM schemes for time series were proposed by Yashchin (1993) and Schmid (1997b). An extensive comparison of several control schemes was given in Schmid (1997b). He assumed that the target process is an autoregressive process of order 1 (AR(1)). As a measure for the performance the average run length (ARL) is used.

Indeed, for independent processes the ARL of the Shewhart and the CUSUM chart is independent of the time where the process changes its distribution [cf. Siegmund (1985)]. Therefore it is usually assumed that the change is at the first position. The definition of the ARL is based on this assumption. For dependent processes however, this property is no longer valid. The process may concluded to be out-of-control although no change has arisen. Thus the use of the ARL for time series is questionable.

Here we consider criteria which are based on the average time between the occurrence of a change and its detection. We make use of a change point model. The observed process $\{X_t\}$ and the target process $\{Y_t\}$ are defined according to

$$X_t = Y_t + a\mathbf{1}_{\{q,q+1,\ldots\}}(t) \tag{23.1}$$

with $a \in \mathbb{R}$ and $q \in \mathbb{N}$. If $a \neq 0$ then it is said that a change point arises at position q with size a. After a change the process is called to be out-of-control, else in-control.

Other performance criteria are based on the average delay of a stopping time N which is defined according to

$$D(N; a, q) \quad = \quad \mathrm{E}_{a,q}(N - q + 1 | N \geq q)$$

provided that $E_{a,q}(N) < \infty$. Here $\mathrm{E}_{a,q}$ denotes the expectation with respect to the change point model (1). Moreover, E_0 means that there is no deviation, i.e. that $a = 0$. Roberts (1966) proposed the following performance criterion

$$D_{\mathrm{lim}}(N; a) \quad = \quad \lim_{q \to \infty} D(N; a, q).$$

For independent samples this criterion was considered by Pollak and Siegmund (1985) in the context of optimal stopping rules if the distributions of the in- and out-of-control process are known. Also in the case of estimated shifts Pollak and Siegmund (1991) considered D_{lim} as $\mathrm{E}_0(N) \to \infty$. Gordon and Pollak (1994) introduced a nonparametric method for detecting a shift and used the limit of the average delay as a quality criterion.

However, D_{lim} is not a worst-case criterion. Therefore, we are in favor of the following criterion proposed by Pollak and Siegmund (1975)

$$D_{\mathrm{sup}}(N; a) \quad = \quad \sup_{q \geq 1} D(N; a, q).$$

For the Shewhart chart and the CUSUM chart all the proposed criteria are equal to the average run length if the target process is independent. For EWMA charts this result is not valid. In 1985, Pollak proved that a control scheme proposed by Shiryayev (1963) and Roberts (1966) is optimal with respect to D_{sup} as $E_0(N) \to \infty$. Srivastava and Wu (1993) and Wu (1996) use D_{sup} as they show that the EWMA procedure is slightly less efficient than the CUSUM or the Shiryayev-Roberts procedure as $E_0(N) \to \infty$. But for fixed $E_0(N)$ Lucas and Saccucci (1990) show that there is not much difference between the EWMA and CUSUM procedure.

Note that all the authors mentioned above assume the target processes to be independent. They focused on certain notions for the optimality of control schemes and compared those schemes by means of the proposed quality criteria. Mainly the asymptotic behaviour of the considered quality criteria was of interest as $E_0(N) \to \infty$. For practical purposes, however, the properties of a criterion for given $E_0(N) < \infty$ are of importance.

In Section 23.2 we give several inequalities for the average delay of a Shewhart scheme for time series. It is shown that in the out-of control case the average delay is less then or equal to the in-control average delay (Theorem 23.2.1). Moreover, for autocorrelated processes the in-control average delay is proved to be greater than or equal to the in-control average delay for independent variables (Theorem 23.2.2). Upper bounds for the in-control average delay are given in Theorems 23.2.3 and 23.2.4. The behaviour for exchangeable variables is discussed in Theorem 23.2.5. The results of an extensive computer study are presented in Section 23.3. We compare several EWMA and CUSUM procedures for dependent observations. As measures for the performance the ARL and D_{sup} are used. The target process is assumed to be an autoregressive process of order 1. It turns out that for both performance criteria the same ranking between the charts is obtained. This result justifies the use of the ARL for time series. Since the practical calculation of the maximal average delay is more time consuming the ARL must be favoured.

23.2 On the Average Delay of a Generalized Shewhart Chart

Schmid (1995) proved several results about the average run length of the modified Shewhart chart. Here similar results are derived for the average delay and a more general control scheme.

The average delay can be calculated as follows

$$D(N; a, q) = \frac{\sum_{k=q-1}^{\infty} P_{a,q}\big(N > k\big)}{P_{a,q}\big(N \geq q\big)} = \sum_{k=q-1}^{\infty} P_{a,q}\big(N > k \big| N \geq q\big) \quad (23.2)$$

for all a and q with $P_{a,q}(N \geq q) > 0$. Especially,

$$D(N;0,q) = \frac{E_0(N) - \sum_{k=0}^{q-2} P_0(N > k)}{P_0(N \geq q)}. \qquad (23.3)$$

Note that the average delay in the in-control state is a function of q although no change is present. However, as shown in Section 23.2.1 (cf. Table 23.1), the influence of q is small if for instance the target process is an AR(1) process.

For the modified Shewhart chart [cf. Schmid (1995)] it is concluded that the process is in the out-of-control state at time t if $|X_t - \mu_0| > c\sqrt{\gamma_0}$. Here $c > 0$ denotes a constant. μ_0 and $\gamma_0 > 0$ are the expectation and the variance of the target process $\{Y_t\}$. Usually μ_0 is called the target value. Thus it is always assumed that the moments of the target process do not change over time. The following results however are also valid if the expectation as well as the variance are a function of t. Let $\mu_t = E(Y_t)$ and $\gamma_t = \text{Var}(Y_t) > 0$ be known quantities. Then the run length of this Shewhart type scheme is given by

$$N_s(c) = \inf\{t \in \mathbb{N} : |X_t - \mu_t| > c\sqrt{\gamma_t}\}.$$

According to Schmid (1995, p. 115) we obtain that $D(N_s; a, q) = D(N_s; -a, q)$ if the random vectors $(Y_q - \mu_q, \ldots, Y_k - \mu_k)$ and $-(Y_q - \mu_q, \ldots, Y_k - \mu_k)$ have the same distribution for all $k \geq q$. Furthermore the Corollary to Theorem 4.1 of Schmid (1995) can easily be extended to the present situation. The same argumentation leads to the result that for an AR(1) process with coefficient α_1 the delay $D(N_s; 0, q)$ is symmetric in α_1.

23.2.1 Bounds for the average delay

Theorem 23.2.1 states that for a large family of stochastic processes which are of practical interest the average delay for $|a| > 0$ is always less than or equal to the average delay for $a = 0$.

Theorem 23.2.1 *Let* $\mathbf{Y} = (Y_1, \ldots, Y_k)'$ *be a k-dimensional random variable with mean vector* $\boldsymbol{\mu}$. *If* $\mathbf{Y} - \boldsymbol{\mu}$ *has a continuous density f which is symmetric about the origin and A-unimodal [Tong (1990, p. 71)], then we obtain that for all a and q*

$$P_{a,q}(N_s > k | N_s \geq q) \leq P_0(N_s > k | N_s \geq q).$$

Moreover, $P_{a,q}(N_s > k | N_s \geq q)$ *is a nonincreasing function in* $|a|$ *for fixed q.*

PROOF.   Analogously to Schmid (1995, proof of Theorem 2.3) it follows that for $|a_1| \geq |a_2|$

$$P_{a_1,q}(N_s > k) \leq P_{a_2,q}(N_s > k).$$

Since $P_{a,q}(N_s \geq q) = P_0(N_s \geq q)$ we readily obtain the desired result. ∎

Assuming that the conditions of Theorem 23.2.1 are satisfied for all k we obtain that $D(N_s; a, q) \leq D(N_s; 0, q)$ and that $D(N_s; a, q)$ is a nonincreasing function in $|a|$ for fixed q.

It is important to note that Theorem 23.2.1 is still valid for the more general relationship between the observed process and the target process

$$X_t \;=\; Y_t + a f_t(q) \mathbf{1}_{\{q, q+1, \ldots\}}(t).$$

Here $f_t(q)$ denotes a known function. By setting $f_t(q) \equiv 1$ the change point model is obtained and by $f_t(q) = t - q + 1$ a linear drift can be modelled. The extension of Theorem 23.2.1 to this case is obvious.

For the rest of this section only the in-control average delay is considered. At first a lower bound is derived. Theorem 23.2.2 can be regarded as the main result of this paper. In Corollary 23.2.1 it is proved that for a Gaussian process the in-control average delay is always greater than or equal to the in-control average delay of an independent random process. The superscript "iid" means that independent and identically distributed random variables are considered.

Theorem 23.2.2 *Let* $\mathbf{Y} = (Y_1, \ldots, Y_k)'$ *be a* k-*dimensional normal random variable. Then*

$$P_0\left(N_s > k \middle| N_s \geq q\right) \;\geq\; P_0^{iid}(N_s > k \middle| N_s \geq q) = \left(\Phi(c) - \Phi(-c)\right)^{k-q+1}$$

for $k \geq q - 1$.

PROOF. First we observe that

$$P_0\left(N_s > k \middle| N_s \geq q\right)$$
$$= P_0\left(|Y_i - \mu_i| \leq c\sqrt{\gamma_i} \, \forall i = 1, \ldots, k \, \middle\| Y_j - \mu_j| \leq c\sqrt{\gamma_j} \, \forall j = 1, \ldots, q-1\right)$$
$$= \frac{P_0\left(|Y_i - \mu_i| \leq c\sqrt{\gamma_i} \, \forall i = 1, \ldots, k\right)}{P_0\left(|Y_i - \mu_i| \leq c\sqrt{\gamma_i} \, \forall i = 1, \ldots, q-1\right)}.$$

By applying the Corollary to Theorem 2.2.2 from Tong (1980) successively we obtain

$$P_0\left(|Y_i - \mu_i| \leq c\sqrt{\gamma_i} \, \forall i = 1, \ldots, k\right)$$
$$\geq \prod_{i=q}^{k} P_0\left(|Y_i - \mu_i| \leq c\sqrt{\gamma_i}\right) \cdot P_0\left(|Y_i - \mu_i| \leq c\sqrt{\gamma_i} \, \forall i = 1, \ldots, q-1\right).$$

Therefore

$$P_0\Big(N_s > k\Big|N_s \geq q\Big) \;\geq\; \prod_{i=q}^{k} P_0\Big(|Y_i - \mu_i| \leq c\sqrt{\gamma_i}\Big)$$

$$= \;\Big(\Phi(c) - \Phi(-c)\Big)^{k-q+1} = P_0^{iid}\Big(N_s > k\Big|N_s \geq q\Big).$$

∎

Corollary 23.2.1 *Let $\{Y_t\}$ be a Gaussian process then*

$$D(N_s;0,q) \;\geq\; D^{iid}(N_s;0,q) \;=\; \mathrm{E}_0^{iid}(N_s).$$

PROOF.   Follows immediately from Theorem 23.2.2 and equation (23.2).   ∎

In Corollary 23.2.1 $\{Y_t\}$ is not demanded to be stationary. Especially, the corollary holds for causal ARMA processes with normally distributed white noise. Here causality means that the process is obtained from the white noise by application of a causal linear filter [cf. Brockwell and Davis (1987, p. 83)]. Therefore, Corollary 23.2.1 can be applied to many processes of practical interest.

The question arises whether the average delay is considerably greater for dependent processes and whether there are big differences to the ARL. For a stationary AR(1) process with coefficient α_1 Table 23.1 shows values of the average delay. The values were computed by a modified Markov chain approach [Brook and Evans (1972)] which was modified. A detailed description can be found in Kramer (1997). The value c is taken such that the in-control ARL in the i.i.d. case would achieve 500, i.e. $c = 3.09023$.

Table 23.1: $D(N_s;0,q)$ of the Shewhart chart ($c= 3.09023$, $\mu_i=\mu_0=0$) determined via Markov chain approach (401 subintervals) for AR(1) processes with different α_1

α_1	0.0	0.1	0.2	0.3	0.4	0.5	0.6	0.7	0.8	0.9
$q = 1$	500.00	500.56	502.56	506.93	515.44	531.17	560.00	614.90	732.85	1088.01
$q = 2$	500.00	500.56	502.57	506.95	515.47	531.24	560.12	615.13	733.32	1089.18
$q = 5$	500.00	500.56	502.57	506.95	515.48	531.25	560.14	615.18	733.45	1089.59
$q = 10$	500.00	500.56	502.57	506.95	515.48	531.25	560.14	615.19	733.48	1089.73
$q = 20$	500.00	500.56	502.57	506.95	515.48	531.25	560.14	615.19	733.48	1089.77

Table 23.1 gives as a first impression that the average delay is quite stable for $|\alpha_1| \leq 0.5$. The differences of the in-control average delay are diminishingly small for different values of q. Only for large values of α_1 the differences are remarkable. Noting that for $q = 1$ the in-control ARL equals $D(N_s;0,1)$ we find that the in-control average delay is only slightly greater than the in-control ARL. This is an important result since the computation of the average delay is

complicate and rather elaborate, however, especially for computing the critical values c. Table 23.1 however indicates that the differences between the ARL and the average delay are negligibly small. Therefore in our simulations we have chosen the critical value c such that the in-control ARL achieves a certain value even if the average delay is considered. And indeed, the error of this procedure was unnoticeably small in every case.

Eventually, Table 23.1 gives rise to the conclusion that the average delay seems to be monotonically increasing in q. For exchangeable variable we are able to prove this result in Theorem 23.2.5. We believe that this is also true for AR(1) processes but the proof of this conjecture is still open.

Our next aim is to derive an upper bound for the in-control average delay. The first bound (equation (23.4)) is valid for arbitrary normal random variables whereas the second one (equations (23.5) and (23.6)) requires the process $\{Y_t\}$ to be an AR(1) process. If the in-control ARL is known the upper bounds are easily computed. The disadvantage of the upper bounds is that they are increasing without bound in q whereas the average delay is bounded. However, as shown in Table 23.2 the bounds are quite accurate for a wide range of q.

Theorem 23.2.3 *Let $\{Y_t\}$ be a Gaussian process then*

$$D(N_s; 0, q)$$

$$\leq \frac{E_0(N_s)}{\Big(\Phi(c) - \Phi(-c)\Big)^{q-1}} - \frac{1}{\Big(\Phi(c) - \Phi(-c)\Big)^{q-1}} \frac{1 - \Big(\Phi(c) - \Phi(-c)\Big)^{q-1}}{1 - \Big(\Phi(c) - \Phi(-c)\Big)}.$$

$$(23.4)$$

If moreover $\{Y_t\}$ is a (weakly) stationary AR(1) process then

$$D(N_s; 0, q) \leq \frac{E_0(N_s)}{\zeta^{\frac{q-1}{2}}} - \frac{1 + \Phi(c) - \Phi(-c)}{\zeta^{\frac{q-1}{2}}} \frac{1 - \zeta^{\frac{q-1}{2}}}{1 - \zeta} \quad \textit{for } q \textit{ odd}$$

$$(23.5)$$

and

$$D(N_s; 0, q) \leq \frac{E_0(N_s)}{\Big(\Phi(c) - \Phi(-c)\Big)\zeta^{\frac{q-2}{2}}} - \frac{1}{\Big(\Phi(c) - \Phi(-c)\Big)}$$

$$- \frac{1 + (\Phi(c) - \Phi(-c))}{\Big(\Phi(c) - \Phi(-c)\Big)\zeta^{\frac{q-2}{2}}} \frac{1 - \zeta^{\frac{q-2}{2}}}{1 - \zeta} \quad \textit{for } q \textit{ even}$$

$$(23.6)$$

where $\zeta = P_0\Big(|Y_1 - \mu_0| \leq c\sqrt{\gamma_0} \cap |Y_2 - \mu_0| \leq c\sqrt{\gamma_0}\Big)$.

PROOF. By equation (23.3) we obtain

$$D(N_s; 0, q) = \frac{E_0(N_s) - \sum_{k=1}^{q-1} P_0\left(|Y_i - \mu_i| \le c\sqrt{\gamma_i} \; \forall i = 1, ..., k-1\right)}{P_0\left(|Y_i - \mu_i| \le c\sqrt{\gamma_i} \; \forall i = 1, ..., q-1\right)}$$

$$\le \frac{E_0(N_s) - \sum_{k=1}^{q-1} \prod_{i=1}^{k-1} P_0\left(|Y_i - \mu_i| \le c\sqrt{\gamma_i}\right)}{\prod_{i=1}^{q-1} P_0\left(|Y_i - \mu_i| \le c\sqrt{\gamma_i}\right)}$$

$$= \frac{E_0(N_s)}{\left(\Phi(c) - \Phi(-c)\right)^{q-1}}$$

$$- \frac{1}{\left(\Phi(c) - \Phi(-c)\right)^{q-1}} \frac{1 - \left(\Phi(c) - \Phi(-c)\right)^{q-1}}{1 - \left(\Phi(c) - \Phi(-c)\right)}.$$

by Theorem 5.1.2 of Tong (1990). Thus, the first part of the theorem is proved.

For an AR(1) process we have $\mu_i = \mu_0$ and $\gamma_i = \gamma_0 = \sigma^2/(1 - \alpha_1^2)$. Since $\{Y_t\}$ is also strictly stationary this bound can be improved by considering $\zeta = P_0\left(|Y_i - \mu_0| \le c\sqrt{\gamma_0} \cap |Y_{i+1} - \mu_0| \le c\sqrt{\gamma_0}\right)$ instead of $P_0\left(|Y_i - \mu_0| \le c\sqrt{\gamma_0}\right)$. Therefore we split

$$D(N_s; 0, q) = \left(P_0\left(|Y_i - \mu_0| \le c\sqrt{\gamma_0} \; \forall i = 1, ..., q-1\right)\right)^{-1}$$

$$\left(E_0(N_s) - \sum_{\substack{k=1 \\ k \text{ odd}}}^{q-1} P_0\left(|Y_i - \mu_0| \le c\sqrt{\gamma_0} \; \forall i = 1, ..., k-1\right)\right.$$

$$\left. - \sum_{\substack{k=1 \\ k \text{ even}}}^{q-1} P_0\left(|Y_i - \mu_0| \le c\sqrt{\gamma_0} \; \forall i = 1, ..., k-1\right)\right).$$

Note that $\left(\frac{Y_1 - \mu_0}{\sqrt{\gamma_0}}, ..., \frac{Y_{k-1} - \mu_0}{\sqrt{\gamma_0}}\right)'$ is a $(k-1)$-dimensional normal random variable with mean 0 and covariance matrix $\left(\alpha_1^{|i-j|}\right)_{i,j=1,...,k-1}$. The inverse of $\left(\alpha_1^{|i-j|}\right)_{i,j=1,...,k-1}$ is an M-matrix [cf. Tong (1990)]. An explicit formula for the inverse of the covariance matrix can be found in Graybill (1969, p. 182). Thus the inverses of

$$\text{diag}\left(\begin{pmatrix} 1 & \alpha_1 \\ \alpha_1 & 1 \end{pmatrix}, ..., \begin{pmatrix} 1 & \alpha_1 \\ \alpha_1 & 1 \end{pmatrix}\right) \qquad \text{and}$$

$$\text{diag}\left(1, \begin{pmatrix} 1 & \alpha_1 \\ \alpha_1 & 1 \end{pmatrix}, ..., \begin{pmatrix} 1 & \alpha_1 \\ \alpha_1 & 1 \end{pmatrix}\right)$$

are M-matrices. Hence, the assumptions of Theorem 5.1.6 in Tong (1990) are fulfilled and we obtain for q odd

$$D(N_s; 0, q) \leq \left(\prod_{\substack{i=1 \\ i \text{ odd}}}^{q-2} P_0\left(|Y_i - \mu_0| \leq c\sqrt{\gamma_0} \cap |Y_{i+1} - \mu_0| \leq c\sqrt{\gamma_0} \right) \right)^{-1}$$

$$\left(E_0(N_s) - \sum_{\substack{k=1 \\ k \text{ odd}}}^{q-1} \prod_{\substack{i=1 \\ i \text{ odd}}}^{k-2} P_0\left(|Y_i - \mu_0| \leq c\sqrt{\gamma_0} \cap |Y_{i+1} - \mu_0| \leq c\sqrt{\gamma_0} \right) \right.$$

$$- \sum_{\substack{k=1 \\ k \text{ even}}}^{q-1} P_0\left(|Y_1 - \mu_0| \leq c\sqrt{\gamma_0} \right) \times$$

$$\left. \times \prod_{\substack{i=1 \\ i \text{ even}}}^{k-2} P_0\left(|Y_i - \mu_0| \leq c\sqrt{\gamma_0} \cap |Y_{i+1} - \mu_0| \leq c\sqrt{\gamma_0} \right) \right)$$

$$= \zeta^{-\frac{q-1}{2}} \left(E_0(N_s) - \sum_{\substack{k=1 \\ k \text{ odd}}}^{q-1} \zeta^{\frac{k-1}{2}} - \sum_{\substack{k=1 \\ k \text{ even}}}^{q-1} \left(\Phi(c) - \Phi(-c) \right) \zeta^{\frac{k-2}{2}} \right)$$

$$= \frac{E_0(N_s)}{\zeta^{\frac{q-1}{2}}} - \frac{1 + \Phi(c) - \Phi(-c)}{\zeta^{\frac{q-1}{2}}} \frac{1 - \zeta^{\frac{q-1}{2}}}{1 - \zeta} \qquad \text{for } q \text{ odd.}$$

The same reasoning yields equation (23.6). ∎

Table 23.2: Upper bounds for $D(N_s; 0, q)$ for AR(1) processes according to inequalities (23.5) and (23.6) and inequality (23.4) ($c = 3.09023$)

	q	1	2	3	4	5	10	100	200
$\alpha_1 = 0.3$	$D(N_s; 0, q)$	506.934	506.948	506.949	506.949	506.949	506.949	506.949	506.949
	(5)&(6)	506.934	506.948	506.949	506.963	506.964	507.007	507.738	508.718
	(4)	506.934	506.948	506.948	506.962	506.990	507.061	508.455	510.328
$\alpha_1 = 0.6$	$D(N_s; 0, q)$	560.000	560.120	560.136	560.141	560.142	560.143	560.143	560.143
	(5)&(6)	560.000	560.120	560.136	560.257	560.273	560.669	567.455	576.497
	(4)	560.000	560.120	560.241	560.361	560.482	561.091	573.152	589.366

In Theorem 23.2.4 we derive a uniform upper bound for $P_0(N_s > k | N_s \geq q)$ for Gaussian processes. Let $\Gamma_k = (Corr(Y_i, Y_j))_{i,j=1,\ldots,k}$ and $\delta_k^2 = \min_{q \leq i \leq k} \frac{|\Gamma_i|}{|\Gamma_{i-1}|}$ and $|\Gamma_0| := 1$.

Theorem 23.2.4 *Let* **Y** *be a k-dimensional normal random variable and* Γ_k *be nonsingular. Then for* $k \geq q - 1$

$$P_0\left(N_s > k \Big| N_s \geq q \right) \leq \left(1 - \left(\Phi\left(\frac{c}{\delta_k} \right) - \Phi\left(-\frac{c}{\delta_k} \right) \right) \right)^{k-q+1}. \qquad (23.7)$$

PROOF. First we apply a direct extension of Theorem 3.2 of Das Gupta et al. (1972). We obtain for $k \geq q - 1$

$$P_0(N_s > k) \leq P_0\left(|Y_1 - \mu_1| \leq c\sqrt{\gamma_1}, ..., |Y_{q-1} - \mu_{q-1}| \leq c\sqrt{\gamma_{q-1}}\right)$$

$$\times \prod_{i=q}^{k} P_0\left(|Y_i - \mu_i| \leq \frac{c\sqrt{\gamma_i}}{\delta_k}\right).$$

A detailed proof can be found in Kramer (1997, Lemma 4.1). Therefore

$$P_0\left(N_s > k \,\middle|\, N_s \geq q\right) \leq \prod_{i=q}^{k} P_0\left(|Y_i - \mu_i| \leq \frac{c\sqrt{\gamma_i}}{\delta_k}\right)$$

$$\leq \prod_{i=q}^{k} P_0\left(|Y_i - \mu_i| \leq \frac{c\sqrt{\gamma_i}}{\delta_k}\right)$$

$$= \left(\Phi\left(\frac{c}{\delta_k}\right) - \Phi\left(-\frac{c}{\delta_k}\right)\right)^{k-q+1}.$$

∎

Thus, it follows for a Gaussian process satisfying the conditions of Theorem 23.2.4

$$D_{\sup}(N_s(c); 0) \leq \left(1 - \left(\Phi\left(\frac{c}{\delta}\right) - \Phi\left(-\frac{c}{\delta}\right)\right)\right)^{-1} = D_{\sup}^{iid}\left(N_s\left(\frac{c}{\delta}\right); 0\right)$$

(23.8)

with $\delta := \inf\{\delta_k : k \geq 1\}$.

For the case of independent and identically distributed random variables $D_{\sup}(N(c); 0)$ equals the upper bound in Theorem 23.2.4 since $\delta = 1$. For a stationary process $\{Y_t\}$ with $E(Y_t) = \mu_0$ and autocovariance function converging to 0 as the lag tends to infinity. Schmid (1995) proved that $\{\delta_\nu\}$ is nonincreasing and $\lim_{\nu \to \infty} \delta_\nu^2 = \sigma^2$ where σ^2 is the variance of the prediction error for the best linear one-step predictor. Thus, we may choose $\delta = \sigma$ in Theorem 23.2.4.

23.2.2   The average delay for exchangeable variables

Up to now explicit expressions for the ARL of control charts for dependent variables are known for exchangeable normal variables only [cf. Schmid (1995)]. For a Gaussian process $\{Y_t\}$ with $E(Y_t) = \mu_0$, $\mathrm{Var}(Y_t) = \gamma_0 \in (0, \infty)$ and $\mathrm{Corr}(Y_t, Y_s) = \varrho \in [0, 1)$ for all $t \neq s$, Kramer (1997) proved that

$$D(N_s; a, q) = \frac{\displaystyle\int_{-\infty}^{\infty} \frac{\zeta_0(z, \varrho)^{q-1}}{1 - \zeta_a(z, \varrho)} \phi(z)\, dz}{\displaystyle\int_{-\infty}^{\infty} \zeta_0(z, \varrho)^{q-1} \phi(z)\, dz}$$

with

$$\zeta_a(z, \varrho) = \Phi\left(\frac{c - \frac{a}{\sqrt{\gamma_0}} - \sqrt{\varrho}z}{\sqrt{1 - \varrho}}\right) - \Phi\left(\frac{-c - \frac{a}{\sqrt{\gamma_0}} - \sqrt{\varrho}z}{\sqrt{1 - \varrho}}\right)$$

where ϕ denotes the density of the standard normal distribution.

In Theorem 23.2.5 it is shown that for exchangeable variables the in-control average delay is monotonically increasing in q. Note that this result is valid without any assumption on the distribution. However, it does not hold in the out-of-control situation.

Theorem 23.2.5 *Let $\{(Y_t - \mu_t)/\gamma_t, t \in I\!\!N\}$ be an infinite sequence of exchangeable random variable where $E(Y_t) = \mu_t$ and $Var(Y_t) = \gamma_t \in (0, \infty)$. Then $D(N_s; 0, q)$ is monotonically increasing in q.*

PROOF. According to Theorem 5.3.2. of Tong (1980) the joint distribution of $Y_1, ..., Y_n$ can be written as a mixture with a common marginal distribution. Thus, Theorem 5.3.3 of Tong (1980) can be applied to

$$\beta(r) = P_0\left(\bigcap_{i=1}^{r}\{|Y_i - \mu_i| \leq c\sqrt{\gamma_i}\}\right) = P_0\left(N_s > r\right).$$

By setting $r = q + \nu + 1$ and $k = 2q + \nu + 2$ (note that $r \geq k/2$) in Theorem 5.3.3 of Tong (1980) we obtain (note that $r \geq k/2$)

$$P_0\left(N_s > q + \nu - 1\right)P_0\left(N_s \geq q\right) \geq P_0\left(N_s > q + \nu - 2\right)P_0\left(N_s \geq q + 1\right)$$

or

$$\frac{P_0\left(N_s > q + \nu\right)}{P_0\left(N_s \geq q + 1\right)} \geq \frac{P\left(N_s > q + \nu - 1\right)}{P_0\left(N_s \geq q\right)}$$

for all $\nu \in I\!\!N$. Therefore

$$D(N_s; 0, q + 1) = 1 + \frac{\displaystyle\sum_{k=q+1}^{\infty} P_0\left(N_s > k\right)}{P_0\left(N_s \geq q + 1\right)}$$

$$\geq 1 + \frac{\displaystyle\sum_{k=q}^{\infty} P_0\left(N_s > k\right)}{P_0\left(N_s \geq q\right)}$$

$$= D(N_s; 0, q).$$

Hence, $D(N_s; 0, q)$ is monotonically increasing in q. ∎

23.3   A Comparison of Several Control Charts

Schmid (1997b) analysed several EWMA and CUSUM charts for time series. As a measure for the performance he used the average run length. In our simulation study we have additionally compared the charts using the average delay. Furthermore, we have made several improvements in the definitions of the charts. The residual charts have been defined in a way allowing to detect a shift already at the first position. Furthermore, the definition of the modified CUSUM chart is based on the variance of the process which allows the reference value k in most cases to be set to the half of the expected shift just as in the independent case. Additionally different values of α_1 and more values of the EWMA parameter r are used here. In the out-of-control case shifts in relation to the root of the variance are considered.

Again, the target process is assumed to be a causal Gaussian AR(1) process with $|\alpha_1| < 1$, and $\sigma = 1$. Thus $\mathrm{Var}(Y_t) = 1/(1 - \alpha_1^2)$. The EWMA charts used in this study are the modified EWMA chart (EWMAmod) introduced by Schmid (1997a) and the EWMA residual chart (EWMAres) defined in Kramer and Schmid (1997) for multivariate observations.

Yashchin (1993) and Schmid (1997b) generalized the CUSUM procedure to dependent target processes. Schmid (1997b) introduced a recursive procedure for the calculation of the exact sequential probability ratio. The corresponding scheme is denoted as modified CUSUM chart (CUSUMmod). The CUSUM residual chart (CUSUMres) is based on the residuals $\Delta_t = X_t - \alpha_1 X_{t-1}$ for $t \geq 2$ of the AR(1) process. As motivated in Kramer and Schmid (1997), it is natural to set $\Delta_1 = \sqrt{1 - \alpha_1^2} X_1$. Thus, run lengths of less than 2 can be obtained since shifts can be detected at the beginning. Yashchin (1993) proposed the classical CUSUM control chart for independent observations to be also used in the case of dependent observations. The control statistic of the classical CUSUM chart is directly applied to dependent observations but the control values h and the reference values k have to be appropriately adapted. This is the third CUSUM chart to be considered in this study.

First, the critical values c for the EWMA charts as well as the values of h for the different CUSUM charts were determined via simulations using an iteration algorithm. The results of our simulations are based on 100 000 realizations. In Tables 23.6 and 23.7 the range of q has to be restricted in order to compute D_{sup}, here to $1 \leq q \leq 20$ (Beyond $q = 20$ $D_{a,q}$ has already approached D_{lim} within an acceptable accuracy).

In the out-of-control state the supremum of the average delay D_{sup} is a worst case criterion and is required to be small, whereas for $a = 0$ this would be the best case to consider since here the value of the quality criterion should be as large as possible. Another reason for not using D_{sup} in the in-control state is the definition of the average delay. Depending on q this criterion only makes sense for a point of time q where a shift of size $a \neq 0$ actually occurs. Therefore the critical values are to be set such that a specific in-control average run length is obtained. This procedure is also recommended by Siegmund (1985) who denotes the in-control case by $q = \infty$.

Therefore for the study of the in-control case the critical values c and h respectively have been determined such that an in-control ARL of 500 was achieved. Then for shifts $a/\sqrt{\gamma_0}$ from 0.5 to 4.0 both criteria, the ARL and D_{sup}, were compared with each other. In both cases the "minimal" ARL and D_{sup} respectively have been computed from the considered values of the smoothing parameter $r \in \{0.01, 0.025, 0.05, 0.1, 0.2, 0.4, 0.6, 0.8, 1.0\}$ for the EWMA charts, the reference value $k \in \{0.0, 0.5, 1.0, 1.5, 2.0, 2.5, 3.0, 3.5, 4.0\}$ for the modified CUSUM chart and the classical CUSUM chart with modified limits, and $k\sqrt{1 - \alpha_1^2} \in \{0.0, 0.5, 1.0, 1.5, 2.0, 2.5, 3.0, 3.5, 4.0\}$ for the CUSUM residual chart.

For the modified EWMA chart and a specified shift $a/\sqrt{\gamma_0}$ the out-of-control ARL increases as α_1 increases. The smaller α_1 the more sensibly the EWMA chart reacts towards the choice of r (cf. Tables 23.3 and 23.4). Although D_{sup} is greater than or equal to the ARL, the differences are small. Thus, D_{sup} shows the same behaviour as the ARL and almost always the same r has to be taken to achieve the "minimal" value of the quality criterion.

Table 23.3: Out-of-control ARL of the modified EWMA chart for $\alpha_1 = -0.8$

		$\frac{a}{\sqrt{\gamma_0}}$ 0.5	1.0	1.5	2.0	2.5	3.0	3.5	4.0
$k=0.000$	($K=0.000, h=11.132$)	22.148	11.514	7.902	6.041	5.010	4.136	3.730	3.202
$k=0.250$	($K=0.417, h=2.774$)	9.381	3.939	2.662	1.958	1.704	1.511	1.318	1.162
$k=0.500$	($K=0.833, h=2.467$)	32.359	4.782	2.854	2.007	1.683	1.489	1.299	1.152
$k=0.750$	($K=1.250, h=2.215$)	163.747	7.134	3.193	2.124	1.695	1.487	1.294	1.150
$k=1.000$	($K=1.667, h=1.965$)	184.361	18.558	3.796	2.279	1.718	1.486	1.298	1.151
$k=1.250$	($K=2.083, h=1.715$)	185.124	49.047	5.214	2.491	1.761	1.489	1.298	1.150
$k=1.500$	($K=2.500, h=1.465$)	185.152	54.156	9.934	2.828	1.816	1.492	1.297	1.149
$k=1.750$	($K=2.917, h=1.215$)	185.152	54.422	16.912	3.522	1.908	1.501	1.296	1.150
$k=2.000$	($K=3.333, h=.965$)	185.152	54.433	18.367	4.981	2.050	1.508	1.296	1.151
	minimal ARL	9.381	3.939	2.662	1.958	1.683	1.486	1.294	1.149
	at $k=$	0.250	0.250	0.250	0.250	0.500	1.000	0.750	1.500

Table 23.4: Out-of-control ARL of the modified EWMA chart for $\alpha_1 = 0.8$

$\frac{a}{\sqrt{\gamma_0}}$	0.5	1.0	1.5	2.0	2.5	3.0	3.5	4.0
$r=0.010$ $(c=1.788)$	119.548	50.791	31.157	22.288	17.340	14.220	12.047	10.492
$r=0.025$ $(c=2.116)$	126.555	47.448	26.886	18.400	13.917	11.198	9.392	8.095
$r=0.050$ $(c=2.316)$	143.605	48.325	25.018	16.070	11.628	9.126	7.520	6.426
$r=0.100$ $(c=2.487)$	170.670	54.185	25.009	14.619	9.883	7.377	5.897	4.947
$r=0.200$ $(c=2.638)$	200.671	63.965	27.225	14.292	8.765	6.000	4.549	3.677
$r=0.400$ $(c=2.779)$	225.624	74.792	30.618	14.816	8.153	5.051	3.450	2.647
$r=0.600$ $(c=2.859)$	236.053	80.187	32.450	15.113	8.008	4.596	2.904	2.052
$r=0.800$ $(c=2.917)$	242.451	83.471	33.653	15.409	7.890	4.251	2.506	1.654
$r=1.000$ $(c=2.963)$	247.699	85.779	34.192	15.395	7.484	3.844	2.176	1.431
minimal ARL	119.548	47.448	25.009	14.292	7.484	3.844	2.176	1.431
at $r=$	0.010	0.025	0.100	0.200	1.000	1.000	1.000	1.000

For the EWMA residual chart the average delay almost always achieves its supremum at $q = 1$. Therefore, the ARL and D_{sup} of an EWMA residual chart behave identically. The value of r has to be chosen smaller for greater α_1 and same shifts $a/\sqrt{\gamma_0}$. For $\alpha_1 < 0$ r is larger and the ARL for small shifts is smaller than for the modified EWMA chart. For $\alpha_1 > 0$ the modified EWMA chart is always better than the residual chart and the EWMA parameter r is smaller for the residual chart. For $\alpha_1 < 0$ the EWMA residual chart reacts more sensibly towards the choice of r compared to the modified chart, whereas for $\alpha_1 > 0$ also the opposite behaviour is shown.

Now, we first compare the CUSUM charts by means of their average run lengths. For $\alpha_1 < 0$ the modified CUSUM chart compares favourable with the other CUSUM charts. Only for some moderate shifts the CUSUM residual chart is better for negative α_1. For positive α_1 in almost all cases the classical CUSUM chart with modified limits out performs the other CUSUM charts. Only for $a/\sqrt{\gamma_0} = 4.0$ the CUSUM residual chart is better than the classical CUSUM chart.

For CUSUM charts with independent observations it is optimal to set the reference value k to half of the size of the expected shift. This is also true for the modified chart except for $|\alpha_1| \geq 0.6$ and great shifts. For the CUSUM residual chart this is not true anymore. For negative α_1 and moderate shifts k should be chosen greater than for the modified chart and for all other cases smaller. For the classical CUSUM chart with modified limits and negative α_1 k is less than or equal to the k of the modified chart, and for $\alpha_1 < 0$ vice versa.

As already mentioned D_{sup} of the CUSUM residual chart equals the average run length. But this is not necessarily true for the modified CUSUM chart and the classical CUSUM chart with modified limits D_{sup}. Indeed the supremum of the average delay sometimes differs considerably from the ARL (cf. Table 23.5). However for the optimal choice of k, the supremum is almost always attained at $q = 1$. Thus, it seems to be true that for the optimal choice of k the supremum of the average delay D_{sup} is nearly equal to the ARL. This is an astonishing and a remarkable result of this simulation study.

Table 23.5: Out-of-control ARL of the classical CUSUM chart with modified limits for $\alpha_1 = -0.8$

		$\frac{q}{\sqrt{n_0}}$ 0.5	1.0	1.5	2.0	2.5	3.0	3.5	4.0
$k=0.000$	($K=0.000, h=11.132$)	22.148	11.514	7.902	6.041	5.010	4.136	3.730	3.202
$k=0.250$	($K=0.417, h=2.774$)	9.381	3.939	2.662	1.958	1.704	1.511	1.318	1.162
$k=0.500$	($K=0.833, h=2.467$)	32.359	4.782	2.854	2.007	1.683	1.489	1.299	1.152
$k=0.750$	($K=1.250, h=2.215$)	163.747	7.134	3.193	2.124	1.695	1.487	1.294	1.150
$k=1.000$	($K=1.667, h=1.965$)	184.361	18.558	3.796	2.279	1.718	1.486	1.298	1.151
$k=1.250$	($K=2.083, h=1.715$)	185.124	49.047	5.214	2.491	1.761	1.489	1.298	1.150
$k=1.500$	($K=2.500, h=1.465$)	185.152	54.156	9.934	2.828	1.816	1.492	1.297	1.149
$k=1.750$	($K=2.917, h=1.215$)	185.152	54.422	16.912	3.522	1.908	1.501	1.296	1.150
$k=2.000$	($K=3.333, h=.965$)	185.152	54.433	18.367	4.981	2.050	1.508	1.296	1.151
	minimal ARL	9.381	3.939	2.662	1.958	1.683	1.486	1.294	1.149
	at $k=$	0.250	0.250	0.250	0.250	0.500	1.000	0.750	1.500

The same behaviour is also shown for the EWMA charts. Regarding that we choose only a few values of r for the EWMA charts (and especially not necessarily "optimal" choices of r as the choice of k for the CUSUM charts can be regarded) it is surprising how often $D_{\sup}$ for this "optimal" r was achieved at $q = 1$ for the modified EWMA chart. For the EWMA residual chart the supremum of the average delay was almost always attained at $q = 1$. This leads to the conclusion that for the optimal choice of k and r respectively the supremum of the average delay and the average run length seem to be nearly the same.

As a conclusion it is possible to consider the more appropriate criterion $D_{\sup}$. But for the optimal choice of k and r this is identical to the ARL. Thus, a main objection against using the average run length in the context of control charts with dependent observations is no longer valid provided that the parameters of the control charts are appropriately chosen.

For the overall comparison of all considered control charts the smallest out-of-control ARL and $D_{\sup}$ respectively are indicated with bold digits in Tables 23.6 and 23.7. For negative α_1 and small shifts the EWMA residual chart should be preferred while for greater shifts the modified EWMA or the modified CUSUM chart are the best. For positive α_1 and small shifts the modified EWMA chart should be applied, whereas for moderate shifts the classical CUSUM chart with modified limits is the best chart. For large shifts the CUSUM residual chart and the classical CUSUM chart with modified limits are better for $\alpha_1 \leq 0.4$ and for $\alpha_1 \geq 0.6$ the modified EWMA chart performs better for large shifts. The behaviour for both quality criteria is again almost identical.

Table 23.6: A comparison of the ARL of several control charts for "optimal" values of k and r ($\varepsilon_t \sim \Phi$, in-control ARL $= 500$)

$\frac{a}{\sqrt{\gamma_0}}$		0.5	1.0	1.5	2.0	2.5	3.0	3.5	4.0
$\alpha_1 = -0.8$	EWMAmod	7.141	3.193	2.076	1.632	1.407	1.232	1.107	1.041
	EWMAres	**6.185**	**2.649**	2.044	1.967	1.836	1.559	1.246	1.063
	CUSUMmod	6.279	2.679	**1.775**	**1.499**	**1.307**	**1.159**	**1.067**	**1.023**
	CUSUMres	6.352	2.751	2.070	1.968	1.840	1.561	1.248	1.064
	CUSUMcla	9.381	3.939	2.662	1.958	1.683	1.486	1.294	1.149
$\alpha_1 = -0.6$	EWMAmod	11.077	4.392	2.720	**1.939**	1.584	1.358	1.193	**1.085**
	EWMAres	10.834	4.067	**2.531**	2.038	1.790	1.546	1.305	1.127
	CUSUMmod	11.145	4.127	2.570	**1.939**	1.611	1.396	1.222	1.103
	CUSUMres	11.404	4.123	2.583	2.048	1.786	1.546	1.306	1.131
	CUSUMcla	11.966	4.930	3.049	2.207	1.789	1.530	1.327	1.170
$\alpha_1 = -0.4$	EWMAmod	15.939	5.845	3.455	2.336	**1.824**	1.482	**1.267**	**1.130**
	EWMAres	**15.715**	**5.721**	**3.254**	**2.334**	1.880	1.579	1.331	1.161
	CUSUMmod	16.585	5.805	3.306	2.368	1.863	1.531	1.310	1.157
	CUSUMres	17.069	5.766	3.274	2.336	1.871	1.566	1.333	1.165
	CUSUMcla	16.889	6.131	3.535	2.476	1.916	1.580	1.346	1.180
$\alpha_1 = -0.2$	EWMAmod	21.780	7.759	4.366	2.847	2.115	1.648	**1.359**	**1.171**
	EWMAres	**21.752**	**7.671**	4.258	2.836	2.110	1.670	1.379	1.182
	CUSUMmod	23.044	7.875	4.230	2.816	2.094	**1.642**	1.364	1.181
	CUSUMres	23.168	7.891	**4.221**	**2.793**	**2.076**	1.647	1.368	1.183
	CUSUMcla	23.163	7.937	4.307	2.862	2.105	1.660	1.371	1.188
$\alpha_1 = 0.0$	EWMAmod	**28.731**	**10.314**	5.498	3.517	2.517	1.874	1.449	1.217
	EWMAres	**28.731**	**10.314**	5.498	3.517	2.517	1.874	1.449	1.217
	CUSUMmod	31.016	10.503	**5.432**	3.411	2.388	1.791	1.425	1.201
	CUSUMres	31.016	10.503	**5.432**	3.411	2.388	1.791	1.425	1.201
	CUSUMcla	31.016	10.503	**5.432**	3.411	2.388	1.791	1.425	1.201
$\alpha_1 = 0.2$	EWMAmod	**38.483**	**13.562**	7.125	4.496	3.068	2.166	1.583	1.248
	EWMAres	38.513	13.640	7.232	4.623	3.100	2.205	1.623	1.260
	CUSUMmod	41.868	14.332	7.276	4.399	2.926	2.067	1.552	1.257
	CUSUMres	42.540	14.625	7.405	4.418	2.909	2.038	1.520	1.230
	CUSUMcla	41.492	14.060	**7.103**	**4.280**	**2.827**	**1.983**	**1.500**	**1.229**
$\alpha_1 = 0.4$	EWMAmod	**51.378**	**18.756**	9.771	5.967	3.852	2.545	1.692	1.283
	EWMAres	51.589	19.103	10.042	6.140	4.239	2.847	1.892	1.358
	CUSUMmod	57.630	20.211	10.247	6.032	3.797	2.513	1.768	1.347
	CUSUMres	62.496	20.360	10.289	6.158	3.853	2.500	1.694	**1.278**
	CUSUMcla	56.522	19.351	**9.681**	**5.626**	**3.528**	**2.298**	**1.627**	1.293
$\alpha_1 = 0.6$	EWMAmod	**74.156**	**27.309**	**14.159**	8.413	5.236	3.005	1.858	1.340
	EWMAres	74.495	27.834	14.846	9.260	6.092	4.260	2.621	1.551
	CUSUMmod	84.254	31.118	16.042	9.234	5.468	3.295	2.124	1.483
	CUSUMres	105.180	31.073	15.885	9.335	5.794	3.579	2.097	1.377
	CUSUMcla	81.124	28.631	14.228	**7.992**	**4.710**	**2.846**	1.916	1.443
$\alpha_1 = 0.8$	EWMAmod	**119.548**	**47.448**	25.009	14.292	7.484	**3.844**	**2.176**	**1.431**
	EWMAres	120.984	49.042	26.793	17.067	11.312	8.059	5.519	2.104
	CUSUMmod	143.452	60.245	32.727	16.878	8.913	4.798	2.736	1.708
	CUSUMres	171.465	75.462	33.311	18.078	11.423	7.816	4.079	1.665
	CUSUMcla	133.053	50.085	**24.847**	**13.479**	**7.277**	4.162	2.600	1.824

Table 23.7: A comparison of the $D_{\sup}$ of several control charts for "optimal" values of k and r ($\varepsilon_t \sim \Phi$, in-control ARL = 500)

		$\frac{a}{\sqrt{70}}$ 0.5	1.0	1.5	2.0	2.5	3.0	3.5	4.0
$\alpha_1 = -0.8$	EWMAmod	7.153	3.211	2.275	1.714	1.407	1.232	1.107	1.041
	EWMAres	6.185	2.649	2.044	1.967	1.836	1.559	1.246	1.063
	CUSUMmod	6.279	2.679	1.793	1.504	1.307	1.159	1.067	1.023
	CUSUMres	6.352	2.751	2.070	1.968	1.840	1.561	1.248	1.064
	CUSUMcla	9.381	3.939	2.662	1.958	1.683	1.486	1.294	1.149
$\alpha_1 = -0.6$	EWMAmod	11.082	4.392	2.793	2.016	1.610	1.358	1.193	1.085
	EWMAres	10.834	4.067	2.531	2.038	1.790	1.546	1.305	1.127
	CUSUMmod	11.145	4.127	2.570	1.939	1.611	1.396	1.222	1.103
	CUSUMres	11.404	4.123	2.583	2.048	1.786	1.546	1.306	1.131
	CUSUMcla	11.966	4.930	3.049	2.207	1.789	1.530	1.327	1.170
$\alpha_1 = -0.4$	EWMAmod	15.939	5.845	3.474	2.366	1.848	1.482	1.267	1.130
	EWMAres	15.715	5.721	3.254	2.334	1.880	1.579	1.331	1.161
	CUSUMmod	16.585	5.805	3.306	2.368	1.863	1.531	1.310	1.157
	CUSUMres	17.069	5.766	3.274	2.336	1.871	1.566	1.333	1.165
	CUSUMcla	16.889	6.131	3.535	2.476	1.916	1.580	1.346	1.180
$\alpha_1 = -0.2$	EWMAmod	21.780	7.759	4.366	2.845	2.129	1.653	1.359	1.171
	EWMAres	21.752	7.671	4.258	2.836	2.110	1.670	1.379	1.182
	CUSUMmod	23.044	7.875	4.230	2.816	2.094	1.642	1.364	1.181
	CUSUMres	23.168	7.891	4.221	2.793	2.076	1.647	1.368	1.183
	CUSUMcla	23.163	7.937	4.307	2.862	2.105	1.660	1.371	1.188
$\alpha_1 = 0.0$	EWMAmod	28.731	10.314	5.498	3.517	2.517	1.874	1.461	1.225
	EWMAres	28.731	10.314	5.498	3.517	2.517	1.874	1.461	1.225
	CUSUMmod	31.016	10.503	5.432	3.411	2.388	1.791	1.425	1.201
	CUSUMres	31.016	10.503	5.432	3.411	2.388	1.791	1.425	1.201
	CUSUMcla	31.016	10.503	5.432	3.411	2.388	1.791	1.425	1.201
$\alpha_1 = 0.2$	EWMAmod	38.483	13.562	7.125	4.496	3.068	2.166	1.595	1.248
	EWMAres	38.513	13.640	7.232	4.623	3.100	2.205	1.623	1.260
	CUSUMmod	41.868	14.332	7.276	4.399	2.926	2.067	1.552	1.257
	CUSUMres	42.540	14.625	7.405	4.418	2.909	2.038	1.520	1.230
	CUSUMcla	41.492	14.060	7.103	4.280	2.827	1.983	1.500	1.232
$\alpha_1 = 0.4$	EWMAmod	51.378	18.756	9.771	5.967	3.852	2.545	1.705	1.285
	EWMAres	51.589	19.103	10.042	6.140	4.239	2.847	1.892	1.358
	CUSUMmod	57.630	20.211	10.247	6.032	3.797	2.513	1.768	1.347
	CUSUMres	62.496	20.360	10.289	6.158	3.853	2.500	1.694	1.278
	CUSUMcla	56.522	19.351	9.681	5.626	3.528	2.305	1.637	1.294
$\alpha_1 = 0.6$	EWMAmod	74.156	27.309	14.159	8.413	5.236	3.021	1.858	1.340
	EWMAres	74.495	27.834	14.846	9.260	6.092	4.260	2.621	1.551
	CUSUMmod	84.254	31.118	16.042	9.234	5.468	3.295	2.124	1.483
	CUSUMres	105.180	31.073	15.885	9.335	5.794	3.579	2.097	1.377
	CUSUMcla	81.124	28.631	14.228	7.992	4.741	2.866	1.930	1.446
$\alpha_1 = 0.8$	EWMAmod	119.666	47.448	25.009	14.292	7.556	3.844	2.183	1.431
	EWMAres	120.984	49.042	26.793	17.067	11.312	8.059	5.519	2.104
	CUSUMmod	143.452	60.245	32.727	16.878	8.913	4.798	2.736	1.728
	CUSUMres	171.465	75.462	33.311	18.078	11.423	7.816	4.079	1.665
	CUSUMcla	133.053	50.085	24.847	13.479	7.318	4.163	2.613	1.825

23.4   Conclusions

In this paper we show that important properties of the average run length are also valid for an alternative quality criterion, the average delay. Furthermore, we give approximations for the average delay. Particularly, the average delay for dependent processes is compared with the average delay for independent processes and with the average run length. For exchangeable variables it is shown that the average delay is monotonically increasing in the time of the change in distribution. Finally, in a simulation study the behaviour of several EWMA and CUSUM charts is considered when using the average delay instead of the average run length. In principle, the same ranking of the control designs is obtained as already known for the ARL. In a certain sense this justifies the use of the ARL as a performance criterion for dependent data.

References

1. Alwan, L. C. and Roberts, H. V. (1988). Time-series modeling for statistical process control. *Journal of Business and Economic Statistics*, **6**, 87–95.

2. Brockwell, P. J. and Davis, R. A. (1987). *Time Series: Theory and Methods*, New York: Springer-Verlag.

3. Brook, D. and Evans, D. A. (1972). An approach to the probability distributions of CUSUM run length, *Biometrika*, **59**, 539–549.

4. Das Gupta, S., Eaton, M. L., Olkin, I., Perlman, M., Savage, L. J. and Sobel, M. (1972). Inequalities on the probability content of convex regions for elliptically contoured distributions, In *Sixth Berkeley Symposium*, Vol. II, p. 241–265, London: Cambridge University Press.

5. Gordon, L. and Pollak, M. (1994). An efficient sequential nonparametric scheme for detecting a change of distribution, *The Annals of Statistics*, **22**, 763–804.

6. Graybill, F. A. (1969). *Introduction to Matrices with Applications in Statistics*, Belmont, CA: Wadsworth.

7. Harris, T. J. and Ross, W. H. (1991). Statistical process control procedures for correlated observations, *Canadian Journal of Chemical Engineering*, **69**, 48–57.

8. Kramer, H. G. (1997). On control charts for time series, *Dissertation*, Universität Ulm, Ulm, Germany.

9. Kramer, H. G. and Schmid, W. (1997). EWMA charts for multivariate time series, *Sequential Analysis*, **16**, 131–154.

10. Maragah, H. D. and Woodall, W. D. (1992). The effect of autocorrelation on the retrospective X-chart, *Journal of Statistical Computation and Simulation*, **40**, 29–42.

11. Lucas, J. M. and Saccucci, M. S. (1992). Exponentially weighted moving average control schemes: properties and enhancements, *Technometrics*, **32**, 1–29.

12. Montgomery, D. C. and Mastrangelo, C. M. (1991). Some statistical process control methods for autocorrelated data, *Journal of Quality Technology*, **23**, 179–204.

13. Pollak, M. (1985). Optimal detection of a change in distribution, *The Annals of Statistics*, **13**, 206–227.

14. Pollak, M. and Siegmund, D. (1975). Approximations to the expected sample size of certain sequential tests, *The Annals of Statistics*, **3**, 1267–1282.

15. Pollak, M. and Siegmund, D. (1985). A diffusion process and its applications to detecting a change in the drift of Brownian motion, *Biometrika*, **72**, 267–280.

16. Pollak, M. and Siegmund, D. (1991). Sequential detection of a change in a normal mean when the initial value is unknown, *The Annals of Statistics*, **19**, 394–416.

17. Roberts, S. W. (1966). A comparison of some control chart procedures, *Technometrics*, **8**, 411–430.

18. Schmid, W. (1995). On the run length of a Shewhart chart for correlated data, *Statistical Papers*, **36**, 111–130.

19. Schmid, W. (1997a). On EWMA charts for time series, In *Frontiers in Statistical Quality Control* (Eds., H.-J. Lenz and P.-Th. Wilrich), pp. 115–137, Heidelberg, Germany: Physica-Verlag.

20. Schmid, W. (1997b). CUSUM control schemes for Gaussian processes, *Statistical Papers*, **38**, 191–217.

21. Shiryayev, A. N. (1963). On optimum methods in quickest detection problems, *Theory of Probability and its Applications*, **13**, 22–46.

22. Siegmund, D. (1985). *Sequential Analysis*, New York: Springer-Verlag.

23. Srivastava, M. S. and Wu, Y. (1993). Comparison of EWMA, CUSUM and Shiryayev-Roberts procedures for detecting a shift in the mean, *The Annals of Statistics*, **21**, 645–670.

24. Tong, Y. L. (1980). *Probability Inequalities in Multivariate Distributions*, New York: Academic Press.

25. Tong, Y. L. (1990). *The Multivariate Normal Distribution*, New York: Springer-Verlag.

26. Vasilopoulos, A. V. and Stamboulis, A. P. (1978). Modification of control chart limits in the presence of data correlation, *Journal of Quality Technology*, **10**, 20–30.

27. Wu, Y. (1996). A less sensitive linear detector for the change point based on kernel smoothing method, *Metrika*, **43**, 43–55.

28. Yashchin, E. (1993). Performance of CUSUM control schemes for serially correlated observations, *Technometrics*, **35**, 35–52.

24

Tolerance Bounds and C_{pk} Confidence Bounds Under Batch Effects

Fritz Scholz and Mark Vangel

Boeing Shared Services Group, Seattle, WA, U.S.A.
National Institute of Standards and Technology, Gaithersburg, MD, U.S.A.

Abstract: The capability index C_{pk} for a process, that produces parts with normally distributed characteristic X, is defined as $C_{pk} = \min(U - \mu, \mu - L)/(3\sigma) = (T - |\mu - \nu|)/(3\sigma)$, where U and L are upper and lower specification limits for X, μ and σ are process mean and standard deviation, and $\nu = (U + L)/2$, $T = (U - L)/2$. Using a sample $X_1, \ldots, X_n$ of independent observations from $\mathcal{N}(\mu, \sigma^2)$ Chou et al. (1990) [with clarification by Kushler and Hurley (1992)] showed how to get lower confidence bounds for C_{pk}. Here we extend this methodology to cover the situation where samples come in batches and the intra batch correlation reduces the amount of independent information. In parallel we also apply this extension to the closely related tolerance bounds or confidence bounds for quantiles. Introducing the simple trick of effective sample size these problems are linked quite successfully to existing tables for tolerance bounds or C_{pk} confidence bounds. The basic idea is to "approximate" the complicated data situation with an i.i.d. scenario with reduced overall sample size. The approximation is anchored by analysis to the two extreme situations where the within batch correlation is zero or one. For the in-between cases the effective sample size is chosen on a simple heuristic basis, namely by matching the variances of the sample mean under the batch effect model and its i.i.d. approximation. The coverage properties of the resulting method, examined by simulation, were found to be reasonably accurate near the extreme cases and mildly conservative in-between.

Keywords and phrases: Tolerance bounds, capability index C_{pk}, effective sample size, batch to batch variation, noncentral t-distribution

24.1   Introduction and Overview

It is assumed that we deal with data from a normal population $\mathcal{N}(\mu, \sigma^2)$ with mean μ and standard deviation σ. For i.i.d. samples it has long been known how to construct tolerance bounds or confidence bounds for normal p-quantiles $x_p + \sigma z_p$ based on the noncentral t-distribution. The earliest reference we found was Jennett and Welch (1939), but also see Johnson and Welch (1940), Owen (1968, 1985), and Odeh and Owen (1980) for extensive tables.

Closely related to such quantiles is the process capability index C_{pk}, introduced by Kane (1986), and defined as

$$C_{pk} = \min\left\{\frac{U - \mu}{3\sigma}, \frac{\mu - L}{3\sigma}\right\} = \frac{\frac{1}{2}(U - L) - |\mu - \frac{1}{2}(U + L)|}{3\sigma},$$

where U and L are given upper and lower product specification limits. Confidence bounds for C_{pk}, again for the i.i.d. case, were given by Chou et. al (1990) with clarification by Kushler and Hurley (1992). For a comprehensive overview of capability indices see Kotz and Johnson (1993).

Often the data of a production process arrive in batches with significant within batch correlation. A popular model for such batch data is $\{X_{ij}, j = 1, \ldots, n_i, i = 1, \ldots, B\}$, where B is the number of batches and n_i is the size of the ith batch. It is then assumed that $X_{ij} = \mu + b_i + e_{ij}$, where b_i is normal with mean zero and variance σ_b^2 and e_{ij} is normal with mean zero and variance σ_e^2. The effects b_i and $\{e_{ij}\}$ are assumed to be mutually independent. Hence X_{ij} is normally distributed with mean μ and variance $\sigma^2 = \sigma_b^2 + \sigma_e^2$. The correlation of two different observations within the same batch is $\rho = \sigma_b^2/(\sigma_b^2 + \sigma_e^2)$ which can range anywhere within $[0, 1]$. Under such a scenario one usually still wants to characterize aspects of the overall $\mathcal{N}(\mu, \sigma^2)$ population and not of individual batches. Hence it is desirable to extend the methodology for constructing tolerance bounds or C_{pk} confidence bounds to such batch data.

Although this sampling model reflects greater realism of the industrial data experience, it also makes it impossible to construct exact confidence bounds for x_p and C_{pk}. For the latter we are aware of no attempts. For tolerance bounds several attempts have been made, with various degrees of numerical complexity, see Seeger and Thorsson (1972), Mee and Owen (1983), and Vangel (1995) who also treats additional regression covariates.

Our intent here is to "reduce" the problem to the i.i.d. case by the simple device of *effective sample size*. As with other methods we can only hope for achieved confidence levels that are approximate. The validity of this approximation is checked via simulations and contrasted with the treatment that ignores batch effects altogether. The appeal of this method is its conceptual simplicity and the reduction to a methodology with available tables and that already is widely spread in the industrial quality assurance practice.

We start out by giving the rationale for the *effective sample size*, which depends on the within batch correlation ρ, and show how to estimate it in straightforward fashion. This is followed by confidence bound construction for x_p, either exactly or approximately, for the two extreme cases: $(\sigma_e > 0, \sigma_b = 0)$ or $\rho = 0$ and $(\sigma_e = 0, \sigma_b > 0)$ or $\rho = 1$. The resulting bounds are further simplified so that they only differ in one parameter which can be identified with the effective sample size $N^\star$. The cases between these two extremes can then be interpolated using the effective sample size and using the existing tables from the i.i.d. case. This process is repeated, but more from a testing perspective, for C_{pk}. For this latter case we present some simulation results for validation and give a sample calculation using a composite material strength data set.

24.2 Effective Sample Size and its Estimation

The extreme case $(\sigma_e > 0,\ \sigma_b = 0)$ or $\rho = 0$ reduces the assumed batch data structure to $N = n_1 + \ldots + n_B$ i.i.d. observations, i.e., the effective sample size is $N^\star = N$. The other extreme case $(\sigma_e = 0,\ \sigma_b > 0)$ or $\rho = 1$ leaves us with effectively $N^\star = B$ i.i.d. observations $X_{11}, X_{21}, \ldots, X_{B1}$, since the remaining observations are just copies of those in this independent set and are of no use.

This suggests that we use an *effective sample size* $N^\star \in [B, N]$ for the intermediate cases $0 < \rho < 1$ in the following sense. We aim to approximate the given batch data set by a fictitious i.i.d. data set $X_1^\star, \ldots, X_{N^\star}^\star$, with $X_i^\star \sim \mathcal{N}(\mu, \sigma^2)$, that in some sense carries the same amount of information. Hence each individual observation in either sample has the same distribution but whereas $\{X_{ij}\}$ has sample size N with complex batch structure, the fictitious sample has the simple i.i.d. structure but with effective sample size $N^\star$.

The above vague notion of "carrying the same amount of information" could be made precise in several different ways. Here we choose $N^\star$ to match the variances of $\bar{X} = \sum_{i=1}^{B} \sum_{j=1}^{n_i} X_{ij}/N$ and $\bar{X}^\star = \sum_{i=1}^{N^\star} X_i^\star/N^\star$, i.e., find $N^\star$ such that

$$\operatorname{var}\left(\bar{X}\right) = \sigma_b^2 \sum_{i=1}^{B} \left(\frac{n_i}{N}\right)^2 + \sigma_e^2 \frac{1}{N} = \operatorname{var}\left(\bar{X}^\star\right) = \frac{\sigma_b^2 + \sigma_e^2}{N^\star}.$$

This leads to the following formula for $N^\star = N^\star(\rho)$

$$N^\star = \left[\frac{\sigma_b^2}{\sigma_b^2 + \sigma_e^2} \sum_{i=1}^{B} \left(\frac{n_i}{N}\right)^2 + \frac{1}{N} \frac{\sigma_e^2}{\sigma_b^2 + \sigma_e^2}\right]^{-1} = \left[\rho \frac{1}{f+1} + (1-\rho) \frac{1}{N}\right]^{-1},$$

where we write $1/(f + 1) = \sum_{i=1}^{B} (n_i/N)^2$ for reasons to become clear later. For $\rho = 0$ this becomes $N^\star = N$ and for $\rho = 1$ we get $N^\star = f + 1$ which matches B when $n_1 = \ldots = n_B$. Thus in the latter case of equal batch sizes this effective sample size formula agrees with our previous notion. We will not bother with the fact that $N^\star$ may not be an integer. An actual fictitious sample $X_1^\star, \ldots, X_{N^\star}^\star$ is never used in our procedure and all calculations are based on the actual batch data $\{X_{ij}\}$.

In practice the within batch correlation ρ is unknown but one may find reasonable estimates from the data as follows. Compute the between batch and error sums of squares

$$SS_b = \sum_{i=1}^{B} n_i (\bar{X}_{i\cdot} - \bar{X})^2 \qquad \text{and} \qquad SS_e = \sum_{i=1}^{B} \sum_{j=1}^{n_i} (X_{ij} - \bar{X}_{i\cdot})^2 .$$

Take $\hat{\sigma}_e^2 = SS_e/(N - B)$ as unbiased estimate of σ_e^2 and $\hat{\kappa}^2 = SS_b/(B - 1)$ as unbiased estimate of

$$\kappa^2 = \sigma_e^2 + \sigma_b^2 \frac{N}{B-1} \left(1 - \sum_{i=1}^{B} \left(\frac{n_i}{N} \right)^2 \right) = \sigma_e^2 + \sigma_b^2 \frac{N}{B-1} \frac{f}{f+1} .$$

Combining these two estimates we get $\hat{\sigma}_b^2 = \left(\hat{\kappa}^2 - \hat{\sigma}_e^2 \right)(B - 1)(f + 1)/(N\ f)$ as unbiased estimate for σ_b^2. Unfortunately, this latter estimate may be negative. If that happens it is suggested to set the estimate to zero. We denote this modification again by $\hat{\sigma}_b^2$ but it will no longer be unbiased. The estimate of ρ is then computed as $\hat{\rho} = \hat{\sigma}_b^2/(\hat{\sigma}_b^2 + \hat{\sigma}_e^2)$. It is this estimate that is used in place of ρ in estimating $N^\star$ by $N^\star = N^\star(\hat{\rho})$.

The notion of "effective sample size" is not new although it is not clear whether we have the earliest references. A recent one is Fisher and Van Belle (1993, p. 828) when interpreting the information loss in the Kaplan-Meier estimate due to censoring. Earlier references, provided kindly by Thomas Lumley, are Kish (1965, p. 162, p. 259) interpreting design effects with simple random sampling and Skinner, et al. (1989) who view the same issue from the perspective of misspecification.

24.3   Tolerance Bounds

Let $x_p = \mu + z_p\,\sigma$ denote the p-quantile of the sampled $\mathcal{N}(\mu, \sigma^2)$ population. Here $z_p = \Phi^{-1}(p)$ is the p-quantile of the standard normal population. It is desired to find lower confidence or lower tolerance bounds for x_p based on the batch data $\{X_{ij}\}$. We will approach this problem by first examining two extreme situations, namely $(\sigma_b = 0, \sigma_e > 0)$, i.e., no between batch variation, and $(\sigma_b > 0, \sigma_e = 0)$, i.e., no within batch variation, and then interpolate all intermediate situations using the effective sample size.

24.3.1   No between batch variation

Here we assume $\sigma_b = 0$ and $\sigma_e > 0$, i.e., $\rho = 0$, and thus all observations X_{ij} are mutually independent. $\bar{X} \sim \mathcal{N}(\mu, \sigma^2/N)$ and $SS_T = SS_b + SS_e \sim \sigma^2 \cdot \chi^2_{N-1}$ and both are independent of each other. In the following let

$$Z = \sqrt{N}\,\frac{\bar{X} - \mu}{\sigma} \quad \text{and} \quad V = \frac{S}{\sigma}, \quad \text{where} \quad S = \sqrt{\frac{SS_T}{N-1}}.$$

We consider $100\gamma\%$ lower tolerance bounds of the form $\bar{X} - k\,S$, where the factor k is determined such that

$$\gamma = P\left(\bar{X} - k\,S \le x_p\right) = P\left(\frac{Z - z_p\sqrt{N}}{V} \le k\sqrt{N}\right) = P\left(T_{N-1, -z_p\sqrt{N}} \le k\sqrt{N}\right),$$

where $T_{N-1, -z_p\sqrt{N}}$ represents a noncentral Student t random variable with non-centrality parameter $-z_p\sqrt{N}$ and $N-1$ degrees of freedom. This results in the following expression for the factor k:

$$k = k_0(N) = \frac{1}{\sqrt{N}}\, t_{N-1, -z_p\sqrt{N}, \gamma} = \sqrt{\frac{N-1}{N}}\,\frac{1}{\sqrt{N-1}}\, t_{N-1, -z_p\sqrt{N}, \gamma},$$

where $t_{N-1, -z_p\sqrt{N}, \gamma}$ is the γ quantile of $T_{N-1, -z_p\sqrt{N}}$.

24.3.2   No within batch variation

Here we assume $\sigma_b > 0$ and $\sigma_e = 0$, i.e., $\rho = 1$, and thus $\sigma^2 = \sigma_b^2$ and all observations within each batch are identical. Hence $SS_e = 0$, and thus $S^2 = SS_b/(N-1)$. Using Satterthwaite's method we will approximate the distribution of $SS_T = SS_b$ by a chi-square multiple with g degrees of freedom, i.e., $SS_T = SS_b \approx a \cdot \chi^2_g$, where a and g are determined to match the first two moments on either side. This leads to

$$g = \frac{(1 - \sum w_i^2)^2}{\sum w_i^2 - 2\sum w_i^3 + (\sum w_i^2)^2} \quad \text{and} \quad a = \frac{N}{g}\,\sigma_b^2\left(1 - \sum_{i=1}^{B} w_i^2\right),$$

where $w_i = n_i/N$. In the Appendix it is shown that this complicated expression for g can be approximated very well by a much simpler expression, namely by $f = (\sum w_i^2)^{-1} - 1$, and the approximation is exact when the n_i are all the same. We will use this simplification (f replacing g) from now on since it leads to a convenient similarity of the formulas for the factor k in the two cases studied. With this simplification we have $a \approx N\,\sigma_b^2/(f+1)$ and we can treat

$$V^2 = \frac{SS_T}{a\,f} = S^2\,\frac{(N-1)(f+1)}{f\,N\,\sigma_b^2}$$

as an approximate χ_f^2/f random variable. Further, $\bar{X} \sim \mathcal{N}(\mu, \tau^2)$ with $\tau^2 = \sigma_b^2 \cdot \sum_{i=1}^{B} w_i^2 \approx \sigma_b^2/(f+1)$, i.e., $Z = \sqrt{f+1}\,(\bar{X} - \mu)/\sigma_b$ has a standard normal distribution.

Note that when all samples sizes n_i are the same $(= n)$, then the above complicated expressions for f and a (and their approximations) reduce to $f = B-1$ and $a = n\sigma_b^2$. In that case SS_b actually is exactly distributed like $n\sigma_b^2 \cdot \chi_{B-1}^2$ and then $SS_T = SS_b$ is independent of $\bar{X}$. When the samples sizes are not the same, then SS_T is approximately distributed like the above chi-square multiple and the strict independence property no longer holds. We will ignore this latter flaw in our derivation below. The simulations show that this is of no serious consequence.

Again we have

$$\gamma = P\left(\bar{X} - k\,S \leq x_p\right) \;=\; P\left(\frac{Z - z_p\sqrt{f+1}}{V} \leq k\sqrt{\frac{f\,N}{N-1}}\right)$$

$$=\; P\left(T_{f,-z_p\sqrt{f+1}} \leq k\sqrt{\frac{f\,N}{N-1}}\right)$$

leading to

$$k = k_1(N) = \sqrt{\frac{N-1}{N}}\,\frac{1}{\sqrt{f}}\,t_{f,-z_p\sqrt{f+1},\gamma}\,.$$

24.3.3   The interpolation step

We note that the two expressions for $k_0(N)$ and $k_1(N)$ share the common factor $\sqrt{(N-1)/N}$ and the remainder can be matched if we match $f+1$ and N. We propose to use the previously developed estimated effective sample size $\hat{N}^\star$ as a simple interpolation between $f+1$ and N and use as k-factor in the general case

$$k^\star(N) = \sqrt{\frac{N-1}{N}}\,\frac{1}{\sqrt{\hat{N}^\star - 1}}\,t_{\hat{N}^\star-1,-z_p\sqrt{\hat{N}^\star},\gamma}\,.$$

24.4   Confidence Bounds for C_L, C_U and C_{pk}

For lower and upper specification limits L and U define

$$C_L = \frac{\mu - L}{3\sigma}\,, \quad C_U = \frac{U - \mu}{3\sigma} \quad \text{and} \quad C_{pk} = \min(C_L, C_U)\,.$$

These process capability indices are unknown but can be estimated respectively by

$$\widehat{C}_L = \frac{\bar{X} - L}{3S}, \quad \widehat{C}_U = \frac{U - \bar{X}}{3S} \quad \text{and} \quad \widehat{C}_{pk} = \min\left(\widehat{C}_L, \widehat{C}_U\right).$$

Here S is again the sample standard deviation of all the data, i.e., $S^2 = (SS_b + SS_e)/(N-1)$. We want to use these estimates $\widehat{C}_L$, $\widehat{C}_U$, and $\widehat{C}_{pk}$ in order to decide whether the corresponding population parameters exceed a given threshold C_0. This can be accomplished either by constructing lower confidence bounds based on these estimates or by testing of appropriate hypotheses. Since the available tables so far favor the testing framework we will stay with that preference, but we will indicate confidence bounds at the appropriate places.

We focus on C_L (C_U is handled the same way) and then combine the results for C_{pk}. Consider the problem of testing the hypothesis $H_L(C_0) : C_L \leq C_0$ against the alternative $K_L(C_0) : C_L > C_0$. We will reject $H_L(C_0)$ at level α whenever $\widehat{C}_L \geq C_\star$, where $C_\star = C_\star(\alpha, C_0)$ is determined such that the maximal chance of $\widehat{C}_L$ exceeding $C_\star$ is α when the hypothesis is true. Clearly $C_\star(\alpha, C_0)$ is an increasing function of C_0 and thus has an inverse $C_\star^{-1}(\alpha, \cdot)$. Solving $C_\star(\alpha, C_0) = \widehat{C}_L$ for $C_0 = C_\star^{-1}(\alpha, \widehat{C}_L)$ will give us a $100(1-\alpha)\%$ lower confidence bound $\widehat{C}_L(1-\alpha) = C_\star^{-1}(\alpha, \widehat{C}_L)$ for C_L. By this construction $\widehat{C}_L(1-\alpha) > C_0$ means that we should reject $H_L(C_0)$. Similarly $\widehat{C}_U(1-\alpha) = C_\star^{-1}(\alpha, \widehat{C}_U)$ is a $100(1-\alpha)\%$ lower confidence bound for C_U and $\widehat{C}_{pk}(1-\alpha) = \min(\widehat{C}_L(1-\alpha), \widehat{C}_U(1-\alpha))$ is a $100(1-\alpha)\%$ lower confidence bound for C_{pk}. The latter is easily seen by letting σ get arbitrarily small so that the two-sided problem reduces to the one-sided one, see also Kushler and Hurley (1992).

The main problem now is to find the proper critical value $C_\star$. We will do this again by examining the two extreme situations ($\sigma_b > 0, \sigma_e = 0$) and ($\sigma_b = 0, \sigma_e > 0$). All other situations will then be dealt with by a simple interpolation scheme. Finally, the resulting procedure is examined via simulations.

24.4.1 No between batch variation

Here we assume again ($\sigma_b = 0$, $\sigma_e > 0$). Thus $\sigma = \sigma_e$ and all X_{ij} are mutually independent. $\bar{X}$ is normally distributed with mean μ and variance σ^2/N, SS_T is distributed as $\sigma^2 \cdot \chi^2_{N-1}$ and both are independent of each other. Adopting the notation that P_{C_0} denotes a probability distribution under (μ, σ) with $C_L = C_0$ we find $C_\star$ by solving

$$\alpha = P_{C_0}\left(\widehat{C}_L \geq C_\star\right) = P_{C_0}\left(\frac{\bar{X} - L}{3S} \geq C_\star\right) = P\left(T_{N-1, 3C_0\sqrt{N}} \geq 3C_\star\sqrt{N}\right)$$

which yields

$$C_\star = \frac{1}{3\sqrt{N}} t_{N-1, 3C_0\sqrt{N}, 1-\alpha} = \sqrt{\frac{N-1}{N}} \frac{1}{3\sqrt{N-1}} t_{N-1, 3C_0\sqrt{N}, 1-\alpha}.$$

24.4.2   No within batch variation

Here we assume again ($\sigma_b > 0$, $\sigma_e = 0$) and use the same notation and approximations developed in the corresponding section on tolerance bounds. The α requirement on $C_\star$ leads to

$$\alpha = P_{C_0}\left(\widehat{C}_L \geq C_\star\right) = P_{C_0}\left(\frac{\bar{X} - L}{3S} \geq C_\star\right)$$

$$= P\left(\frac{Z + \delta}{V} \geq 3C_\star\sqrt{N/(N-1)}\sqrt{f}\right) \approx P\left(T_{f,\delta} \geq 3C_\star\sqrt{N/(N-1)}\sqrt{f}\right),$$

where $T_{f,\delta}$ is a noncentral Student t random variable with f degrees of freedom and noncentrality parameter $\delta = 3C_0\sqrt{f+1}$. This yields the following expression for $C_\star$

$$C_\star = \sqrt{\frac{N-1}{N}}\,\frac{1}{3\sqrt{f}}\,t_{f,3C_0\sqrt{f+1},1-\alpha}\,,$$

where $t_{f,\delta,1-\alpha}$ represents the $1 - \alpha$ percentile of that noncentral Student t distribution.

24.4.3   The interpolation step

Note that the two formulas for $C_\star$, developed for the two extreme cases, share the factor $\sqrt{(N-1)/N}$ and the remainder can be matched if we match $f + 1$ and N. We propose to use the previously developed effective sample size N^* as a simple interpolation between $f + 1$ and N, namely

$$N^\star = \left[\hat{\rho}\,\frac{1}{f+1} + (1 - \hat{\rho})\,\frac{1}{N}\right]^{-1}$$

and use as critical point in the general case

$$C_\star = \sqrt{\frac{N-1}{N}}\,\frac{1}{3\sqrt{N^\star - 1}}\,t_{N^\star-1,3C_0\sqrt{N^\star},1-\alpha}\,.$$

Table 3 of Chou et al. (1990) gives the value of

$$C_{\text{Table}\star}(N) = \sqrt{\frac{N-1}{N}}\,\frac{1}{3\sqrt{N-1}}t_{N-1,3C_0\sqrt{N},1-\alpha} \quad \text{for } \alpha = .05,$$

for various values of $C_0 = .7, .8, \ldots, 2.0$ and $N = 10, 20, \ldots, 50, 75, 100, 125, 150,$ $200, 300, 350, 400$. These tabled values are correct when $\sigma_b = 0$, i.e., in the i.i.d. case, which was addressed by Chou et al. and then clarified by Kushler and Hurley (1992). The same table, covering a somewhat different grid, and additional tables for $\alpha = .20, .10, .01$ are given in Tables 24.1–24.4.

Table 24.1: $\alpha = .20$ or 80% confidence

sample size	Critical Values $C_\star$ $\alpha = .20$ C_0											
	1.00	1.10	1.20	1.30	1.33	1.40	1.50	1.60	1.70	1.80	1.90	2.00
500	1.03	1.13	1.24	1.34	1.37	1.44	1.54	1.65	1.75	1.85	1.95	2.06
300	1.04	1.14	1.25	1.35	1.38	1.45	1.56	1.66	1.76	1.87	1.97	2.08
250	1.04	1.15	1.25	1.36	1.39	1.46	1.56	1.67	1.77	1.87	1.98	2.08
200	1.05	1.15	1.26	1.36	1.40	1.47	1.57	1.68	1.78	1.88	1.99	2.09
175	1.05	1.16	1.26	1.37	1.40	1.47	1.58	1.68	1.79	1.89	2.00	2.10
150	1.06	1.16	1.27	1.37	1.41	1.48	1.58	1.69	1.79	1.90	2.00	2.11
125	1.06	1.17	1.28	1.38	1.42	1.49	1.59	1.70	1.80	1.91	2.01	2.12
100	1.07	1.18	1.29	1.39	1.43	1.50	1.60	1.71	1.82	1.92	2.03	2.14
90	1.08	1.18	1.29	1.40	1.43	1.50	1.61	1.72	1.82	1.93	2.04	2.14
80	1.08	1.19	1.30	1.40	1.44	1.51	1.62	1.73	1.83	1.94	2.05	2.15
70	1.09	1.20	1.30	1.41	1.45	1.52	1.63	1.74	1.84	1.95	2.06	2.17
60	1.10	1.21	1.31	1.42	1.46	1.53	1.64	1.75	1.86	1.97	2.07	2.18
50	1.11	1.22	1.33	1.44	1.47	1.55	1.66	1.76	1.87	1.98	2.09	2.20
46	1.11	1.22	1.33	1.44	1.48	1.55	1.66	1.77	1.88	1.99	2.10	2.21
42	1.12	1.23	1.34	1.45	1.49	1.56	1.67	1.78	1.89	2.00	2.11	2.23
38	1.13	1.24	1.35	1.46	1.50	1.57	1.68	1.79	1.91	2.02	2.13	2.24
34	1.14	1.25	1.36	1.47	1.51	1.58	1.70	1.81	1.92	2.03	2.14	2.26
30	1.15	1.26	1.37	1.49	1.52	1.60	1.71	1.82	1.94	2.05	2.16	2.28
28	1.15	1.27	1.38	1.49	1.53	1.61	1.72	1.83	1.95	2.06	2.18	2.29
26	1.16	1.27	1.39	1.50	1.54	1.62	1.73	1.85	1.96	2.07	2.19	2.30
24	1.17	1.28	1.40	1.51	1.55	1.63	1.74	1.86	1.97	2.09	2.20	2.32
22	1.18	1.29	1.41	1.53	1.56	1.64	1.76	1.87	1.99	2.11	2.22	2.34
20	1.19	1.31	1.42	1.54	1.58	1.66	1.77	1.89	2.01	2.13	2.24	2.36
19	1.20	1.31	1.43	1.55	1.59	1.67	1.78	1.90	2.02	2.14	2.26	2.37
18	1.20	1.32	1.44	1.56	1.60	1.68	1.79	1.91	2.03	2.15	2.27	2.39
17	1.21	1.33	1.45	1.57	1.61	1.69	1.81	1.93	2.04	2.16	2.28	2.40
16	1.22	1.34	1.46	1.58	1.62	1.70	1.82	1.94	2.06	2.18	2.30	2.42
15	1.23	1.35	1.47	1.59	1.63	1.71	1.83	1.96	2.08	2.20	2.32	2.44
14	1.24	1.36	1.49	1.61	1.65	1.73	1.85	1.97	2.10	2.22	2.34	2.46
13	1.26	1.38	1.50	1.62	1.67	1.75	1.87	1.99	2.12	2.24	2.36	2.49
12	1.27	1.40	1.52	1.64	1.69	1.77	1.89	2.02	2.14	2.27	2.39	2.52
11	1.29	1.42	1.54	1.67	1.71	1.79	1.92	2.05	2.17	2.30	2.43	2.55
10	1.31	1.44	1.57	1.70	1.74	1.82	1.95	2.08	2.21	2.34	2.47	2.60
9	1.34	1.47	1.60	1.73	1.77	1.86	1.99	2.12	2.25	2.39	2.52	2.65
8	1.37	1.51	1.64	1.77	1.82	1.91	2.04	2.18	2.31	2.45	2.58	2.72
7	1.42	1.56	1.70	1.83	1.88	1.97	2.11	2.25	2.39	2.53	2.67	2.81
6	1.48	1.63	1.77	1.92	1.96	2.06	2.21	2.35	2.50	2.64	2.79	2.93
5	1.58	1.73	1.89	2.04	2.09	2.20	2.35	2.51	2.66	2.82	2.97	3.13
4	1.75	1.92	2.09	2.26	2.32	2.44	2.61	2.78	2.95	3.12	3.29	3.47
3	2.14	2.35	2.56	2.77	2.84	2.98	3.19	3.40	3.61	3.82	4.03	4.24
2	3.94	4.34	4.73	5.13	5.26	5.52	5.92	6.31	6.71	7.10	7.50	7.89

Table 24.2: $\alpha = .10$ or 90% confidence

sample size	Critical Values $C_\star$ $\alpha = .10$ C_0											
	1.00	1.10	1.20	1.30	1.33	1.40	1.50	1.60	1.70	1.80	1.90	2.00
500	1.05	1.15	1.25	1.36	1.39	1.46	1.57	1.67	1.78	1.88	1.98	2.09
300	1.06	1.17	1.27	1.38	1.41	1.48	1.59	1.69	1.80	1.90	2.01	2.11
250	1.07	1.17	1.28	1.39	1.42	1.49	1.60	1.70	1.81	1.91	2.02	2.13
200	1.08	1.18	1.29	1.40	1.43	1.50	1.61	1.72	1.82	1.93	2.04	2.14
175	1.08	1.19	1.30	1.40	1.44	1.51	1.62	1.72	1.83	1.94	2.05	2.15
150	1.09	1.20	1.30	1.41	1.45	1.52	1.63	1.74	1.84	1.95	2.06	2.17
125	1.10	1.21	1.32	1.42	1.46	1.53	1.64	1.75	1.86	1.97	2.08	2.19
100	1.11	1.22	1.33	1.44	1.48	1.55	1.66	1.77	1.88	1.99	2.10	2.21
90	1.12	1.23	1.34	1.45	1.49	1.56	1.67	1.78	1.89	2.00	2.11	2.22
80	1.13	1.24	1.35	1.46	1.50	1.57	1.68	1.79	1.90	2.02	2.13	2.24
70	1.14	1.25	1.36	1.47	1.51	1.58	1.70	1.81	1.92	2.03	2.15	2.26
60	1.15	1.26	1.38	1.49	1.53	1.60	1.72	1.83	1.94	2.06	2.17	2.28
50	1.17	1.28	1.40	1.51	1.55	1.63	1.74	1.85	1.97	2.08	2.20	2.31
46	1.18	1.29	1.41	1.52	1.56	1.64	1.75	1.87	1.98	2.10	2.21	2.33
42	1.19	1.30	1.42	1.53	1.57	1.65	1.77	1.88	2.00	2.12	2.23	2.35
38	1.20	1.31	1.43	1.55	1.59	1.67	1.78	1.90	2.02	2.14	2.25	2.37
34	1.21	1.33	1.45	1.57	1.61	1.69	1.80	1.92	2.04	2.16	2.28	2.40
30	1.23	1.35	1.47	1.59	1.63	1.71	1.83	1.95	2.07	2.19	2.31	2.43
28	1.24	1.36	1.48	1.60	1.64	1.72	1.84	1.97	2.09	2.21	2.33	2.45
26	1.25	1.37	1.49	1.62	1.66	1.74	1.86	1.98	2.11	2.23	2.35	2.47
24	1.26	1.39	1.51	1.63	1.67	1.76	1.88	2.00	2.13	2.25	2.38	2.50
22	1.28	1.40	1.53	1.65	1.69	1.78	1.90	2.03	2.15	2.28	2.40	2.53
20	1.30	1.42	1.55	1.68	1.72	1.80	1.93	2.06	2.18	2.31	2.44	2.57
19	1.31	1.44	1.56	1.69	1.73	1.82	1.95	2.07	2.20	2.33	2.46	2.59
18	1.32	1.45	1.58	1.71	1.75	1.83	1.96	2.09	2.22	2.35	2.48	2.61
17	1.33	1.46	1.59	1.72	1.77	1.85	1.98	2.11	2.24	2.37	2.50	2.63
16	1.35	1.48	1.61	1.74	1.78	1.87	2.00	2.14	2.27	2.40	2.53	2.66
15	1.37	1.50	1.63	1.76	1.81	1.89	2.03	2.16	2.29	2.43	2.56	2.69
14	1.38	1.52	1.65	1.79	1.83	1.92	2.06	2.19	2.33	2.46	2.60	2.73
13	1.41	1.54	1.68	1.81	1.86	1.95	2.09	2.22	2.36	2.50	2.64	2.77
12	1.43	1.57	1.71	1.85	1.89	1.99	2.13	2.26	2.40	2.54	2.68	2.82
11	1.46	1.60	1.74	1.89	1.93	2.03	2.17	2.31	2.45	2.60	2.74	2.88
10	1.50	1.64	1.79	1.93	1.98	2.08	2.22	2.37	2.52	2.66	2.81	2.95
9	1.55	1.69	1.84	1.99	2.04	2.14	2.29	2.44	2.59	2.74	2.89	3.04
8	1.61	1.76	1.91	2.07	2.12	2.22	2.38	2.54	2.69	2.85	3.00	3.16
7	1.69	1.85	2.01	2.17	2.23	2.34	2.50	2.66	2.83	2.99	3.15	3.32
6	1.80	1.97	2.15	2.32	2.38	2.49	2.67	2.84	3.02	3.19	3.37	3.54
5	1.98	2.17	2.36	2.55	2.62	2.74	2.94	3.13	3.32	3.51	3.71	3.90
4	2.31	2.53	2.76	2.98	3.05	3.20	3.43	3.65	3.88	4.10	4.33	4.55
3	3.13	3.43	3.73	4.04	4.14	4.35	4.65	4.96	5.26	5.57	5.88	6.18
2	7.95	8.75	9.55	10.3	10.6	11.1	11.9	12.7	13.5	14.3	15.1	15.9

Table 24.3: $\alpha = .05$ or 95% confidence

sample size	Critical Values $C_\star$											
	$\alpha = .05$											
	C_0											
	1.00	1.10	1.20	1.30	1.33	1.40	1.50	1.60	1.70	1.80	1.90	2.00
500	1.06	1.17	1.27	1.38	1.41	1.48	1.59	1.69	1.80	1.90	2.01	2.11
300	1.08	1.19	1.29	1.40	1.44	1.51	1.61	1.72	1.83	1.93	2.04	2.15
250	1.09	1.20	1.30	1.41	1.45	1.52	1.63	1.73	1.84	1.95	2.06	2.16
200	1.10	1.21	1.32	1.42	1.46	1.53	1.64	1.75	1.86	1.97	2.08	2.19
175	1.11	1.22	1.32	1.43	1.47	1.54	1.65	1.76	1.87	1.98	2.09	2.20
150	1.12	1.23	1.34	1.45	1.48	1.56	1.67	1.78	1.89	2.00	2.11	2.22
125	1.13	1.24	1.35	1.46	1.50	1.57	1.68	1.80	1.91	2.02	2.13	2.24
100	1.15	1.26	1.37	1.48	1.52	1.60	1.71	1.82	1.93	2.05	2.16	2.27
90	1.16	1.27	1.38	1.49	1.53	1.61	1.72	1.84	1.95	2.06	2.18	2.29
80	1.17	1.28	1.39	1.51	1.55	1.62	1.74	1.85	1.97	2.08	2.20	2.31
70	1.18	1.29	1.41	1.53	1.56	1.64	1.76	1.87	1.99	2.10	2.22	2.34
60	1.20	1.31	1.43	1.55	1.59	1.66	1.78	1.90	2.02	2.13	2.25	2.37
50	1.22	1.34	1.46	1.58	1.62	1.70	1.81	1.93	2.05	2.17	2.29	2.41
46	1.23	1.35	1.47	1.59	1.63	1.71	1.83	1.95	2.07	2.19	2.31	2.43
42	1.24	1.37	1.49	1.61	1.65	1.73	1.85	1.97	2.09	2.22	2.34	2.46
38	1.26	1.38	1.50	1.63	1.67	1.75	1.87	2.00	2.12	2.24	2.37	2.49
34	1.28	1.40	1.53	1.65	1.69	1.78	1.90	2.03	2.15	2.28	2.40	2.53
30	1.30	1.43	1.56	1.68	1.72	1.81	1.94	2.06	2.19	2.32	2.44	2.57
28	1.32	1.44	1.57	1.70	1.74	1.83	1.96	2.08	2.21	2.34	2.47	2.60
26	1.33	1.46	1.59	1.72	1.76	1.85	1.98	2.11	2.24	2.37	2.50	2.63
24	1.35	1.48	1.61	1.74	1.79	1.87	2.01	2.14	2.27	2.40	2.53	2.66
22	1.37	1.51	1.64	1.77	1.82	1.90	2.04	2.17	2.30	2.44	2.57	2.71
20	1.40	1.53	1.67	1.80	1.85	1.94	2.08	2.21	2.35	2.48	2.62	2.76
19	1.41	1.55	1.69	1.82	1.87	1.96	2.10	2.23	2.37	2.51	2.65	2.78
18	1.43	1.57	1.71	1.84	1.89	1.98	2.12	2.26	2.40	2.54	2.68	2.82
17	1.45	1.59	1.73	1.87	1.91	2.01	2.15	2.29	2.43	2.57	2.71	2.85
16	1.47	1.61	1.75	1.89	1.94	2.04	2.18	2.32	2.46	2.61	2.75	2.89
15	1.49	1.64	1.78	1.92	1.97	2.07	2.21	2.36	2.50	2.65	2.79	2.94
14	1.52	1.67	1.81	1.96	2.01	2.10	2.25	2.40	2.55	2.69	2.84	2.99
13	1.55	1.70	1.85	2.00	2.05	2.15	2.30	2.45	2.60	2.75	2.90	3.05
12	1.59	1.74	1.89	2.04	2.10	2.20	2.35	2.50	2.66	2.81	2.97	3.12
11	1.63	1.79	1.94	2.10	2.15	2.26	2.42	2.57	2.73	2.89	3.05	3.21
10	1.69	1.85	2.01	2.17	2.22	2.33	2.50	2.66	2.82	2.98	3.15	3.31
9	1.75	1.92	2.09	2.26	2.31	2.43	2.60	2.77	2.93	3.10	3.27	3.44
8	1.84	2.02	2.20	2.37	2.43	2.55	2.73	2.90	3.08	3.26	3.44	3.62
7	1.96	2.15	2.34	2.53	2.59	2.72	2.91	3.10	3.29	3.48	3.67	3.86
6	2.14	2.35	2.55	2.76	2.83	2.96	3.17	3.38	3.58	3.79	4.00	4.21
5	2.43	2.66	2.90	3.13	3.21	3.36	3.60	3.83	4.07	4.30	4.54	4.77
4	2.99	3.27	3.56	3.85	3.94	4.14	4.42	4.71	5.00	5.29	5.58	5.87
3	4.49	4.92	5.36	5.80	5.94	6.23	6.67	7.11	7.55	7.99	8.43	8.87
2	15.9	17.5	19.1	20.7	21.3	22.3	23.9	25.5	27.1	28.7	30.3	31.9

Table 24.4: $\alpha = .01$ or 99% confidence

sample size	Critical Values $C_\star$ $\alpha = .01$ C_0											
	1.00	1.10	1.20	1.30	1.33	1.40	1.50	1.60	1.70	1.80	1.90	2.00
500	1.09	1.19	1.30	1.41	1.44	1.52	1.62	1.73	1.84	1.95	2.05	2.16
300	1.11	1.22	1.33	1.44	1.48	1.55	1.66	1.77	1.88	1.99	2.10	2.21
250	1.13	1.24	1.35	1.46	1.50	1.57	1.68	1.79	1.90	2.01	2.13	2.24
200	1.14	1.26	1.37	1.48	1.52	1.59	1.71	1.82	1.93	2.04	2.16	2.27
175	1.15	1.27	1.38	1.49	1.53	1.61	1.72	1.83	1.95	2.06	2.18	2.29
150	1.17	1.28	1.40	1.51	1.55	1.63	1.74	1.86	1.97	2.09	2.20	2.32
125	1.19	1.30	1.42	1.54	1.57	1.65	1.77	1.89	2.00	2.12	2.23	2.35
100	1.21	1.33	1.45	1.57	1.61	1.69	1.81	1.93	2.04	2.16	2.28	2.40
90	1.23	1.35	1.47	1.59	1.63	1.71	1.83	1.95	2.07	2.19	2.31	2.43
80	1.24	1.37	1.49	1.61	1.65	1.73	1.85	1.97	2.09	2.22	2.34	2.46
70	1.27	1.39	1.51	1.63	1.67	1.76	1.88	2.00	2.13	2.25	2.37	2.50
60	1.29	1.42	1.54	1.67	1.71	1.79	1.92	2.04	2.17	2.30	2.42	2.55
50	1.33	1.46	1.58	1.71	1.75	1.84	1.97	2.10	2.23	2.36	2.49	2.62
46	1.35	1.47	1.60	1.73	1.78	1.87	2.00	2.13	2.26	2.39	2.52	2.65
42	1.37	1.50	1.63	1.76	1.81	1.89	2.03	2.16	2.29	2.42	2.56	2.69
38	1.39	1.53	1.66	1.79	1.84	1.93	2.06	2.20	2.33	2.47	2.60	2.74
34	1.42	1.56	1.70	1.83	1.88	1.97	2.11	2.25	2.38	2.52	2.66	2.80
30	1.46	1.60	1.74	1.88	1.93	2.02	2.16	2.30	2.45	2.59	2.73	2.87
28	1.48	1.63	1.77	1.91	1.96	2.05	2.20	2.34	2.48	2.63	2.77	2.91
26	1.51	1.65	1.80	1.94	1.99	2.09	2.23	2.38	2.53	2.67	2.82	2.97
24	1.54	1.69	1.84	1.98	2.03	2.13	2.28	2.43	2.58	2.73	2.87	3.02
22	1.58	1.73	1.88	2.03	2.08	2.18	2.33	2.48	2.64	2.79	2.94	3.09
20	1.62	1.77	1.93	2.08	2.14	2.24	2.40	2.55	2.71	2.86	3.02	3.18
19	1.65	1.80	1.96	2.12	2.17	2.28	2.43	2.59	2.75	2.91	3.07	3.23
18	1.67	1.83	1.99	2.15	2.21	2.31	2.47	2.64	2.80	2.96	3.12	3.28
17	1.71	1.87	2.03	2.19	2.25	2.36	2.52	2.69	2.85	3.01	3.18	3.34
16	1.74	1.91	2.07	2.24	2.30	2.41	2.57	2.74	2.91	3.08	3.24	3.41
15	1.78	1.95	2.12	2.29	2.35	2.46	2.63	2.81	2.98	3.15	3.32	3.49
14	1.83	2.01	2.18	2.35	2.41	2.53	2.70	2.88	3.06	3.23	3.41	3.59
13	1.89	2.07	2.25	2.43	2.49	2.61	2.79	2.97	3.15	3.33	3.51	3.70
12	1.96	2.14	2.33	2.51	2.57	2.70	2.89	3.07	3.26	3.45	3.64	3.83
11	2.04	2.23	2.42	2.62	2.68	2.81	3.01	3.20	3.40	3.59	3.79	3.99
10	2.14	2.34	2.55	2.75	2.82	2.95	3.16	3.36	3.57	3.77	3.98	4.19
9	2.27	2.49	2.70	2.92	2.99	3.14	3.35	3.57	3.79	4.01	4.23	4.44
8	2.45	2.68	2.92	3.15	3.23	3.38	3.62	3.85	4.09	4.32	4.56	4.79
7	2.71	2.96	3.22	3.47	3.56	3.73	3.99	4.25	4.51	4.77	5.03	5.29
6	3.10	3.39	3.68	3.98	4.07	4.27	4.57	4.86	5.16	5.46	5.76	6.05
5	3.78	4.13	4.49	4.85	4.97	5.21	5.58	5.94	6.30	6.67	7.03	7.39
4	5.24	5.74	6.24	6.75	6.91	7.25	7.76	8.26	8.77	9.27	9.78	10.3
3	10.1	11.1	12.1	13.1	13.4	14.1	15.1	16.1	17.1	18.1	19.1	20.0
2	79.8	87.8	95.7	103.	106.	112.	120.	128.	136.	144.	152.	160.

To allow for the possible batch effect we should, according to the above derivation, use instead the adjusted critical value

$$C_{\text{Adj}\star}(N) = \sqrt{\frac{N-1}{N}} \sqrt{\frac{N^\star}{N^\star - 1}} \, C_{\text{Table}\star}(N^\star).$$

This concludes the derivation of the critical point $C_\star$ for our hypothesis testing problem concerning C_L. The same $C_\star$ in conjunction with $\widehat{C}_U$ works for the testing the hypothesis $H_U : C_U \leq C_0$ against the alternative $K_U : C_U > C_0$.

To combine these two procedures into one for testing the corresponding hypothesis for C_{pk}, namely $H : C_{pk} \leq C_0$ versus $K : C_{pk} > C_0$, we simply reject H when $\widehat{C}_{pk} \geq C_\star$ with $C_\star = C_{\text{Adj}\star}(N)$ as developed previously. Upon rejection of H we can be at least $100(1 - \alpha)\%$ confident that $C_{pk} > C_0$. When $C_L = C_U$ the confidence will be slightly higher than the target of $100(1 - \alpha)\%$ but when C_L and C_U are quite different, the confidence will be approximately equal to $100(1 - \alpha)\%$. Not knowing the actual values of C_L and C_U and wanting to use the simple estimated value $\widehat{C}_{pk}$ as a decision criterion, this procedure should serve its purpose reasonably well.

24.5   Validation

For $\alpha = .10$ the above procedure was validated through simulation as follows. Normal batch random samples were generated according to the assumed model for various values of ρ, using $\mu = 0$, $\sigma^2 = 1 = \rho + (1 - \rho)$, i.e., $\sigma_b^2 = \rho$ and $\sigma_e^2 = 1 - \rho$. The known value of C_L was assumed to be 1, i.e., $L = -3$. For each generated collection of B batch samples we computed the observed $\widehat{C}_{pk}$ and compared it against the value $C_{\text{Adj}\star}(N)$ and also against the value $C_\star$ that would be correct if $\sigma_b = 0$ were correct. Repeating this 1000 times for each configuration of ρ, B and $(n_1, \ldots, n_B)$ we recorded the observed rates of exceeding the respective critical points. This was done for $B = 10, 20, 30, 40$ batches of same size $n = 2, 3, 5$ each for $\rho = 0, .2, .4, .6, .8, 1$. To study the sample size imbalance effect we also simulated $B = 10, 20, 30, 40$ batches, half of the batch samples of size n_1 and half of size n_2 for $(n_1, n_2) = (2, 3), (2, 5)$ and $(3, 5)$.

The resulting observed confidence levels are summarized graphically in Figures 24.1 and 24.2. The three horizontal lines centered on $1 - \alpha = .9$ represent the target confidence for this simulation and 95% uncertainty limits for 1000 replications. The dotted curves represent the observed confidence for the proposed procedure whereas the dashed curves represent the observed confidence of the procedure that ignores batch effects, i.e., treats all observations as mutually independent.

It is quite obvious from the figures that the degradation of the latter procedure can be quite serious even with moderate batch effects, especially for few batches ($B = 10$) of "large" size ($n = 5$). The proposed procedure appears to hold its intended confidence level quite well, being slightly conservative when it is off by a small amount. Also given in these plots are the observed average effective sample sizes for each ρ.

24.6 Sample Calculation

The data Table 24.5 represent data on 21 batches of some composite material property data with lower specification value $L = 45$. From the data in this table we obtain:

$$\bar{X} = 49.638 \,, \quad S = 1.320 \,, \quad \text{thus} \quad \hat{C}_L = 1.17 \,.$$

Ignoring the batch effects and assuming that we deal with $N = 63$ independent observations we obtain from our Table 24.3 (using the column corresponding to $C_0 = 1$ and interpolating at the row corresponding to $N = 63$) the critical value 1.147, i.e., we would then conclude with 90% confidence that the true C_{pk} is at least 1 since $1.17 > 1.147$.

Table 24.5: Example batch data

batch	n_i	sample data	sample average
1	1	50.5	50.5
2	1	50.2	50.2
3	4	50.7, 50.8, 51.4, 51.3	51.05
4	1	49.3	49.3
5	3	51.0, 51.2, 53.4	51.867
6	3	50.9, 51.6, 51.8	51.433
7	1	49.3	49.3
8	3	48.6, 48.2, 46.6	47.8
9	2	50.4, 49.9	50.15
10	2	48.2, 47.5	47.85
11	3	50.5, 48.2, 49.5	49.4
12	3	49.7, 51.4, 50.6	50.567
13	4	49.6, 51.1, 51.1, 52.5	51.075
14	4	48.4, 50.2, 48.8, 49.1	49.125
15	4	48.8, 49.8, 50.0, 50.5	49.775
16	5	49.3, 50.2, 49.8, 48.9, 48.7	49.38
17	4	49.3, 47.5, 49.4, 48.4	48.65
18	4	47.8, 47.7, 48.8, 49.9	48.55
19	3	50.0, 49.5, 49.3	49.6
20	4	48.5, 49.2, 48.3, 47.8	48.45
21	4	47.9, 49.6, 49.8, 49.0	49.075

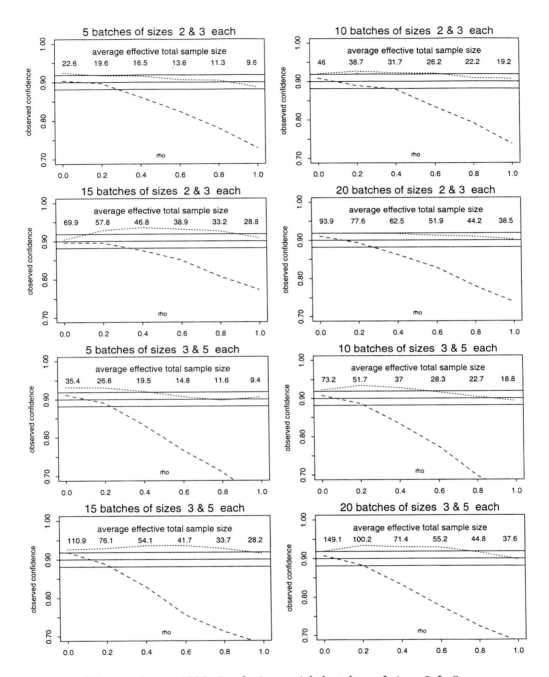

Figure 24.1: 1000 simulations with batches of sizes 2 & 5

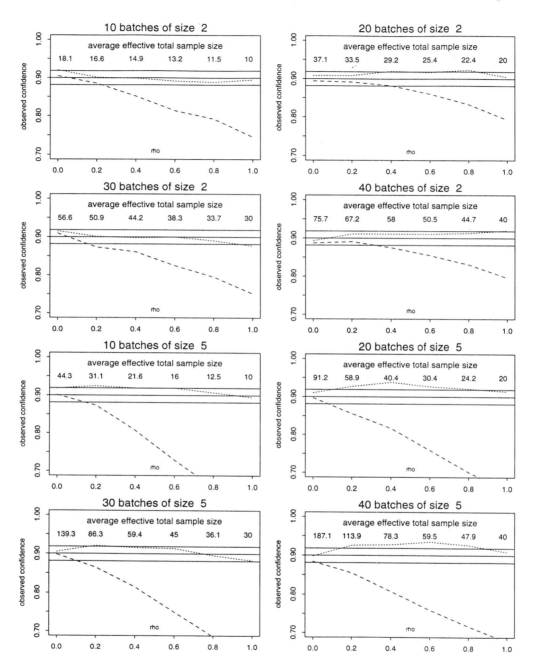

Figure 24.2: 1000 simulations with batches of size 2 & 3 and 3 & 5

However, the given data show strong batch effects and the above conclusion is not warranted as we will see when adjusting by the "effective" sample size. For the data above we obtain

$$SS_b = 78.921 \,, \quad SS_e = 29.148 \,, \quad f = 17.123 \,, \quad \hat{\sigma}_e^2 = .6939 \,, \quad \hat{\sigma}_b^2 = 1.093$$

and thus $\hat{\rho} = .6116$ and $N^\star = 25.056$. Again interpolating from Table 24.3, this time at the row corresponding to $N^\star = 25.056$, we obtain

$$C_{\text{Table}\star}(25.056) = 1.255 \quad \text{and thus} \quad C_{\text{Adj}\star}(63) = 1.0122 \cdot 1.255 = 1.27$$

as the appropriately adjusted critical value. Since $1.17 < 1.27$ we cannot conclude with 90% confidence that the true C_{pk} is at least 1.

24.7 Concluding Remarks

The *effective sample size* device, in the two situations of tolerance and C_{pk} bounds examined here, provides a simple way of "approximating" the complex batch effect scenario by the better understood i.i.d. case. For C_{pk} confidence bounds the solution provided is apparently the first such treatment. Practitioners, who tend to deal more often than not with such complications in data and who understand the statistical analysis solutions for the idealistic i.i.d. case, will welcome the simple modifications required by the approach presented here. From the simulations it appears that any error in coverage performance is on the conservative side. Whether this error can be reduced while maintaining the simplicity is not clear. It is more tempting to systematically examine many other data scenarios, well understood in the i.i.d. case but difficult in the context of random or batch effects, in the light of similar modifications using the *effective sample size* device. Our scheme of finding the effective sample size was based on matching the variances of sample means under the given data scenario and under the i.i.d. approximation. Other criteria for matching, such as information indices, may be examined. Much more broadly but also vaguely at this point, one could contemplate how much inferences of any type under some i.i.d. approximation may differ from the corresponding inferences under the given data situation. One simple but deficient approximation is to randomly select one representative from each batch. What could be gained by resampling this process?

Acknowledgement. We wish to thank the referee for constructive comments improving the presentation of this article.

Appendix

Here we present the rationale for the approximation $g \approx f$. Let $w_i = n_i/N$ and observe $\sum_{i=1}^{B} w_i = 1$. Further let

$$A = \sum_{i=1}^{B} w_i^2 \quad \text{and} \quad U = \sum_{i=1}^{B} w_i (w_i - A)^2 = \sum_{i=1}^{B} w_i^3 - A^2 .$$

Then

$$\frac{(1 - \sum w_i^2)^2}{\sum w_i^2 - 2\sum w_i^3 + (\sum w_i^2)^2} = \frac{1-A}{A} \frac{1-A}{1-A-2U/A} \approx \frac{1-A}{A} ,$$

where in the approximation step we assume that

$$\frac{U}{A} = \sum_{i=1}^{B} w_i \left(\frac{w_i}{A} - 1 \right)^2 A \ll 1 , \quad \text{since} \quad w_i \approx \frac{1}{B} , \quad A \approx \frac{1}{B} , \quad \frac{w_i}{A} \approx 1 .$$

Note that $U/A = 0$, when the n_i are all the same. In that case the above approximation is exact.

References

1. Chou, Y. M., Owen, D. B. and Borego, S. A. (1990). Lower confidence limits on process capability indices, *Journal of Quality Technology*, **22**, 223–229.

2. Fisher, L. and Van Belle, G. (1993). *Biostatistics*, New York: John Wiley & Sons.

3. Jennett, W. J. and Welch, B. L. (1939). The control of proportion defective as judged by a single quality characteristic varying on a continuous scale, *Journal of the Royal Statistical Society–Supplement*, **6**, 80–88.

4. Johnson, N. L. and Welch, B. L. (1940). Applications of the noncentral *t*-distribution, *Biometrika*, **31**, 362–389.

5. Kane, V. E. (1986). Process capability indices, *Journal of Quality Technology*, **18**, 41–52.

6. Kish, L. (1965). *Survey Sampling*, New York: John Wiley & Sons.

7. Kotz, S. and Johnson, N. L. (1993). *Process Capability Indices*, London: Chapman & Hall.

8. Kushler, R. and Hurley, P. (1992). Confidence bounds for capability indices, *Journal of Quality Technology*, **24**, 188–195.

9. Mee, R. W. and Owen, D. B. (1983). Improved factors for one-sided tolerance limits for balanced one-way ANOVA random model, *Journal of the American Statistical Association*, **78**, 901–905.

10. Odeh, R. E. and Owen, D. B. (1980). *Tables for Normal Tolerance Limits, Sampling Plans, and Screening*, New York: Marcel Dekker.

11. Owen, D. B. (1968). A survey of properties and applications of the non-central t-distribution, *Technometrics*, **10**, 445–478.

12. Owen, D. B. (1985). Noncentral t-distribution, *Encyclopedia of Statistical Sciences, Vol. 6*, New York: John Wiley & Sons.

13. Seeger, P. and Thorsson, U. (1972). Two-sided tolerance limits with two-stage sampling from normal populations – Monte Carlo studies of the distribution of coverages, *Applied Statistics*, **22**, 292–300.

14. Skinner, C. J., Holt, D. and Smith, T. M. F. (Eds.) (1989). *Analysis of Complex Surveys*, New York: John Wiley & Sons.

15. Vangel, M. G. (1995). Design allowables from regression models using data from several batches, *Composite Materials: Testing and Design*, Twelfth Volume (Eds., R. B. Deo and C. R. Saff), pp. 358–370, ASTM STP 1274, American Society for Testing and Materials, Philadelphia.

Subject Index